Wireless and Mobile Device Security

SECOND EDITION

Jim Doherty

JONES & BARTLETT
LEARNING

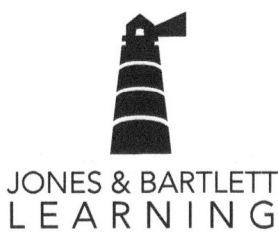

World Headquarters
Jones & Bartlett Learning
25 Mall Road
Burlington, MA 01803
978-443-5000
info@jblearning.com
www.jblearning.com

Jones & Bartlett Learning books and products are available through most bookstores and online booksellers. To contact Jones & Bartlett Learning directly, call 800-832-0034, fax 978-443-8000, or visit our website, www.jblearning.com.

Substantial discounts on bulk quantities of Jones & Bartlett Learning publications are available to corporations, professional associations, and other qualified organizations. For details and specific discount information, contact the special sales department at Jones & Bartlett Learning via the above contact information or send an email to specialsales@jblearning.com.

Copyright © 2022 by Jones & Bartlett Learning, LLC, an Ascend Learning Company

All rights reserved. No part of the material protected by this copyright may be reproduced or utilized in any form, electronic or mechanical, including photocopying, recording, or by any information storage and retrieval system, without written permission from the copyright owner.

The content, statements, views, and opinions herein are the sole expression of the respective authors and not that of Jones & Bartlett Learning, LLC. Reference herein to any specific commercial product, process, or service by trade name, trademark, manufacturer, or otherwise does not constitute or imply its endorsement or recommendation by Jones & Bartlett Learning, LLC and such reference shall not be used for advertising or product endorsement purposes. All trademarks displayed are the trademarks of the parties noted herein. *Wireless and Mobile Device Security, 2nd Edition* is an independent publication and has not been authorized, sponsored, or otherwise approved by the owners of the trademarks or service marks referenced in this product.

There may be images in this book that feature models; these models do not necessarily endorse, represent, or participate in the activities represented in the images. Any screenshots in this product are for educational and instructive purposes only. Any individuals and scenarios featured in the case studies throughout this product may be real or fictitious, but are used for instructional purposes only.

Production Credits

Director of Product Management: Laura Pagluica
Product Manager: Edward Hinman
Content Strategist: Melissa Duffy
Content Coordinator: Paula-Yuan Gregory
Manager, Project Management: Jessica deMartin
Project Specialist: Allie Koo
Senior Digital Project Specialist: Angela Dooley
Digital Project Specialist: Rachel DiMaggio
Marketing Manager: Michael Sullivan
Product Fulfillment Manager: Wendy Kilborn
Technical Editor: Justin Hensley
Composition: S4Carlisle Publishing Services
Project Management: S4Carlisle Publishing Services
Cover Design: Briana Yates
Text Design: Briana Yates
Media Development Editor: Faith Brosnan
Rights Specialist: James Fortney
Cover Image (Title Page, Part Opener,
    Chapter Opener): © Cherezoff/Shutterstock
Printing and Binding: Gasch Printing

Library of Congress Cataloging-in-Publication Data
Names: Doherty, Jim, 1968- author.
Title: Wireless and mobile device security / Jim Doherty, Professor,
Description: Second edition. | Burlington, Massachusetts : Jones & Bartlett
    Learning, [2022] | Series: Issa | Includes bibliographical references
    and index.
Identifiers: LCCN 2020043841 | ISBN 9781284211726 (paperback)
Subjects: LCSH: Wireless communication systems--Security measures. | Mobile
    communication systems--Security measures. | Wireless LANs--Security
    measures. | Mobile computing--Security measures.
Classification: LCC TK5105.78 .D64 2022 | DDC 005.8--dc23
LC record available at https://lccn.loc.gov/2020043841

6048

Printed in the United States of America
25 24 23 22 21   10 9 8 7 6 5 4 3 2

To Katie, Samantha, and Conor

Contents

Preface xvii
Acknowledgments xix
About the Author xxi

PART I Introduction to Wireless and Mobile Networks 1

CHAPTER 1 The Evolution of Data and Wireless Networks 3

The Dawn of Data Communication 4
Early Data Networks 5
The Internet Revolution 7
Advances in Personal Computers 7

Networking and the Open Systems Interconnection Reference Model 8
The Seven Layers of the OSI Reference Model 9
Communicating over a Network 11
IP Addressing 11
Data Link Layer 12
Physical Layer 13

From Wired to Wireless 14

Business Challenges Addressed by Wireless Networking 16
The Economic Impact of Wireless Networking 16
Wireless Networking and the Way People Work 16

The Wi-Fi Market 19
IP Mobility 20

The Internet of Things 22

CHAPTER SUMMARY 23
KEY CONCEPTS AND TERMS 24
CHAPTER 1 ASSESSMENT 24

CHAPTER 2 The Mobile Revolution 27

Introduction to Cellular (Mobile Communication) 28
Cellular Coverage Maps 28
Cellular Handoff 32

The Evolution of Mobile Networks 33

AMPS 1G 33
GSM and CDMA 2G 34
GPRS and EDGE 2G+ 35
3G Technology 35
4G and LTE 36
5G 36

The BlackBerry Effect and the BYOD Revolution 37

The Economic Impact of Mobile IP 38

The Business Impact of Mobility 40

Business Use Cases 40

CHAPTER SUMMARY 42
KEY CONCEPTS AND TERMS 42
CHAPTER 2 ASSESSMENT 43

CHAPTER 3 Anywhere, Anytime, on Anything: "There's an App for That!" 45

Anywhere, Anytime, on Anything 46

Convenience Trumps Security 47
Always Connected, Always On 47

The Rise of the Mobile Workforce 48

From Castle-and-Moat toward Zero Trust 50

The Mobile Cloud 50

Mobile Cloud Computing 51
Cloud Apps versus Native Mobile Apps 51

Deploying Wireless: Different Strokes for Different Folks 51

The Industrial Internet of Things 52

IoT Wireless Technologies 53

Wireless Communication Technologies 54

Bluetooth Low Energy 55
Zigbee IP 55
Z-Wave 55
RFID 56
NFC 56
Thread 56
6LoWPAN 56

Cloud VPNS, WANs, and Interconnects 56

Free Space Optics 57
WiMAX 57
vSAT 57

Contents

SD-WAN   57

WAN Technologies for IoT   58

Sigfox   58
LoRaWAN   58
Low-Power Wi-Fi (HaLow)   59
Millimeter Radio   59

Private LTE Networks   59

Wireless Network Security   60

Lingering Security Issues   62

Mobile IP Security   62

CHAPTER SUMMARY   62

KEY CONCEPTS AND TERMS   63

CHAPTER 3 ASSESSMENT   63

CHAPTER 4   Security Threats Overview: Wired, Wireless, and Mobile   65

What to Protect?   66

General Threat Categories   67

Confidentiality   68
Integrity   68
Availability   69
Accountability   69
Nonrepudiation   70

Threats to Wireless and Mobile Devices   70

Data Theft Threats   71
Device Control Threats   72
System Access Threats   73

Risk Mitigation   74

Mitigating the Risk of BYOD   75
BYOD for Small-to-Medium Businesses   78

Defense in Depth   78

Authorization and Access Control   80

AAA   80

Information Security Standards   82

ISO/IEC 27001:2013   82
ISO/IEC 27002:2013   83
NIST SP 800-53   83

Regulatory Compliance   84

The Sarbanes–Oxley Act   84
The Gramm–Leach–Bliley Act   84

The Health Insurance Portability and Accountability Act
 and the Health Information Technology for Economic and Clinical Health Act   85
The Payment Card Industry Data Security Standard   85
GDPR & CCPA   86
Detrimental Effects of Regulations   86

CHAPTER SUMMARY   88

KEY CONCEPTS AND TERMS   88

CHAPTER 4 ASSESSMENT   89

PART II — WLAN Security   91

CHAPTER 5 — How Do WLANs Work?   93

WLAN Topologies   94

ISM Unlicensed Spectrum   94

WLAN Anatomy   96
Wireless Client Devices   96
802.11 Service Sets   97

The 802.11 Standards   98

New Wi-Fi Alliance Naming System   99

802.11 Unlicensed Bands   102

Narrowband and Spread Spectrum   103
Multipath   103
Frequency Hopping Spread Spectrum   104
Direct Sequence Spread Spectrum   104

Wireless Access Points   105

How Does a WAP Work?   105
WAP Architecture   108

Wireless Bridges   109

Wireless Workgroup Bridges   109
Residential Gateways   110
Enterprise Gateways   111

Wireless Antennas   111

Omnidirectional Antennas   112
Semi-Directional Antennas   112
Highly Directional Antennas   113
MIMO Antennas   114
Determining Coverage Area   118

Site Surveys   118

Spectrum and Protocol Analysis   120

CHAPTER SUMMARY   122

KEY CONCEPTS AND TERMS   123

CHAPTER 5 ASSESSMENT   123

Contents

CHAPTER 6 **WLAN and IP Networking Threat and Vulnerability Analysis** 125

Types of Attackers 126
Skilled versus Unskilled Attackers 127
Insiders versus Outsiders 127

Targets of Opportunity versus Specific Targets 128

Scouting for a Targeted Attack 129

Physical Security and Wireless Networks 130

Social Engineering 131

Wardriving 133

Rogue Access Points 134
Rogue Access Point Vulnerabilities 134

Evil Twins 135

Bluetooth Vulnerabilities and Threats 137
Bluetooth Versions 137
Revisions Compared 138
Bluetooth Pairing 139
Bluejacking 140
Bluesnarfing 141
Bluebugging 142
Is Bluetooth Vulnerable? 143

Packet Analysis 143

Wireless Networks and Information Theft 144

Malicious Data Insertion on Wireless Networks 146

Denial of Service Attacks 147

Peer-to-Peer Hacking over Ad Hoc Networks 148

When an Attacker Gains Unauthorized Control 149

CHAPTER SUMMARY 149
KEY CONCEPTS AND TERMS 150
CHAPTER 6 ASSESSMENT 150

CHAPTER 7 **Basic WLAN Security Measures** 153

Design and Implementation Considerations for Basic Security 154
Radio Frequency Design 154
Equipment Configuration and Placement 155
Interoperability and Layering 156
Security Management 157
Basic Security Best Practices 158

Authentication and Access Restriction   158

SSID Obfuscation   159
MAC Filters   160
Authentication and Association   160
VPN over Wireless   161
Virtual Local Area Networks   162

Data Protection   163

Wired Equivalent Privacy   164
Wi-Fi Protected Access   165
Wi-Fi Protected Access 2   165
WPA2 with AES   166
WPA2 with CCMP   167
Order of Preference for Wi-Fi Data Protection   167
WPA3   168

Ongoing Management Security Considerations   169

Firmware Upgrades   169
Physical Security   170
Periodic Inventory   170
Identifying Rogue WLANs/Wireless Access Points   171

CHAPTER SUMMARY   171
KEY CONCEPTS AND TERMS   172
CHAPTER 7 ASSESSMENT   172

CHAPTER 8   Advanced WLAN Security Measures   175

Establishing and Enforcing a Comprehensive Security Policy   176

Centralized versus Distributed Design and Management   176
Remote Access Policies   177
Guest Policies   177
Quarantining   178
Compliance Considerations   178
Employee Training and Education   179

Implementing Authentication and Access Control   179

Extensible Authentication Protocol   180
Remote Authentication Dial-In User Service   180
Intrusion Detection Systems and Intrusion Prevention Systems   182
Protocol Filtering   182
Authenticated Dynamic Host Configuration Protocol   183

Data Protection   184

WPA2 Personal and Enterprise Modes   184
WPA3   184
Internet Protocol Security   185
Virtual Private Networks   186
Malware and Application Security   186

User Segmentation   187

Contents   xi

        Virtual Local Area Networks   187
        Guest Access and Passwords   188
        Demilitarized Zone Segmentation   188

    Managing Network and User Devices   189

        Simple Network Management Protocol Version 3   189
        Discovery Protocols   190
        IP Services   190
        Coverage Area and Wi-Fi Roaming   191
        Client Security Outside the Perimeter   192
        Device Management and User Logons   193
        Hard Drive Encryption   194
        Quarantining   194
        Wi-Fi as a Service   195

    CHAPTER SUMMARY   196

    KEY CONCEPTS AND TERMS   197

    CHAPTER 8 ASSESSMENT   197

CHAPTER 9   WLAN Auditing Tools   199

    WLAN Discovery Tools   200

        Enterprise Wi-Fi Audit Tools   201
        HeatMapper   202

    Penetration Testing Tools   204

        Metasploit   204
        Security Auditor's Research Assistant   204

    Password-Capture and Decryption Tools   205

        Network Enumerators   208

    Network Management and Control Tools   208

        Wireless Protocol Analyzers   208
        Aircrack-ng   209
        Airshark   209
        Network Management System   210

    WLAN Hardware Audit Tools and Antennas   211

        Hardware Audit Tools   211
        Antennas   211

    Attack Tools and Techniques   212

        Radio Frequency Jamming   212
        Denial of Service   213
        Hijacking Devices   213
        Hijacking a Session   214

    Network Utilities   216

    CHAPTER SUMMARY   217

    KEY CONCEPTS AND TERMS   218

    CHAPTER 9 ASSESSMENT   218

CHAPTER 10 WLAN and IP Network Risk Assessment 221

Risk Assessment 222

Risk Assessment on WLANs 224
Other Types of Risk Assessment 225

IT Security Management 225

Methodology 225
Legal Requirements 226
Other Justifications for Risk Assessments 226

Security Risk Assessment Stages 226

Planning 227
Information Gathering 228
Risk Analysis 229
Identifying and Implementing Controls 234
Monitoring 235

Security Audits 235

CHAPTER SUMMARY 236
KEY CONCEPTS AND TERMS 236
CHAPTER 10 ASSESSMENT 236

PART III Mobile Security 239

CHAPTER 11 Mobile Communication Security Challenges 241

Mobile Phone Threats and Vulnerabilities 242

Exploits, Tools, and Techniques 244

Google Android Security Challenges 244

Criticism of Android 247
Android Exploitation Tools 248
Android Security Architecture 249
Android Application Architecture 249
Google Play 250

Apple iOS Security Challenges 251

Apple iOS Exploits 252
Apple iOS Architecture 254
The App Store 254

Windows Phone Security Challenges 255

Windows Phone OS Exploits 255
Windows Phone Security Architecture 255
Windows Phone Architecture 256
Windows Store 256

CHAPTER SUMMARY 257
KEY CONCEPTS AND TERMS 257
CHAPTER 11 ASSESSMENT 257

Contents   xiii

CHAPTER 12   Mobile Device Security Models   259

Google Android Security   260

The Android Security Model   260
The Android Sandbox   261
File-System Permissions   261
Android SDK Security Features   261
Rooting and Unlocking Devices   262
Android Permission Model   262

Apple iOS Security   263

The Apple Security Model   263
Application Provenance   264
iOS Sandbox   264
Security Concerns   265
Permission-Based Access   265
Encryption   266
Jailbreaking iOS   266

Windows Phone 8 Security   267

Platform Application Security   267
Security Features   267

iOS and Android Evolution   267

Android Version Evolution   268
Apple iOS   269

Security Challenges of Handoff-Type Features   270

BYOD and Security   271

Security Using Enterprise Mobility Management   272

Mobile Device Management   273
Mobile Application Management   273

CHAPTER SUMMARY   275
KEY CONCEPTS AND TERMS   276
CHAPTER 12 ASSESSMENT   276

CHAPTER 13   Mobile Wireless Attacks and Remediation   279

Scanning the Corporate Network for Mobile Attacks   280

Security Awareness   280

Scanning the Network: What to Look For   281
Scanning for Vulnerabilities   282

The Kali Linux Security Platform   283

Scanning with Airodump-ng   284

Client and Infrastructure Exploits   285

Client-Side Exploits   285

Other USB Exploits   **286**

Network Impersonation   286

Network Security Protocol Exploits   **287**

RADIUS Impersonation   287
Public Certificate Authority Exploits   288
Developer Digital Certificates   289
Browser Application and Phishing Exploits   289
Drive-By Browser Exploits   290

Mobile Software Exploits and Remediation   **290**

Weak Server-Side Security   291
Unsecure Data Storage   291
Insufficient Transport Layer Protection   292
Data Leakage   292
Poor Authorization and Authentication   293
Broken Cryptography   293
Client-Side Injection   293
Security Decisions via Untrusted Inputs   294
Improper Session Handling   294
Lack of Binary Protections   295

CHAPTER SUMMARY   **295**

KEY CONCEPTS AND TERMS   **295**

CHAPTER 13 ASSESSMENT   **296**

CHAPTER 14   Fingerprinting Mobile Devices   297

Is Fingerprinting a Bad or a Good Thing?   **298**

Types of Fingerprinting   **299**

Network Access Control
 and Endpoint Fingerprinting   299
Network Scanning and Proximity Fingerprinting   299
Online or Remote Fingerprinting   300

Cookies   **301**

Cross-Site Profiling   **302**

Fingerprinting Methods   **303**

Passive Fingerprinting   303
Examining TCP/IP Headers   304
Application Identification   304
Active Fingerprinting   305

Unique Device Identification   **305**

Apple iOS   305
Android   306

HTTP Headers   306
New Methods of Mobile Fingerprinting   307
JavaScript   307
Fingerprinting Users   308
Fingerprinting Users via Biometrics   308
Spyware for Mobile Devices   309
Spy Software   310
Spy Cells: Stingray   311
Fingerprinting on Modern Cellular Networks   312
MNmap   313
Man-in-the-Middle Attack   313
CHAPTER SUMMARY   314
KEY CONCEPTS AND TERMS   314
CHAPTER 14 ASSESSMENT   315

CHAPTER 15 Mobile Malware and Application-Based Threats   317

Malware on Android Devices   318
Software Fragmentation   319
Criminal and Developer Collaboration   320
Madware   323
Excessive Application Permissions   323
Malware on Apple iOS Devices   325
Mobile Malware Delivery Methods   326
Mobile Malware and Social Engineering   327
Captive Portals   327
Drive-By Attacks   327
Clickjacking   328
Likejacking   328
Plug-and-Play Scripts   328
Mitigating Mobile Browser Attacks   328
Mobile Application Attacks   329
Mobile Malware Defense   330
Mobile Device Management   330
Penetration Testing and Smartphones   331
CHAPTER SUMMARY   332
KEY CONCEPTS AND TERMS   332
CHAPTER 15 ASSESSMENT   333

APPENDIX A	**Answer Key** 335
APPENDIX B	**Standard Acronyms** 337

Glossary of Key Terms 339

References 351

Index 363

Preface

Purpose of This Book

This book is part of the Information Systems Security & Assurance Series from Jones & Bartlett Learning (www.jblearning.com). Designed for courses and curriculums in IT Security, Cybersecurity, Information Assurance, and Information Systems Security, this series features a comprehensive, consistent treatment of the most current thinking and trends in this critical subject area. These titles deliver fundamental information security principles packed with real-world applications and examples. Authored by professionals experienced in information systems security, they deliver comprehensive information on all aspects of the topic. Reviewed word for word by leading technical experts in the field, these books are not just current but forward-thinking—putting you in the position to solve current cybersecurity challenges and future ones, as well.

Part I of the text reviews the history of wireless and mobile networks and the evolution of wired and wireless networking—from Alexander Graham Bell to the present bring-your-own-device (BYOD) phenomenon. You'll read about the mobile revolution that took users from clunky analog phones to "smart" devices people can't live without and about the implications of the always on, ever-present aspect of these devices. Although most people view the resulting changes as a net positive, both wireless and mobile networking have introduced significant security vulnerabilities to networking in general. You'll get an overview of network security threats and considerations, with a particular emphasis on wireless and mobile devices.

Part II focuses on wireless local area network (WLAN) security. You'll read about WLAN design and the operation and behavior of wireless in general, particularly on 802.11 WLANs. You'll review the threats and vulnerabilities directly associated with 802.11 wireless networks, their various topologies, and devices. The text will discuss basic security measures that satisfy the needs of small office/home office (SOHO) networks, as well as more advanced concepts in wireless security unique to the needs of larger organizations. You'll learn about the need to audit and monitor a WLAN and the tools available for doing so. Finally, you'll review risk assessment procedures as applied to WLAN and Internet Protocol mobility.

Part III discusses security solutions to the risks and vulnerabilities of wireless networks and mobile devices. You'll read about the three major mobile operating systems and the vulnerabilities of each. Then you'll review the security models of these operating systems and explore how IT organizations manage the security and control of smart devices on a large scale. The text will look at the risks mobile clients present to corporate networks, as well as the tools and techniques used to mitigate these risks. You'll also learn about the issues surrounding fingerprinting of mobile devices. Finally, you'll review the mobile malware landscape and mitigation strategies to prevent malware from finding its way into an organization's information security resources.

Learning Features

The writing style of this book is practical and conversational. Step-by-step examples of information security concepts and procedures are presented throughout the text. Each chapter begins with a statement of learning objectives. Illustrations are used both to clarify the material and to vary the presentation. The text is sprinkled with notes, tips, FYIs, warnings, and sidebars to alert the reader to additional helpful information related to the subject under discussion. Chapter assessments appear at the end of each chapter with solutions provided in the back of the book. Chapter summaries are included in the text to provide a rapid review or preview of the material and to help students understand the relative importance of the concepts presented.

Audience

The material is suitable for undergraduate or graduate computer science majors or information science majors, students at a 2-year technical college or community college who have a basic technical background, or readers who have a basic understanding of IT security and want to expand their knowledge.

Cloud Labs

This text is accompanied by Cybersecurity Cloud Labs. These hands-on virtual labs provide immersive mock IT infrastructures where students can learn and practice foundational cybersecurity skills as an extension of the lessons in this textbook. For more information or to purchase the labs, visit go.jblearning.com/doherty2e

New to this Edition

Quite a lot has changed in the 5 years between the 1st and 2nd edition. We've dedicated an entire chapter to the "Always On, Ever Present" phenomenon and put a significant focus on the Internet of Things (IOT), which has resulted in both an explosion of devices on the wireless and mobile networks and in a secondary explosion of vulnerabilities. To make room for these new topics the two chapters dedicated to the evolution of data and mobile networks have been consolidated into a single chapter.

Additionally, this second edition includes the latest Wi-Fi standards (renamed since the original version of this text) and updated version descriptions for both the Android OS and Apple iOS. We've dropped all but passing mentions of the Windows phone operating system and have added up-to-date information on 5G. Finally, we've revised this version to include the latest wireless and mobile security vulnerabilities and remediations. As with any book on network and mobile security, we do our best to keep up with this never-ending game of cat and mouse.

Acknowledgments

I may be the fortunate one with my name on the cover, but there are several people who were instrumental in the creation of this book. Without them I would not have been able to take on, much less finish, this project.

- First and foremost on this list is Alasdair Gilchrist, my researcher, writing assistant, and technical bodyguard. His knowledge skill, effort, and technical expertise were indispensable in all aspects of creating both the manuscript for this book and the associated labs. One of the best things about all of this new technology is its ability to connect people from across the world. In this case, two guys who never met in person—one in North Carolina and one in Bangkok—were able to collaborate while both going about work and life using many of the tools and technologies discussed in this book (in a very secure way, of course). Alasdair is a true professional, and I'm very fortunate to have him on my team.
- Justin Hensley, my technical reviewer/editor and the director of Information Security and Infrastructure at the University of Cumberlands. Justin kept us sharp and did his level best not only to ensure the information in this book was accurate but also that it was accessible to the reader. This book is better and more readable than it would have otherwise been without his keen eye.
- My content strategist Melissa Duffy kept this project moving with unwavering attention and good cheer. Through multiple revisions, technical and editorial reviews, and a couple of nearly missed deadlines, Melissa competently moved the project forward. Writing a book is easy in comparison to getting a book published. Melissa did the heavy lifting.
- The production team at JB Learning included production specialist Allie Koo; media development editor Faith Brosnan; rights specialist James Fortney; and project manager (as S4Carlisle) Manjusha Chandrasekaran who had to fix all my grammatical, punctuation, and formatting errors—no small task! Thanks to this entire team and to all the many others who provided production assistance.

Thank you all. I am truly grateful.
Jim Doherty

About the Author

Jim Doherty, has more than 20 years of engineering, marketing, and sales experience across a broad range of networking, security, and technology companies. Focusing on technology strategy, product positioning, and marketing execution, he has held leadership positions for CommScope, Cisco Systems, Certes Networks, Ixia, and Ericsson Mobile.

Doherty is also the coauthor of the Networking Simplified series of books, which includes *Cisco Networking Simplified*, *Home Networking Simplified*, and several other titles. He is a former U.S. Marine Corps sergeant and holds a bachelor's degree in electrical engineering from North Carolina State University and a Master of Business Administration degree from Duke University.

PART I

Introduction to Wireless and Mobile Networks

CHAPTER 1	The Evolution of Data and Wireless Networks 3
CHAPTER 2	The Mobile Revolution 27
CHAPTER 3	Anywhere, Anytime, on Anything: "There's an App for That!" 45
CHAPTER 4	Security Threats Overview: Wired, Wireless, and Mobile 65

CHAPTER 1

The Evolution of Data and Wireless Networks

IN THE SPACE OF JUST 25 YEARS, network security, which had already gone through several evolutions, has once again been turned on its head due to the confluence of two phenomena: the untethering of network connectivity and the proliferation of devices that are always on, always connected, and often moving. The ability to log on to both the Internet and corporate networks without having to physically connect a computer to the network via an Ethernet cable has radically altered the culture and has greatly blurred the line between work life and personal life. Further, the willingness (or obsessive need, perhaps) to be connected to all things directly (via computers and smartphones) or indirectly (via smart devices) has redefined privacy and the security of personal data.

Although most people view the resulting changes as a net positive, both wireless and mobile networking have introduced significant security vulnerabilities to networking in general and company and personal information in particular. These vulnerabilities, along with prevention and detection methods, are the focus of this text. However, before jumping into the details of wireless and mobile network security, let's take a look at how these profound changes came about.

Chapter 1 Topics

This chapter covers the following concepts and topics:

- How early forms of data communication worked
- How computers went mobile
- How mobile and data networks converged
- What the origin, purpose, and function of the OSI Reference Model are
- What the origins of wireless technology are
- What the economic impact of wireless networking is
- How wireless networking has changed the way people work

Chapter 1 Goals

When you complete this chapter, you will be able to:

- Describe the evolutionary history of networking
- Describe the function of the OSI Reference Model
- Understand and describe the functions of each layer in the OSI Reference Model
- Describe IP addressing and the key differences between IPv4 and IPv6
- Describe MAC addresses and how they differ from IP addresses
- Provide examples of how wireless networking is used in health care, warehousing, and retail

The Dawn of Data Communication

Data communication and networking have a long history going back to 1837, when Samuel Morse developed the first practical telegraph system. In 1844, Morse sent his first long-distance message, "What hath God wrought!" encoded in Morse code, from Washington, D.C. to Baltimore, Maryland. By 1850, more than 12,000 miles of telegraph lines traversed the country, run by more than 20 different commercial operators. **Telegraphy**, as it was known, used start and stop signals of dots and dashes transmitted over copper wires. It was a one-way message protocol that evolved to support two, and then four, channels. Telegraphy monopolized electronic communication until 1877, when the first telephone networks started to appear.

Despite reservations that telephones would be too technical for the common man, **telephony** was quickly adopted. Indeed, the telephone system quickly usurped telegraphy in terms of traffic carried, revenue generated, and network coverage. Telephony, however, was initially limited to voice. Therefore, despite telegraphy losing out to the telephone as the popular means for interpersonal communications, it remained an effective medium for carrying digital data traffic.

In 1923, the first teletypewriter services came into being, serving the need for true and accurate communications. By 1935, the introduction of rotary dial telex services emerged.

By the 1950s, the **public switched telephone network (PSTN)** had become ubiquitous and affordable, in large part due to broad interconnectivity, creating a network effect. The PSTN could interconnect telephones from anywhere in the community, the country, or even internationally over its network of exchanges. To accomplish this, a hierarchy of networks connected local exchange carriers (LECs) with regional, national, and international carriers via interexchange carriers (IXCs). It was this national and international reach that made the PSTN such an inviting medium when it became necessary to network large business computer mainframes.

> **NOTE**
>
> The **network effect** is a phenomenon in which a technology becomes more valuable as the number of users or units increases. A common example is the fax machine. The first fax machine was useless, but the second fax machine made the first one useful. As more were added, all fax machines had greater utility.

Early Data Networks

By the late 1950s, there was a demand to network the growing number of business computers being deployed by large companies. These computers were large standalone machines that operated independently. IBM accomplished the first successful interface between two digital devices over the analog PSTN using acoustic couplers and telephone sets. These couplers operated over the PSTN at 300 bits per second (bps). At this point, voice networks and the burgeoning data network began to merge.

> **FYI**
>
> Until the appearance of the personal computer (PC), computers were huge mainframes that often occupied an entire room. Access to these computers was achieved through a "dumb terminal," which offered a simple text display (often called a *green screen* due to the green color of the font on the black screen), as shown in **FIGURE 1-1**. These displays had no computing power; rather, they were simple readout displays.

FIGURE 1-1

A dumb terminal or green screen.

Courtesy of U.S. Department of Defense.

Another significant point in the history of data communications was the transmission of the first fax over the standard PSTN in 1962. This was possible due to the modulation of data into sound by devices called *modems*, which were attached to either end of the analog telephone lines. A **modem**, short for modulator/demodulator, was required to transfer digital communications over the analog PSTN for the several decades that followed. Modems convert digital data into an analog signal for transport over the wire. At the other end, the analog signal is demodulated to recover the original digital signal. By using modems, computers that had access to a telephone line could communicate over the analog PSTN.

Soon, however, telephone companies saw the obvious benefits of digital technology and began upgrading their networks. Digital communication was accepted as technically superior to analog. Furthermore, digital technology had become both cheaper and more reliable, which made it suitable for transmitting voice communications.

Digital communications have the following advantages over analog:

- More efficient use of bandwidth
- Greater utilization
- Improved error rates (that is, fewer errors)
- Less susceptibility to noise and interference
- Increased throughput
- Support for additional services (such as caller ID, auto-forwarding, and call waiting)

As telecom providers rolled out new digital networks, high-speed digital communication became a widely available service.

The innovation that enabled the technological leap in long-distance digital communication was **packet switching**, used in lieu of **circuit switching**. In circuit switching, a physical connection was made between two phones using a series of telephony switches, creating an electric circuit. While the circuit was in use, no other phones could use the wires connecting the two phones. This was very inefficient because conversations—even among chatty people—are in fact about 50 percent silence when you account for the pauses between words and between speakers. It was also very expensive, especially on long-distance calls, because the callers had to "rent" the exclusive use of a circuit that was almost always in demand.

There were also concerns with circuit switching regarding the resilience of the message path. Circuit switching restricted communications to a preprovisioned point-to-point circuit for the duration of the call. Should any intermediary exchange along the message path fail, the circuit was lost and had to be reprovisioned. Ideally, the call would be automatically rerouted over a different path. However, that required multiple paths to any given destination and an awareness of alternative routes to the destination, which greatly increased the user cost.

In packet switching, the voice signal is first digitized and then chopped up into a series of packets. These packets contain the voice information along with the source and destination. The packets are then forwarded from the source to the destination. Taking advantage of the silent gaps, packets from multiple conversations can share the same circuit, making packet switching much more efficient. Additional efficiencies were created through the development of digital compression techniques so that many of today's conversations exist on the same wires.

Packet switching is also much more resilient to circuit switching. Packets can take multiple paths from source to destination, so there is no dependence on a single circuit. Also, because each single packet is such a small fragment of the speech signal, many packets can be lost or dropped without noticeably affecting the quality of the call. With packet switching, if any one circuit or exchange fails, the packets are rerouted. Any dropped packets are simply ignored. As it turned out, packet switching was also key to modern data communication.

Not surprisingly, the military was very much involved in developing packet switching. Its interests lay in the possibilities of high-speed failover and resilient data communications under battlefield conditions.

The Internet Revolution

This interest led to the U.S. government's creation of the Advanced Research Projects Agency (ARPA) to research and develop computer networks. The results of the ARPA project were the design, creation, and development of the ARPANET, the first computer network based on packet switching.

The ARPANET project was the predecessor of the modern Internet. However, during the 1970s and 1980s, it was a noncommercial network developed and used by universities and research institutions. During this period, despite being little known and seldom used, the ARPANET project developed several key protocols—one of the most important being the **Transmission Control Protocol/Internet Protocol (TCP/IP)**. TCP/IP would become the protocol of the Internet in the early 1990s.

During that same period, many **local area network (LAN)** technologies vied for supremacy. Several proprietary networking protocols were prevalent and vied for market dominance. The problem was that the proprietary protocols were not compatible. That is, LANs using different protocols could not be connected to each other. The challenge was to connect all these different operating systems and protocols into one heterogeneous network. By the early 1990s, a clear winner had emerged in the LAN technology war: the Ethernet protocol. It became the ubiquitous LAN network standard protocol across the globe. Thirty years later, Ethernet is still the dominant LAN protocol.

ARPA was more than LANs

ARPA also went on to create what would become the standard methods of connecting business computers and networks over long distances. These wide area networks (WANs), as they are known, were point-to-point or point-to-multipoint topologies, which enabled companies to connect networks and computers in cities and countries across the globe at high speeds and high throughput.

Advances in Personal Computers

Digital communications were not the only technology growing by leaps and bounds. Within businesses themselves, a revolution was occurring that spelled the end of the road for the huge mainframes and their dumb terminals.

The IBM PC was launched in the early 1980s. It immediately caught the attention of businesses and became popular for running standalone word processing and accounting packages. But because most of the business data resided on the mainframe, both a PC and a mainframe terminal (the "dumb" screen connected to the mainframe) were required on the desktop. Not only was that inconvenient but there was also no easy way to transfer information from the mainframe to the PC applications for local processing. The solution was to connect each PC to the mainframe by networking them over a LAN and harnessing the growing processing power of PCs by just connecting them directly to each other over the LAN. This made the dumb terminal redundant as PC sales skyrocketed and the computer networking industry exploded. Up to this point, however, if you wanted to connect a device to a network or to the Internet, you had to physically connect the device via an Ethernet connection or via a modem connected to the PSTN. However, all of this was about to change with the advent of wireless networking.

Networking and the Open Systems Interconnection Reference Model

Before examining how wired networks evolved into wireless networks, it's important to have a fundamental grasp of basic networking. The logical place to start is the **Open Systems Interconnection (OSI) Reference Model**. Initially proposed in 1984, the OSI Reference Model was the industry's response to the issues created by proprietary data networking and equipment development at the dawn of the networking era. Before this, most equipment manufacturers had developed their own proprietary methods for data networking. These included unique communication protocols, connection interfaces, and procedures for data storage and retrieval, to name a few. In some cases, vendors did this because much of the development in this area was new. But many insisted on going their own way in an attempt to gain a competitive advantage.

This may have seemed like a good idea for the companies breaking into this newly developing market. But it was bad for customers because it locked them into a single vendor solution. Concerns over the limited scope of solutions and clients' natural hesitance to remain locked into a single vendor ultimately stalled the market. This, of course, was not only bad for customers, but it was also bad for vendors because it put an artificial limitation on opportunities.

The answer was to develop a model that defined standards for communication protocols as well as for the physical and logical interfaces between machines and subsystems.

> **FYI**
>
> In many cases, the term *stack* is used in reference to a particular protocol, such as IP. However, you will often see cases in which the phrase "moving up or down the stack" refers to the OSI layers. For example, firewalls were originally Layer 3/4 (IP) devices; however, application firewall vendors are now said to have gone "up the stack" to Layer 7. In cases such as this, the context of how and where the term is used helps identify which meaning of the term applies.

CHAPTER 1 | The Evolution of Data and Wireless Networks

Published by the International Organization for Standardization, the OSI Reference Model defines seven layers (referred to as a **stack**) that describe standards from the physical wire all the way up to the application interfaces on computers. Further, the standard was written such that each layer has a specific function, common among all devices, and a specific way of communicating with the layer above and the layer below in the stack.

As a result of this standardization, different companies were able to specialize in specific solutions or products, confident that their products would be compatible with products from other manufacturers, even if there were no up-front collaboration. This led to a great deal of innovation. It also greatly improved the number of available solutions, because the barriers to entering the market were reduced. It's much easier to start a company based on a niche product than to create an entire end-to-end solution.

More importantly, the OSI Reference Model proved to be an enormous benefit to consumers—businesses, universities, and governments—who enjoyed greater choice and the opportunity to choose "best in breed" products rather than be forced to standardize around a single vendor for their entire network. Most economists and network historians agree that it was this willingness to adopt the OSI Reference Model that helped propel the networking industry into the economic power that it has become.

The Seven Layers of the OSI Reference Model

As noted, the OSI Reference Model has seven layers, each with a specific function and a standard way of communicating with the adjacent layers. From the bottom up, the layers are as follows:

- **Layer 1 (Physical Layer)**—The **Physical Layer** is the signal path over which data is transmitted. This can include copper wires; optical fibers; radio signals such as **Wi-Fi, Worldwide Interoperability for Microwave Access (WiMAX)**, and **Bluetooth**; and any other transmission path. The units of information at this layer are bits or bytes.
- **Layer 2 (Data Link Layer)**—The **Data Link Layer** helps establish the communication path by specifying the **Media Access Control (MAC) address** of each device. Layer 2 is viewed as the "switching" layer because it is the layer where switching paths are determined in LANs. Ethernet is a Layer 2 protocol. The unit of information at Layer 2 is the data frame.
- **Layer 3 (Network Layer)**—The **Network Layer** is typically viewed as the routing or IP layer, although over the years, the line between routing and switching has blurred. This layer handles communication paths between LANs. The IP exists at Layer 3. As such, communication paths are defined in terms of their IP addresses. The unit of information at Layer 3 is the packet.
- **Layer 4 (Transport Layer)**—Known as the **Transport Layer**, Layer 4 is the bridge between the network and the application-processing software on devices. This is where data from applications is broken down into small chunks, or packets, that are suitable for transport from the sending device, and then reassembled on the receiving device.
- **Layer 5 (Session Layer)**—The **Session Layer** defines and manages communications between applications on separate devices.
- **Layer 6 (Presentation Layer)**—Known as the **Presentation Layer**, Layer 6 formats information sent to and from applications.

> **NOTE**
>
> An easy way to remember the layer names in order is to use a mnemonic. There are dozens of them, ranging from funny, to topical, to outrageous. A good top-to-bottom one that is suitable in the workplace is *All People Seem To Need Data Processing*.

- **Layer 7 (Application Layer)**—At the top of the stack is the **Application Layer**, which provides appropriate protocols for Internet applications (e.g., HTTP, SMTP).

Inter-stack communication is accomplished through the use of headers. As data is created (at the Application Layer) and sent down the stack and across the network to another device, each layer adds its own set of instructions for the corresponding layer on the other machine and for the layer below or above the stack on the same machine.

Although the OSI Reference Model does provide a clear framework, the lines have begun to blur over the years. For example, many people will now reference a new condensed framework called the TCP Model, which lumps Layers 5 through 7 together, referring to them as the "Application Layers." In addition, Layer 3 and Layer 4 are closely associated due the use of TCP/IP, the de facto standard for communication over the Internet. Finally, as previously mentioned, the functional lines between switches and routers have been greatly blurred as innovation and processing power have continued to accelerate, allowing for greater efficiency in multitasking devices. As a result, Layer 2 and Layer 3 are no longer the sole domains of switching and routing, respectively.

An OSI layer can communicate only with the layers immediately above and below it on the stack and with its peer layer on another device. This process of passing instructions to the layer above or below must be used so that information (including data and stack instructions) can be passed down the stack, across the network, and back up the stack on the peer device. **FIGURE 1-2** shows the seven layers in relation to each other along with their basic functions.

FIGURE 1-2

The OSI Reference Model describes how devices communicate with each other over networks.

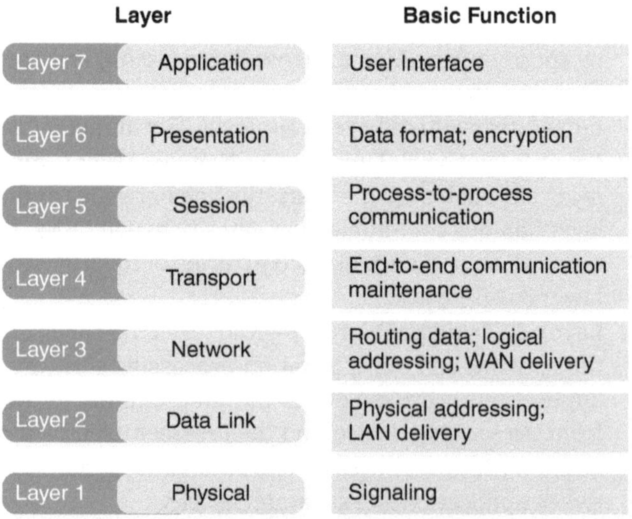

Layer		Basic Function
Layer 7	Application	User Interface
Layer 6	Presentation	Data format; encryption
Layer 5	Session	Process-to-process communication
Layer 4	Transport	End-to-end communication maintenance
Layer 3	Network	Routing data; logical addressing; WAN delivery
Layer 2	Data Link	Physical addressing; LAN delivery
Layer 1	Physical	Signaling

CHAPTER 1 | The Evolution of Data and Wireless Networks

Communicating over a Network

While a detailed look at the fundamentals of data communication is beyond the scope of this text, a brief overview of a few key aspects of network communication is necessary. In particular, we will look at those aspects of networking that weigh heavily in understanding wireless security. At the simplest level, the two main keys to communication over a network are conditioning the data for transport and ensuring that the data reaches the correct destination.

In the context of this simplified definition, the role of the Application Layers is to condition data for transport over the network or to reassemble data for use by the application on the receiving end. While there are many security considerations at these layers, they are largely independent of the method of transport (either wired or wireless), and are beyond the scope of this text. With the advent of network-connected mobile devices, all known issues that exist in fixed settings become moving targets.

Communication at the Network Layer is achieved through a logical addressing scheme that enables routers to move packets across the network to the correct destination. (In this case, "logical" means that the addresses are assigned and can be changed if needed.) This addressing scheme, called *IP addressing*, allows for data communication between (switched) LANs via routed WANs.

IP Addressing

IP addressing enables a network to correctly route packets across a network. For most of the last 30 years, an IP addressing scheme called **Internet Protocol version 4 (IPv4)** was used. This scheme used a format called **dotted decimal**, which consists of four *octets* (groups of eight in binary code) that define the location of a source or destination LAN—much in the same way a home or business address includes the street number, street, town, state or province, zip code, and country. A typical IP address has the format 198.10.249.168, where one or more of the first octets defines the network segment and the remaining octets define the computers or devices (referred to as *hosts*) on that network. Again, using the analogy of a street address, you can think of the network as being like the street (for example, 5th St.) and the hosts as being like houses on that street (for example, 121 5th St., 122 5th St.). Each octet ranges from a minimum of 1 to a maximum of 255. The maximum number is 255 because it is the largest eight-digit binary number (11111111). Certain numbers, such as 0 and 255, are reserved for certain types of communication.

When IPv4 was first developed, it was thought to be sufficient enough to allow for future expansion. After all, the total number of combinations is more than 4 trillion—although, due to disallowed combinations, as noted, the actual number of usable addresses is in the billions. Even so, that seemed like a lot of addresses. But this was before the Internet exploded and every business office, nearly every home in the developed world, and every **smartphone** needed an IP address. As a result, address space quickly became scarce. Technologies such as **network address translation (NAT)**, a method to mask the address of devices on a network from the outside world, and **Dynamic Host Configuration Protocol (DHCP)**, used to automatically assign IP addresses to devices as needed, work to mitigate the address shortage but both proved to be only stopgap measures given the quick adoption of IP-enabled devices. The real answer to the problem was IPv6.

IPv6

Unlike IPv4, whose 32-bit addresses yielded about 4 billion usable addresses, **Internet Protocol version 6 (IPv6)**, a 128-bit address scheme, allows more than 3.5 undecillion addresses. That's a number with 38 zeroes after it. Given this very large number, it's a good bet that address space will no longer be an issue no matter how many network-connected smart **Internet of Things (IoT)** devices come along. There are also other benefits to IPv6, including the following:

- Auto-configuration
- Improved address management
- Built-in security/encryption capability
- Optimized routing

An IPv6 address contains eight fields of four-digit hexadecimal (base 16) numbers (0 to 9 and a to f) as follows:

    2051:0011:13A2:0000:0000:03b2:000a:19aa

The notation also allows for a shorthand method that eliminates leading zeroes, with an all-zero field represented by a single zero:

    2051:11:13a2:0:0:3b2:a:19aa

Further, successive fields of all zeroes can be shortened with a double colon (::) as follows:

    2051:11:13a2::3b2:a:19aa

The changeover to IPv6 has been discussed for many years. Initial predictions in the late 1990s suggested the conversion would need to occur around 2002. However, as of this writing, the changeover is not yet complete, despite an explosion in the use of devices that has far exceeded any predictions in the early part of the century. IPv4 remains in wide use because converting to IPv6 requires massive upgrades that will cost millions and perhaps billions of dollars and because, so far, the workarounds seem to work. The conversion does seem to be slowly taking shape, however, as most new devices are IPv6 enabled. The tipping point of the conversion will likely be when most networking devices have gone through the natural refresh process over a 7 to 10-year period and are replaced with devices that can accommodate the new addressing scheme.

Data Link Layer

Layer 2, the Data Link Layer, has been dominated by the Ethernet protocol for the better part of 30 years. Used primarily in LANs, Ethernet is how switches move data between machines. Over the years, Ethernet has been expanded. It is used in massive data centers and large metro area networks, furthering its hold on Layer 2 just as IP has dominated Layer 3.

The Dominance of Switching

Switches have always been a critical part of networks. Recently, however, switching has claimed a dominant position in networking. Why? One reason is that switch vendors have added functionality to switches, including features such as Power over Ethernet (PoE). This enabled Voice over Internet Protocol (VoIP) phones to work without an additional power cord (they still needed an Ethernet cable, of course), and gave them Layer 3 and even Layer 4 forwarding capabilities. Improved performance drove ever-increasing speeds, from 1 gigabit (Gb) to 10 Gb to 40 Gb and even 100 Gb and 400 Gb Ethernet. These improvements weighed heavily as dedicated data centers came into prominence in enterprise networks. They have greatly accelerated with the advent of virtualization, where fast switching within massive virtualized data centers is a primary design criterion.

Unlike IP, which was meant to connect LANs with each other, Ethernet connects data between machines that are all on the same network. As such, one of the key components of Ethernet deals with preventing data collisions that occur when multiple computers transmit data at the same time on the same segment of the network.

Much in the way that routers are associated with Layer 3, switches rule Layer 2. A switch connects different network segments together via switch ports, each of which can have multiple machines (or other switches in a hierarchical network). When a data frame (the Layer 2 version of a packet) arrives on a switch port, the switch looks up the source and destination MAC address (a unique identifier assigned by manufacturers to any network-connected device) and makes a forwarding decision based on the rule set associated with the destination MAC address.

Unlike IP addresses, MAC addresses are physically assigned to every individual device that can connect to a network by the manufacturer as it rolls off the factory assembly line. MAC addresses are intended to be both unique and permanent. However, it is possible to spoof a MAC address, which—not surprisingly—has significant security implications.

Physical Layer

At the bottom of the stack, the OSI Reference Model specifies standards for the medium over which data is transmitted. While not often discussed in the same way as Layer 2 or Layer 3, the standards specified at Layer 1 are quite complex. This text will emphasize their importance because **wireless local area network (WLAN)** communication standards are specified here.

Layer 1 defines how the actual bits are transmitted between devices, including specifications on the following:

- **Transmission**—Bit-by-bit procedures
- **Electrical**—Signal levels, amplification, and attenuation
- **Mechanical**—Specifications for cables (type, length) and connectors

- **Procedural**—Modulation schemes, synchronization, signaling, and multiplexing
- **Wireless transmissions**—Frequencies, signal strength, and bandwidth
- **Throughput**—Bit rates
- **Topologies**—Bus, ring, mesh, point-to-point, and point-to-multipoint

Protocols defined at Layer 1 include the following:

- Synchronous Optical Networking (SONET)/Synchronous Digital Hierarchy (SDH)
- Digital subscriber line (DSL)
- T1/E1
- Integrated Services for Digital Networks (ISDN)
- Ethernet Physical Layer
- Bluetooth
- 802.11 (WLAN)

From Wired to Wireless

The networking industry began to grow in the 1980s and exploded in the 1990s, with the culmination of affordable PCs and the growing popularity of the Internet and the World Wide Web. Already on an incredible trajectory in the late 1990s, the industry benefited from another boost in the form of wireless networking—and there was another even bigger boost to come.

This "second wave" of networking was initially made possible as a result of a groundbreaking decision by the U.S. regulatory body in charge of telecommunication rules, the **Federal Communications Commission (FCC)**—the opening of several bands (contiguous ranges of radio frequencies) of the radio spectrum for unlicensed use in 1985. This was a big change, given that apart from ham radio, which was valued as a nationwide emergency communication system, the radio spectrum was a tightly controlled government asset that required licensed approval for use. This visionary decision (not a phrase often associated with a government regulatory body) had a profound effect on networking as well as on several other industries.

The frequency bands in question—900 MHz, 2.4 GHz, and 5.8 GHz—had previously been reserved for things such as microwave ovens, among others. The FCC's decision allowed anyone to use these bands (or any company to build a product that used these bands) as long as they managed interference with other devices. This made products such as cordless phones and remote-controlled ceiling fans—and, later, wireless networks—possible.

At first glance, it's hard to see exactly why wireless had such a huge impact. At the time, wireless performance was not that great. In fact, compared to a hard-wired Ethernet connection, it was pretty lousy. As it turned out, though, users were far more interested in convenience than performance—at least initially. Before the advent of WLAN, if you wanted to connect to a network, you had to go to where the computer was tethered to an Ethernet port. Or, if you had a laptop (these were also becoming cheaper), you had to go to where the connection port was. This may not seem like a big deal, but "going to the computer" meant leaving where you were and dropping what you were doing.

Wireless networking changed all that. With WLAN, you brought your computer to where you wanted to be and connected to the network from there. The ability to connect in a meeting room or on your couch far outweighed the slower connection speed, especially since there were very few high-speed network applications at the time. (Streaming media meant waiting 5 or 10 minutes to download a single song, for example.) This convenience factor created a massive surge in WLAN usage. In response, manufacturers poured millions into research and development (R&D), which improved performance, which in turn attracted more users.

The first generation of WLAN operated at about 500 kilobits per second (Kbps) on an unlicensed frequency band. In this case, "unlicensed" meant that anyone could use it. It was not restricted or reserved for commercial or government use as long as the transmission power was kept low. The second-generation boosted performance to 2 megabits per second (Mbps), a 400-percent improvement. (Note that the term "generation" is used here in the generic sense rather than as a name, as it is when describing mobile network technology.)

In 1990, the IEEE established a working group to create a standard for WLANs. In 1997, the IEEE 802.11 standard was ratified, specifying the use of the 2.4 GHz band with data rates of up to 2 Gbps. Different versions of the 802.11 standard were developed in subsequent years and were noted via extensions such as a, b, g, and n. Notationally, this would appear as "802.11b," for example.

Outside the enterprise workplace, mobile data communication was becoming commonplace. The combination of affordable powerful laptop computers combined with the availability of affordable, easy-to-configure WLAN routers conditioned Internet users to expect wireless connectivity. This was reinforced when local hotspots sprang up in shopping malls, cafés, restaurants, bars, airports, and even sports stadiums. Home users rushed to buy WLAN access points and routers, as this enabled them to create a WLAN for all devices to connect throughout the household. The fact that one could easily build a WLAN that covered the entire home without having to run or hide cables was a huge selling point to a world that had begun to expect Internet access. Despite lower data rates as compared to wired access, people were willing to trade data rate for broader connectivity options.

In addition to performance issues, WLANs presented new security risks. It was perhaps not surprising then, that when WLAN vendors tried to push into enterprise markets, information technology (IT) security and network managers were less than excited at the prospect of using them. In fact, most were very much against it. The problem was, users were beginning to demand access from anywhere in the office, especially in conference rooms. When IT managers said they would not support wireless connections, many people simply connected their own WLAN routers to an Ethernet port in conference rooms and other locations creating their own rogue access points. This was a huge problem for IT, which rightly set harsh rules against it.

In the end, however, it was too much to fight. User demand simply overwhelmed IT departments' resolve. Consequently, after 2005, businesses gradually started to roll out WLANs in areas where temporary network connections were a convenience rather than a necessity. These areas—reception areas, meeting rooms, cafeterias, and recreational facilities—could be supplied by wireless access points. From a purely functional perspective, this proved to be an ideal use of the technology, as people using their portable devices in those areas would typically not require high throughput, anyway. Rather, they would more likely than not just be checking email or using an instant message application.

On the horizon, though, an even more disruptive technology was rolling in: 3G Mobile IP broadband. This new technology promised to have an even larger footprint—one that could truly be described as "ubiquitous mobile data access."

Business Challenges Addressed by Wireless Networking

Wireless networking addresses several challenges thanks to its inherent ability to allow network access without the hindrance of cables. The most obvious benefit is that areas considered for WLAN deployment do not require a cable run to each desk or print station. This offered major savings in time and effort. Moving cables and activating the Ethernet ports is a considerable burden in any installation, office move, or reshuffle. With WLAN being predominantly wireless, there is a greatly reduced burden on cabling devices because only the backhaul from the access point to the network may need to be cabled. As a result, you can put desks and network printers wherever you want. There is also significant cost savings, particularly in new installations where fewer Ethernet cables are being run to each desk or work space.

> **NOTE**
> A Wi-Fi device is generally any device based on the Institute of Electrical and Electronics Engineers (IEEE) 802.11 standard. But the term is often used to describe any WLAN-capable device (most of which are 802.11 compatible). The terms Wi-Fi and WLAN are often used interchangeably. The IEEE develops global standards for a wide range of technologies. The 802 standard pertains to LANs and the ".11" extensions define standards for WLAN.

The Economic Impact of Wireless Networking

To understand how wireless became available almost everywhere, you must understand the desire for mobility. It is no longer acceptable to be located at a fixed network point of access. Nowadays, consumers demand the right to move around—to be mobile—so much so that for the average person, it's getting harder to find a networking device that will physically connect at all.

The first wave of the wireless revolution included early adopters who provided fixed-line PC-based Internet services at cafés and shopping malls. Their fee-based services, which were fraught with security issues—typically keyboard sniffers to capture passwords and banking details—soon succumbed to free public wireless broadband offerings.

Soon, it became common to have high-speed asymmetric digital subscriber line (ADSL), and broadband fixed-line networks in residential properties. To reduce churn, Internet service providers (ISPs) often supplied Wi-Fi–enabled gateways to consumers. *Churn* is the movement of a subscriber from one network to another and is one of the **key performance indicators (KPIs)** most relevant to overall performance. These measures helped Wi-Fi gain traction in homes and small businesses, but they alone cannot account for the fact that in 2005, wireless broadband accounted for productivity gains of $28 billion and the trend has continued.

Wireless Networking and the Way People Work

It's somewhat difficult to separate the business impact of Wi-Fi from its economic impact because they are intrinsically tied together. However, this section will focus on one key point: the way in which wireless has changed how companies work. As previously mentioned, wireless networking has changed how people work in the office—particularly **knowledge workers** (that is, professionals whose jobs involve the use and manipulation of data). These were the first wave of employees to use computers and, later, laptops. These workers found it very useful to be able to connect to the network from anywhere in the office, such as meeting rooms, particularly given the collaborative nature of their work.

This was an important change, of course. But there were some industries for which wireless fundamentally changed the way they did business. A few of them are discussed here.

Health Care

One of the first industries to adopt wireless technology was health care, and it has profoundly changed how hospitals work and how doctors and nurses interact with patients. In 2005, productivity improvements from the use of mobile wireless broadband solutions across the U.S. health care industry were valued at almost $6.9 billion. Today, wireless technology productivity is almost impossible to carve out because it's woven into the very fabric of modern health care.

Health care is one of the most labor-intensive industries, as well as one of the most sensitive to keeping personal data private. Nonetheless, health care has managed to use wireless technology to improve communication, as well as provide instant access to information portals for diagnosis and general health information such as diets and lifestyle plans. However, these improvements in efficiency alone don't explain the savings in health care. Over $1 billion of inventory loss avoidance is achieved through the use of wireless tagging or radio-frequency identification (RFID).

One of the obvious improvements that wireless brings to the health care industry is that it allows modern medical devices and monitors to be moved with patients as they travel to different places in the hospital—from their room to a lab, to pre- and post-op, to recovery, and back to their room. Before the availability of wireless, there was a constant need to disconnect, move, and reconnect these devices and monitors, which presented numerous opportunities for errors or breakdowns. Cables, connectors, and ports break down with extended use—and this is compounded when they are used next to hospital beds because of the frequent need to adjust the bed's height and position (from sitting to prone, for example). The result was often pinched or severed cables, many of which went unnoticed for some time, leaving patients unmonitored.

Another benefit to patients, health care professionals, and insurance companies is that wireless technology allows for real-time data at the point of care. This has a couple of significant implications. First, it means the patient can always be monitored, which improves staff reaction time to alerts—a potential lifesaver. Second, it allows for real-time access to patients' files. This is profound. Prior to wireless technology, patients often had a paper chart at the foot of their bed, which a doctor or nurse would look at prior to administering care or running a test. These charts were highly prone to errors—from lost or missing pages, to misinterpreted handwriting, to out-of-date or incomplete information. With wireless, caregivers now have real-time access to patients' digital records, where information is backed up and complete. This also allows for features such as auto-alerts, which help eliminate errors with treatment, incompatible prescriptions, or incorrect dosages. These are lifesaving improvements. Of course, many important security considerations go along with using wireless technology in this manner.

A less dramatic use of wireless in health care—but nonetheless an important one—is that most hospitals offer free Wi-Fi to patients and visitors. In this way, hospitals have greatly improved the satisfaction of patients and visitors alike. For all the drama of a medical emergency, the reality is that the vast majority of the time in a hospital for both visitors and patients is spent sitting around waiting. Offering free access to the Internet provides the perfect distraction.

Warehousing and Logistics

Seemingly at the opposite end of the technology spectrum from health care is the warehousing and logistics industry. Nevertheless, wireless networking has also made a significant impact here.

Prior to the advent of wireless networking, warehouse personnel logged storage locations on paper and used written notes to retrieve them. This method was fraught with errors and inefficiencies, especially in bigger warehouses, which can be as large as 1 million square feet.

Wireless networking allows for much greater efficiency throughout the entire process, from receiving, to shelving, to picking (retrieval), to outbound distribution. Specifically, wireless networking helps with the following:

- **Asset tracking**—With wireless networking, companies can automatically track assets in real time, providing a major productivity gain over manual, semiannual inventory checks.
- **Picking efficiency**—Retrieving inventory for distribution (picking the item off the warehouse floor) used to be a long process. Often, workers walked (or rode) several hundred yards through a warehouse, only to arrive and not find the item. The worker would then have to walk (or ride) all the way back, try to find the error, and repeat the process. With wireless technology, location accuracy is much improved, as is the picking process. This saves a great deal of time and money.
- **Loss control**—Another Wi-Fi automation success has been the savings gained by reducing inventory loss. In the first edition of this text, the authors were able to cite industry studies on the estimated savings through the use of wireless. Since then, "smart inventory" systems (all based on wireless technologies) have become so prevalent that it's difficult to parse out specific benefits of wireless as a standalone aspect any more than one could have estimated the economic impact of Ethernet cables.

Retail

The retail industry has also taken great advantage of wireless technology. As with the back end of the business, discussed in the preceding section, the front end of the business has also changed as a result of wireless networking. Specifically, wireless has resulted in the following key changes:

- **Inventory counts**—Retailers always want the most popular products on hand, but they must carefully manage their inventory to avoid overstocking. Accurate inventory counts and direct front-of-store and point-of-customer interaction can put warehouse orders into motion in real time.
- **Customer satisfaction**—Retailers spend a great deal of money getting shoppers into their stores. Once shoppers are there, retailers must do their best to convert them into

buyers. This can be difficult, however, in an industry with a transient workforce, especially during those critical shopping seasons when a retail business can be made (or ruined). Arming sales staff with wireless devices turns even brand-new employees into product experts by allowing them to verify back-of-store inventory, suggest popular merchandise tie-ins or up-sells, or check with other local outlets for popular items.

General Business and Knowledge Workers

While it's true that specific industries have taken advantage of wireless networking, the biggest impact of wireless networking has been the way it has fundamentally changed how, where, and when people work. Before the wide adoption of wireless networking, people tended to work mostly in the office. When workers did work outside the office, either after hours or while traveling, they were often limited to offline work. Alternatively, if they went through emails, for example, they submitted work in batches.

With wireless, of course, there is greater flexibility with regard to where you can work—whether it's on the deck of a beach house, at a coffee shop, in an airport terminal, or on an airplane traveling 500 miles per hour, 30,000 feet in the air. As a result, worker productivity has gone up, and continues to rise.

This is great for many, but it does come at a cost. Perhaps most significant is the fact that the line between "work" and "not work" has blurred to the point that it's hard to distinguish where work ends and one's personal life begins. More and more, it seems that businesses expect their employees to be available and checking emails late at night, on weekends, and on vacations. More on topic for this discussion is that using a portable work device to access a network via a public Wi-Fi connection opens a vast array of potential security vulnerabilities that must be accounted for.

Nevertheless, Wi-Fi has brought about considerable increases in savings and overall efficiency. Its impact on the economy was considerable—even in its fledgling state, back in the early 2000s, before the advent of high-speed devices and mobile web applications. Even then, those who embraced the technology found that it delivered major benefits to their businesses.

The Wi-Fi Market

In 2017, the worldwide WLAN market was estimated to be $5.9 billion per year and is projected to be as much as $15 billion per year in 2022. About half of that was attributed to growth in the enterprise space, as fewer and fewer office spaces even bother installing Ethernet ports for users because most are connecting to the network over wireless most, if not all, of the time. However, this annual spending on WLAN equipment is only part of the story. As noted, Wi-Fi technology has changed the way many organizations do business. Indeed, it has created whole new business models.

Coffee shops, bookstores, and cafés have embraced wireless connectivity in two waves. The first wave involved providing wireless access to a private LAN that charged for a one-time use or for a subscription. Many such businesses felt that this was a nice add-on feature and an opportunity for revenue. What's interesting, however, is that due to the reduction in price of Wi-Fi equipment, many businesses have converted to providing free Wi-Fi access, choosing to forego the additional revenue in favor of attracting more customers, whom they welcome to come in and stay. In other words, they encourage clients to actually use their stores as an office, for three reasons:

- Those customers tend to buy other goods and services while there
- It creates a regular and loyal customer base
- Customers simply expect it to be available

Hotels have also been transformed. Not surprisingly, providing Internet access to guests has become essential. Perhaps the biggest impact of wireless technology on the hotel industry is that it has drastically lowered the cost of providing Internet access. This is particularly important in situations where retrofitting an older hotel to offer a wired network would be prohibitively expensive. If not for wireless technology, many hotels would have to choose between a very expensive upgrade and the potential loss of revenue. With wireless technology, Internet access can be provided at a much lower cost. This also has an enormous impact on hosting conferences and events, which is a major component of hotel revenues.

Much like cafés, some hotels offer free network access, especially in the lobby, where they encourage users—especially business travelers—to meet. Interestingly, however, some hotels still charge for "high-speed" or "premium" access in rooms but nearly all offer some level of free Wi-Fi to their guests. Wireless connectivity is simply an expectation as much as running water is.

IP Mobility

The number of mobile wireless devices now far exceeds the number of fixed devices. The growth and adoption rates of smartphones and tablets have created a huge demand on mobile operator data networks, with data traffic rates growing upward of 115 percent compounded per year since 2011. The public's and business's adoption of smartphones and tablets has been so fast that manufacturers of fixed PCs and desktop computers have either shifted to laptops, tablets, or servers, or are out of the market entirely.

The shift toward wireless mobile devices has presented businesses with many opportunities and challenges—most notably, the challenge of how to make best use of new mobile wireless technologies. After all, networks have been designed and secured with static devices in mind. When LANs were designed, it was assumed that employees would be at a desk within a department. The network was segmented accordingly via **subnets** to accommodate physically present numbers of employees and allow for future growth. The emergence of WLAN technology was used to address any unexpected growth. However, the growth in the number of IP-capable wireless devices means that employees are now far more mobile and can work from anywhere in the network or even from outside the network—at home or at a client's site.

This has proved to be very productive for business and has created tremendous improvements in employee efficiencies and communications. Laptops, smartphones, and tablets can be used in any location where there is a WLAN or 4G/5G network connection. These devices can also roam around the workplace LAN, connecting to the WLAN wherever there is a signal. If the WLAN is one single subnet, users can maintain application and web browser sessions.

The ability to roam and maintain an IP session is fundamental to true IP mobility. Ideally, the wireless device must not only be usable in any location, but it should also be usable when in transit between locations and even between IP and mobile networks. This presents a significant problem—when moving from one network or subnet to another, the device will require a change of IP address. However, if the IP address of the mobile device changes, all its current sessions will be lost, and applications will hang and crash.

What is required is a method to allow the seamless transfer of an IP address from one network to another without losing IP sessions. Only then will there be true mobility with roaming using IP wireless devices. This is termed *IP mobility*. The International Engineering Task Force (IETF) uses the term **Mobile IP** to describe its standard communications protocol for addressing this problem. It does so by preserving existing sessions as a device moves to a network with a different IP address space. Because this function is performed at the Network Layer of the OSI Reference Model rather than at the Physical Layer, a device can span different types of wireless and wired networks while maintaining connections and application sessions.

Another goal for the Mobile IP standard is for a device to be able to cross not just network boundaries but technologies as well. Ideally, the device should transparently connect to any technology it can support including wired, wireless, and 4G/WiMAX networks.

In a nutshell, with IP mobility, any compatible device that communicates at the Network Layer can roam from a fixed Ethernet to a wireless Ethernet to a mobile (cell) network without any loss of session and only a noticeable change in the access speeds, if that. There is no need to restart or reboot the operating system (OS) because the Network Layer handles it all seamlessly.

Mobile IP handles the change of IP address and maintains current sessions by using certain Mobile IP client stack specific components. These are as follows:

- **Mobile node (MN)**—This is a device (it could be anything) that changes its point of attachment from one subnet or network to another. It does its own move detection and must determine not just the change in access type, if any, but also the change in the subnet.
- **Home address**—This refers to the mobile node's home IP address, which is where it is registered with the home agent. The address can be static or dynamically assigned when registering with the home agent.
- **Home agent (HA)**—This is a router capable of processing and tracking mobile routing IP updates, tracking mobile node registrations, and forwarding traffic to mobile nodes on visited networks through IP tunnels.
- **Care-of-address (CoA)**—This is the new IP address the mobile node has been assigned by the visited network. The mobile node informs the HA of the CoA when registering its movement.
- **Foreign agent (FA)**—This stores all information about mobile nodes that are visiting its network. It advertises CoAs and routing services to the MN while it is visiting its network. If there is no FA present on a network, then the MN itself must handle getting a local address and advertising it.

Mobile IP enables a wireless device to traverse different network types—fixed, wireless, and cellular—while maintaining session and application status. It provides for transparent

FIGURE 1-3

How Mobile IP sessions are maintained as the user moves around.

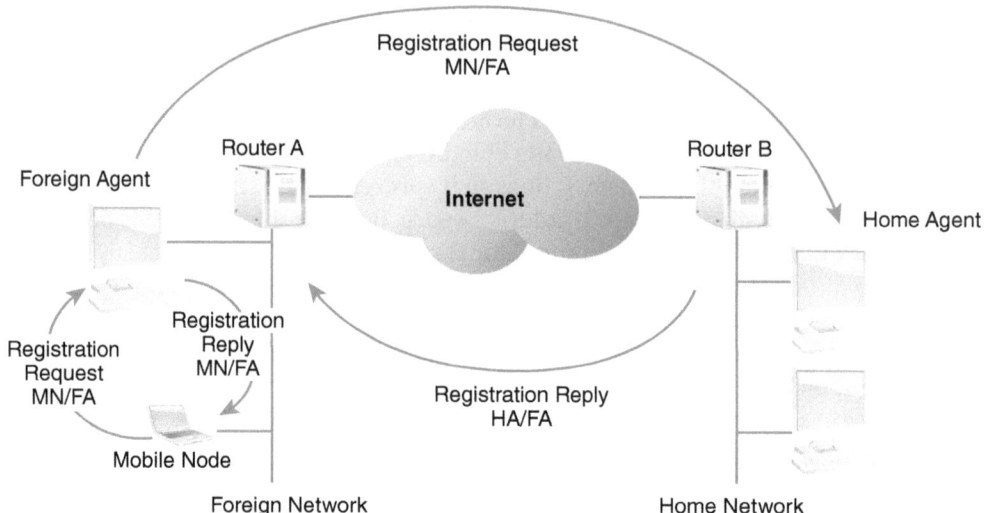

handover and supports different access types and IP subnets through the use of IP tunnels from the home network to visited networks. This not only enables wireless devices to work on different networks but it allows them to be seamlessly accommodated without any drop in service. This is true IP mobility. It facilitates real roaming of wireless devices in which the device reconfigures itself automatically and registers with another network type and IP address while the user works without any interruption. **FIGURE 1-3** shows how Mobile IP sessions are maintained as the user moves around.

The Internet of Things

In 2005, having wireless access in the workplace or home was still something of a novelty. Today, wireless networking is so prevalent that one can obtain wireless access on a transcontinental flight. Indeed, passengers will even complain to the flight attendant if the connection is slow or unstable.

One of the biggest technology trends of today is connecting all manner of the Internet and each other. Known as the Internet of Things (IoT), it is the interconnection of virtually any electronic device that can be controlled and optimized via automation or remotely via an Internet connection. With IoT, most people's interactions will be via smart homes and electronics, allowing energy-saving programming, automated lighting, safety programming, and home entertainment. Unlike obtaining broadband access on a plane, which is likely a novel but infrequent experience for most, IoT is dramatically shaping the way people live and interact with their home environment and electronic devices. IoT will also extend to enterprise and industrial settings, and even into implanted medical devices such as heart monitors.

IoT has long been available through wired networks, but adoption among the general public has been slow and limited to those in high-end homes due to the costs associated with wiring and, in older homes, retrofitting. This key barrier has been removed in recent years due to the reduction in the size and cost of wireless technology.

As IoT continues to ramp over the next several years, people will truly live in a wireless world, where everything is connected and remotely controllable. And although this will have a great many benefits with regard to safety, energy efficiency, and convenience, it will also open a whole new world of security threats and vulnerabilities. It's one thing to have your computer files destroyed due to a virus or other malware, or to have your credit card run up by a cyber thief. It's another thing altogether to have a hacker lock you out of your own home, ransom control of your heating system in the dead of winter, or take control of a moving vehicle (while you are in it). For those in the information security business, this will be another front in the cybersecurity war that sees no end in sight.

CHAPTER SUMMARY

Data communication and networking have a long rich history—the advancements from telegraphy to the near universal use of the PSTN seemed to happen at a steady and measured pace over the course of 60 years or so. However, with the advent of packet switching, which enabled multiple transmissions to share a single circuit and the creation of ARPANET and the Internet, the pace of networking and communication innovation accelerated even beyond the wildest expectations of the most enthusiastic futurist of the 1980s.

Initially, networks were wired; but with advancements in mobile telephony came the development of the WLAN, which could be accessed by mobile users.

Wireless networking has changed the world in many ways. It has altered entire industries and brought productivity gains even to sectors that had seemingly squeezed out as much productivity as possible. More than that, wireless networking has changed not only how and where people work but also the very relationship between employees and employers. Wireless technology has brought work into people's homes in a way that previously did not exist.

In a macro sense, wireless technology has made it easier to bring Internet access to people and places that for too long have been on the wrong side of the digital divide. This extended access improves not only the lives of those directly affected but also the surrounding community and, to some degree, the world at large. As they say, a rising tide lifts all boats.

The rise in the use of mobile devices, such as smartphones and tablets, has fueled demand for wireless networks. It has likewise presented businesses with many opportunities and challenges. Mobility has become a way of life, with smartphones and tablets part of the fabric of modern society. For the generation born in the Internet era, to be denied this mobility is unthinkable.

But all of this easy and near constant access, for all the good it has done, has also exposed and even introduced many security vulnerabilities. Vulnerabilities that impact individuals, companies (both large and small), and even nation states. Fortunately, there are many people like you who are interested in mitigating those vulnerabilities so that the benefits can be fully felt by all.

KEY CONCEPTS AND TERMS

Application Layer
Bluetooth
Circuit switching
Data Link Layer
Dotted decimal
Dynamic Host Configuration Protocol (DHCP)
Federal Communications Commission (FCC)
Internet of Things (IoT)
Internet Protocol version 4 (IPv4)
Internet Protocol version 6 (IPv6)
IP addressing
Key performance indicators (KPIs)
Knowledge workers
Local area network (LAN)
Media Access Control (MAC) address
Mobile IP
Modem
Network address translation (NAT)
Network effect
Network Layer
Open Systems Interconnection (OSI) Reference Model
Packet switching
Physical Layer
Presentation Layer
Public switched telephone network (PSTN)
Session Layer
Smartphone
Stack
Subnets
Telegraphy
Telephony
Transmission Control Protocol/Internet Protocol (TCP/IP)
Transport Layer
Wi-Fi
Wireless local area network (WLAN)
Worldwide Interoperability for Microwave Access (WiMAX)

CHAPTER 1 ASSESSMENT

1. Digital communication offers which of the following advantages?
   A. More efficient use of bandwidth
   B. Greater utilization
   C. Improved error rates
   D. Less susceptibility to noise and interference
   E. All of the above

2. ARPANET was the predecessor of the modern Internet.
   A. True
   B. False

3. Wireless networking was initially supported by IT departments because of the productivity gains it provided.
   A. True
   B. False

4. Mobile IP solves which important problem?
   A. Battery life
   B. Wireless connections to the Internet
   C. Access to app stores
   D. The ability to maintain an IP session while moving

5. Which of the following is *not* a requirement for successful mobility?
   A. Location discovery
   B. Movement detection
   C. Update signaling
   D. Omnidirectional antennas
   E. Path establishment

6. Switches primarily operate at which layer of the OSI Reference Model?
   A. Physical Layer
   B. Data Link Layer
   C. Network Layer
   D. Transport Layer
   E. None of the above

7. Wireless networking standards are defined at the Network Layer.
   A. True
   B. False

8. Layers 4 to 7 are often grouped together and referred to as the "Application Layers."
   A. True
   B. False

9. IP addressing is specified in which layer?
   A. Layer 1
   B. Layer 2
   C. Layer 3
   D. Layer 4
   E. All of the above

10. The Data Link Layer uses a logical addressing scheme to switch data frames.
    A. True
    B. False

11. Which of the following is *not* a use of Wi-Fi in warehousing?
    A. Asset tracking
    B. Loss control
    C. Forklift automation
    D. Picking efficiency

CHAPTER 2

The Mobile Revolution

THIS CHAPTER TAKES A HISTORICAL LOOK AT MOBILE NETWORKS, smartphones, and other mobile devices. With this understanding, you'll be better able to grasp the security issues related or specific to mobile networks and devices.

Over the last 30 years, the advances in mobility have been significant. Evolving from clunky analog phones that were little more than novelty status symbols (and poor communication devices) to business "smart devices" that people can't live without, these devices and the systems that support them have changed how people live, work, and interact. For the security professional, however, mobility represents a new and complex set of challenges.

Chapter 2 Topics

This chapter covers the following concepts and topics:

- How early cellular or mobile devices operated
- How mobile networks evolved
- What the effects of the BlackBerry were
- What the economic impact of mobility has been
- What the business impact of mobility has been
- What some business use cases for mobility are

Chapter 2 Goals

When you complete this chapter, you will be able to:

- Describe basic cellular design
- Provide examples of frequency sharing techniques
- List the main considerations in cellular network design
- Describe the security issues and concerns with both 3G and 4G systems
- Provide examples of business uses for Mobile IP and smart devices

© Cherezoff/Shutterstock

Introduction to Cellular (Mobile Communication)

One of the greatest accomplishments of the 20th century was the rollout of the public switched telephone network (PSTN)—not only because of the technology itself but also because of its ubiquitous reach. With the PSTN, nearly every home in the developed world (and a high percentage of homes even in some undeveloped areas) had a wired communication channel that connected it to the rest of the world, providing a lifeline in times of trouble. The system even provided its own power, keeping the communication channel open when the lights went out. Just imagine the scope and cost of running and maintaining a wired connection in the United States alone, with approximately 125 million homes and 20 million apartments. (This does not even include every business and office.)

In the early 1990s, telephony was extended beyond the limitations of wired connections with the emergence of the first mobile, or **cellular**, phones. Cellular is a generic term for mobile phone systems or devices. It refers to the portioning of frequency coverage maps, discussed shortly. Initially viewed as a perk for high-powered executives and a status symbol for young professionals, cellular phones caught on fast. As their popularity rose, technology companies poured hundreds of millions of dollars into research and development, and the pace of innovation took off.

The first-generation cellular phones in the 1990s had limited range and coverage, short battery life, and poor voice quality. Even so, people clearly saw the benefit of having a phone that could travel with them—although few considered their mobile phone to be their primary phone. Flash forward just 30 years, and mobile phones are now viewed as an essential part of people's lives. In some cases, they are the predominate means by which people interact with the world.

The expansion has been impressive. In the United States, 90 percent of adults now own a mobile phone. Ownership in the 18–29-year-old group is 98 percent. In addition, more and more teens and even preteens have their own smartphones. More impressive is that these statistics transcend gender, race, and income categories. The most amazing aspect of this phenomenon, however, is the growing number of mobile phone users who have disconnected their landlines—something that was unthinkable even 15 years ago.

Mobile phones use all the principles of two-way radio communication that have been around since the early 20th century. Well-known problems such as range, power, signal-to-noise ratios, and interference all come into play. This keeps a lot of radio frequency (RF) engineers gainfully employed. Mobile telephony, however, presents some unique challenges. Indeed, one was so critical to making mobile telephony feasible that the solution to the problem became the name that now describes the entire system: cellular. The problem stems from the fact that in a mobile telephony system, there are far more users than there are available frequency channels over which to communicate. This is a two-part problem; this discussion will begin with the physical distribution of channels, or frequency bands.

Cellular Coverage Maps

One of the limitations of cellular technology is the transmission power of the phone. Because the phone is battery-powered, and because battery life is a big consideration, transmission power must be kept low. (There are health considerations as well. You don't really want a high-powered transmitter pressed to your head for several hours a day!) However, low-transmission power limits the signal range, which means you need to have a receiver nearby.

The solution was to create a coverage map of small geographic sectors, or *cells*, each with its own antenna tower. Two separate teams of engineers from Bell Labs, 20 years apart, conceived and then perfected the idea of using hexagonal (six-sided) cells. This mapping provides the best coverage, leaving no gaps in the coverage plan. This was referred to as a *cellular design* and was so critical to the design of the system that term *cell phone* came into being.

In each cell, there is an antenna array called a **base transceiver station (BTS)**, which communicates directly with the subscriber phones within its coverage area. Usually perched on a tall metal structure, these antenna arrays came to be known as **cell towers**, or simply *towers*. In some places, local ordinances require towers to be camouflaged. As a result, many look like tall trees and are easy to miss if you are not looking closely (which is exactly the point). A mobile phone communicates with the tower. The tower, in turn, communicates over a backhaul circuit either originally on fixed-line **T1/E1** trunks (T1/E1 are the standard digital carrier signals that transmit both voice and data) or on point-to-point microwave links to a **base controller station (BCS)**, which connects to the core network. Today, with 4G/5G networks, the backhaul is typically over high-speed fiber backhaul in urban areas and microwave links in remote locations. Typically, multiple towers will connect to a single BCS. The core network links all BCSs so that calls can be established over the local cellular network. It also has connections via gateways to the PSTN and, more recently, to the Internet.

Bell Labs

For most of its existence in the 20th century, the PSTN was a monopoly service delivered primarily by the Bell Telephone System (which became AT&T), often referred to as "Ma Bell" because it eventually spawned many smaller regional providers called "Baby Bells." One common criticism of monopolies is that they stall innovation due to the lack of competition. However, this did not seem to be the case with Bell; its engineering division, Bell Labs, had a remarkable 70-plus-year run of technology breakthroughs and innovations. These include, among other things, the first operational transistor, the first binary digital computer, the first transatlantic phone call, the development of UNIX operating system, and the development of both the C and C++ programming languages.

Cellular design fixed the phone transmission power problem. However, this created a frequency interference problem. As shown in **FIGURE 2-1**, if each cell uses the same sets of frequencies, then users in two different cells on the same channel interfere with each other.

The solution to the interference problem was to split up the frequencies to prevent interference from adjacent cells. With this pattern, interference is greatly reduced (see **FIGURE 2-2**).

Taking a step back and looking at the repeating pattern, you can see the genius behind the concept of cellular and **frequency reuse** patterns (that is, the practice of assigning multiple users to the same frequency channel, achieved by the physical separation and power

FIGURE 2-1

Adjacent cells using the same frequency will interfere with each other, especially near the cell borders.

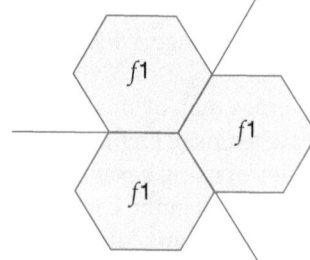

Frequency Reuse 1

FIGURE 2-2

By segmenting frequency use, interference can be greatly reduced or avoided.

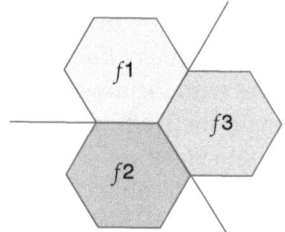

Frequency Reuse 3

FIGURE 2-3

A basic frequency reuse pattern on a large scale. Note that no cell is adjacent to another that uses the same sets of frequencies.

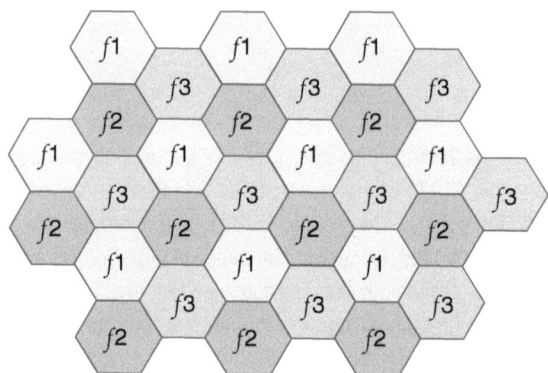

management of the transmission streams). This is a simplified view, however. Radio-frequency planning requires more than just creating areas that roughly correspond to the hexagonal cell pattern, because the distribution and density of potential subscribers is not likely to be uniform. Therefore, large cells called **macrocells** (that is, cells within a mobile system for large coverage areas) are needed for rural areas. **Microcells** (cells within a mobile system for small coverage areas) are needed for urban areas. **Picocells** (small hotspot cells offering Wi-Fi connectivity via a mobile carrier) are needed for dense urban areas. This ensures sufficient capacity per cell or area (see **FIGURE 2-3**). Picocells are now often being used inside buildings and large venues as most cellular communication happens inside of buildings where signals from external cell towers are often attenuated. However, picocells are alternatively deployed at the other end of the scale in remote rural areas when a home or office is out of range of the cellular network.

Frequency Sharing

Another challenge with cellular phones is the limitation of frequency channels. For example, the first cellular system rolled out in the United States had only 830 usable channels—not many at all. This limitation was compounded by the fact that frequency reuse patterns reduced the number of channels in any one cell to about 280 channels per cell. Even in the early days of cellular telephony, this small number of channels was not nearly enough to meet demand. There were solutions to this problem, but all of them were based on the concept of allowing multiple access—either through frequencies, time, or code division.

Frequency Division Multiple Access

Frequency Division Multiple Access (FDMA) is the foundation of cellular coverage maps, but in this case, each channel is split up further so that multiple users can share a common channel without interference. FDMA does not require a great deal of timing synchronization, but it does require very precise transmission and receiving filters. FDMA frequencies are assigned for the length of the communication, the downside being that unused channels sit idle. FDMA is a 1G technology, and is still common in satellite communications (see **FIGURE 2-4**).

Time Division Multiple Access

Time Division Multiple Access (TDMA) allows multiple users on the same frequency channel, each with its own sliver of time. This works well in a voice conversation because a phone conversation between two people is mostly silence. That means there's a lot of "empty space" on a channel even when it's in use.

Channel efficiency was greatly improved through the use of voice-compression techniques. These employed intelligent algorithms that could turn speech into mathematical points on a graph. This allowed speech to be replicated with high fidelity (that is, it sounded like the real person on the receiving end) without ever sending the speech signal. As a result, a lot of conversations could be stuffed on to one frequency channel.

TDMA does not require high-performance filtering as FDMA does, but it does require very tight timing synchronization. TDMA helped bridge 1G technology to 2G and allowed for rapid subscriber expansion from the original analog cell systems to digital without expensive upgrades to the system itself (see **FIGURE 2-5**).

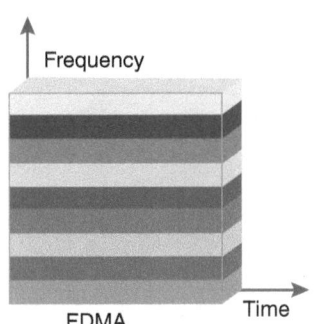

FIGURE 2-4

With FDMA, the frequency spectrum is divided among users.

FIGURE 2-5

With TDMA, each user is assigned a time slot so that packets from different communication sessions can occupy a shared frequency without interference.

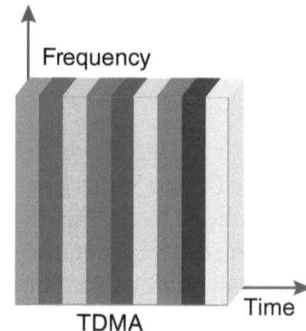

Code Division Multiple Access

Code Division Multiple Access (CDMA) makes it possible for several users to share multiple frequency bands at the same time by spreading the signal out over the frequencies. This spread-spectrum technique uses codes to distinguish between connections. The wide bandwidths and improved power usage greatly reduce interference, and the coding allows multiple users to occupy the same channel at the same time (see **FIGURE 2-6**).

CDMA is a 3G technology that improved the capacity of 1G systems by a factor of 18 and 2G systems by a factor of 6. However, because it relies on lower-powered signals, CDMA suffers from what is known as the *near–far problem*. This is when a receiver locks onto a strong signal from a nearby source, preventing it from detecting a wanted signal from a source that is farther away (and therefore weaker). Because CDMA has multiple signals on the same frequency, the near–far problem creates a frequency jam. This is a potential security issue from an availability standpoint, as would-be **hackers** could prevent communication via jamming. **FIGURE 2-7** shows all three types of basic cellular modulation—FDMA, TDMA, and CDMA—together.

Cellular Handoff

Because mobile phones are—obviously—mobile, cellular networks must be able to accommodate subscribers as they pass out of the range of one transmitter and into the area of another without losing the connection. This requires a controlled **handoff** from one base station to another. This is known as the *handover process*, and it occurs at the point when both neighbor frequency signals are at their lowest, usually at the border between two cells. If

FIGURE 2-6

With CDMA, communication is spread over multiple frequencies at the same time. Coding algorithms are used to spread and then reassemble the transmissions.

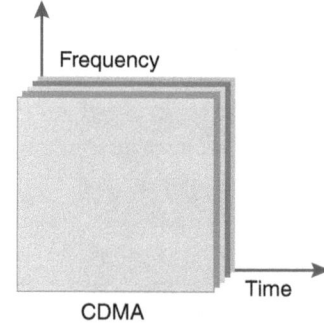

FIGURE 2-7
All three main types of basic cellular modulation—FDMA, TDMA, and CDMA—shown together.

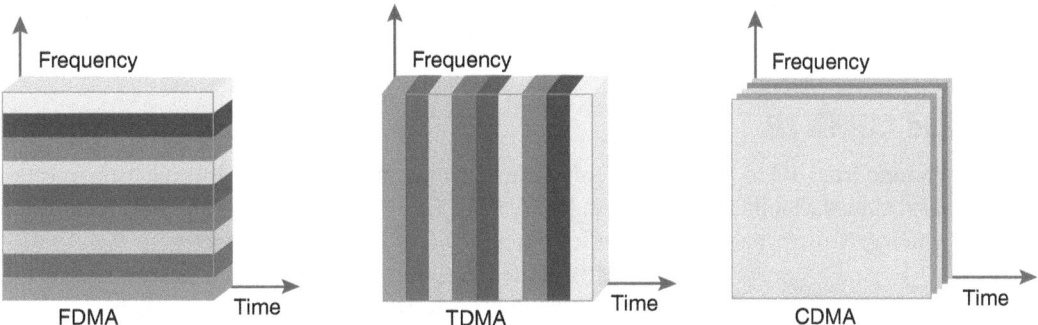

the handover process is designed correctly, the mobile phone can be passed back and forth repeatedly as the user remains on the border of two cells.

As noted, by overlaying a cell pattern on the coverage area, cells of different sizes can be planned to cater for population density and frequency reuse. The tower, in turn, will communicate over a backhaul circuit using either fixed-line T1/E1 trunks or point-to-point microwave links to the base controller station, which connects to the core network. The core network links all BTSs/BCSs so that calls can be established over the local network. It also has connections via gateways to the PSTN and to the Internet.

The Evolution of Mobile Networks

Mobile phone technology has been available to consumers for around 35 years, but there have been some amazing advancements in that time. Since the first limited commercial rollout in 1983, there have been four distinct generations of technology. These have gone from basic radio communication with a limited connection range and poor-quality voice to smartphones capable of managing high-quality voice while taking and sending a 7-megapixel picture with no noticeable drop in quality. This section reviews each generation of cell phone technology, looking at what it was, how it worked, and what the security implications were and are.

AMPS 1G

A commercial cellular system, called the **Advanced Mobile Phone System (AMPS)**, was deployed in North America in 1993. AMPS used analog signals to connect to cell towers, using FDMA for channel assignment. AMPS succeeded where previous attempts to create a commercial cellular service failed because of its ability to reuse frequencies (FDMA) and to hand off calls between cells in a relatively seamless way that did not involve the user.

The AMPS system was a commercial success despite serious performance issues. Call quality and reliability were nowhere near that of the PSTN, which limited its usefulness. In addition, FDMA, while considered a breakthrough, still consumed a lot of bandwidth per channel, which limited capacity. AMPS calls were also unencrypted, making it possible to eavesdrop on a call using a scanner. Finally, AMPS phones were relatively easy to clone, allowing non-subscribers to gain access to the service.

Although much-improved second-generation technology soon came along, carriers continued to support AMPS phones until 2002, when the older technology was finally phased out.

GSM and CDMA 2G

The big change from 1G to 2G was the conversion from analog to digital. Initially referred to as **Digital Advanced Mobile Phone System (D-AMPS)**, 2G cellular phones and networks used TDMA, which greatly improved bandwidth efficiency and subscriber capacity.

Unlike AMPS, which was essentially the same everywhere it was deployed, two distinct systems emerged for D-AMPS. The first of these was a TDMA-based second-generation technology developed in the late 1980s by an industry consortium consisting mostly of European companies. This technology was called **Groupe Spécial Mobile (GSM)**, although its name was later changed to **Global System for Mobile (GSM)**. The use of GSM was mandated throughout Europe to ensure continent-wide compatibility between countries.

> **NOTE**
>
> The "Generation" naming convention, such as 1G, 2G, 3G, and 4G, did not come into vogue until the 3G systems came online in the mid-2000s. Even then, the original technologies were referred to as AMPS and D-AMPS. The 1G and 2G designations have been retroactively assigned.

The second major 2G technology was CDMA, which refers to both the cellular system and the method of subscriber access. CDMA was the dominant 2G system used in the United States. While CDMA and GSM were not compatible, dual-system phones were eventually developed that could operate on either system.

In addition to offering more efficient use of bandwidth, 2G systems also used encryption, which greatly improved security. One of the downsides, however, was that the lower power requirements of digital systems meant that coverage was often poor outside populated areas, which had greater cell density. Another problem with digital was that unlike an analog signal, which degrades in a linear way, digital signals drop off completely when the signal strength falls below a certain threshold. When it's good, digital quality can be very good. But when it's bad, it's essentially unusable.

This 2G technology was the precursor to mobile data networks. The first of these was used for **Short Message Service (SMS)**, which introduced the world to texting. At first, SMS

> **FYI**
>
> One big breakthrough with GSM was the introduction of the **subscriber identity module (SIM)** card. A SIM card is a small, detachable smart card that fits into a standardized card slot on the phone. It contains all the subscriber's information, as well as their contacts list. Not only did SIM cards help to address the 1G system's cloning vulnerability, they also allowed users to switch phones without carrier involvement. This led to the emergence of third-party phone retailers who could (and did) sell phones directly to consumers.

did not seem like a compelling feature. But its use exploded with teens and young adults to the point where many used their phones only for texting. Eventually, subscription plans were created to accommodate these users.

GPRS and EDGE 2G+

Although GSM and CDMA were digital technologies and took advantage of multiple access techniques, both were still circuit-switched technologies, much in the way the PSTN was. **General Packet Radio Service (GPRS)** was the first packet-switching technology method that allowed data sharing over mobile networks. Still considered a 2G technology but often called 2G+ or 2.5G, GPRS allowed access to some websites—although data rates proved to be too slow for what was becoming a growing need and expectation.

EDGE, which AT&T rolled out in 2003, and which other carriers quickly offered, represented an enhancement over GPRS. It offered high data rates through better data encoding and (at that time) viable data access to many websites.

3G Technology

The third generation of mobile technology, called 3G, was the first generation specifically designed to accommodate both voice and data. Based on the **International Mobile Telecommunications-2000 (IMT-2000)** standards set by the **International Telecommunications Union (ITU)**, 3G can accommodate voice, data, and video.

The first 3G system was rolled out in Japan in 2001. In 2002, it was rolled out in many other parts of the world, including the United States and the European Union. However, implementation of 3G took longer than anticipated. This was in large part due to the need for expanded frequency licensing to accommodate higher bandwidth needs and rapidly increasing subscriber rates. By the end of 2007, there were 190 3G systems online in more than 40 countries worldwide.

The most noticeable improvement in 3G was its high-speed data rates. One enhancement to 3G was a mobile protocol called **High Speed Downlink Packet Access (HSDPA)**, which improved data rates to an impressive 14 Mbps. For the first time, the streaming of music and video to mobile devices was supported. Responding to this capability, many content providers created streaming offerings that catered specifically to mobile users.

FYI

The term "3G" started as an industry insider term—a catchall phrase for the many different technologies that adhered to the IMT-2000 standard. It became a common term, largely because it coincided with the explosion of smartphone users—mostly equipped with iPhones and Android phones. In an attempt to capture as much of the new market as possible, carriers invested in massive, aggressive marketing campaigns touting the superior performance and coverage of their "3G data networks." As a result, the term "3G" became widely adopted, even in consumer circles.

In addition to the security benefits of 2G, such as encryption, 3G systems also allowed for network authentication, which ensured that users connected to the correct network. On the negative side, smartphones that attached to 3G networks had far more personal-data capabilities—for example, access to bank accounts—as well as access to corporate systems and applications. With the growth in the number of users and an increase in the types of opportunities to exploit, 3G systems and smartphones soon attracted the attention of cybercriminals.

4G and LTE

Mobile telephony is in its fourth generation, called 4G, and the fifth generation, called 5G, is beginning its initial rollout as of this writing. Among other improvements, 4G is an all-IP network, allowing the use of ultra-broadband and the promise of 1 Gbps data rates. At that throughput level, voice communications can be converted to Voice over IP (VoIP) with high quality, high-definition TV can be streamed to mobile devices, and a host of live interactive gaming applications can be enjoyed.

The two systems currently deployed for 4G are Mobile Worldwide Interoperability for Microwave Access (WiMAX) and **Long-Term Evolution (LTE)**. The standards for 4G were developed by the ITU as the International Mobile Telecommunications Advanced (IMT-Advanced) specification. 4G also supports IPv6, which is especially important given the growth of smart devices.

> **NOTE**
>
> Some of the original WiMAX and LTE systems (as well as some later 3G+ systems) were not fully compatible with the 4G specification, but were allowed to call themselves 4G.

An important change in 4G is the authentication method used. Previous systems used a signaling system called **Signaling System 7 (SS7)** to set up calls and mobile data sessions. In contrast, 4G uses a signaling protocol called **Diameter**. Some critics say Diameter sessions are potentially open to hijacking or having users' personal information exposed, making it a less-than-ideal replacement for SS7. In addition, the fact that 4G is an all-IP network opens it up to all the Internet's known security issues. Given the vast amounts of private, personal information, as well as company information, stored on or captured from mobile devices, this represents a significant security vulnerability for both individuals and businesses.

5G

At the time of this writing, most carriers have begun the rollout of their 5G networks (although their marketing and ad campaigns are greatly overstating the expanse and practical impact). Nevertheless, 5G is here and will soon be ubiquitous (and surely surpassed by what will likely be 6G). First things first though.

The fifth generation of mobile networks is being specifically designed to meet not only the explosion of data traffic per user but also the explosion of the number of data consuming (and generating) devices.

5G promises higher data rates, lower latency, greater connection density, and improved reliability. Two of the highly anticipated applications of 5G are:

- **AR/VR mobile gaming**—One of the promises of 5G is the ability to support mobile augmented reality and virtual reality (AR and VR) gaming. Imagine a group of people with VR

headsets on engaging with each other in a fully rendered virtual three-dimensional (3D) environment in real time with the headsets communicating with the network (and a gaming engine) and each other.
- **Driverless vehicles**—In addition to the massive amount of local computing power required for so-called "autonomous" vehicles, an enormous amount of data must flow from and to a driverless car at all times in order to ensure public safety. It remains to be seen whether or not 5G networks will actually enable driverless vehicles in a commercially significant way, but the very fact that several pilot programs are underway is encouraging.

The BlackBerry Effect and the BYOD Revolution

One could make the argument that the company Research in Motion (RIM) Ltd., later called BlackBerry Limited, first opened the door through which **Bring Your Own Device (BYOD)** charged. BlackBerry got two things right, which led to its meteoric rise. Interestingly, one of those same things led to the company's subsequent decline.

The first thing BlackBerry got right was the development of the **BlackBerry Enterprise Server (BES)** in 1999. The BES enabled BlackBerry devices to receive "push" emails from Microsoft Exchange Server, which meant that users could send and receive emails no matter where they were (assuming they had cell coverage, which by then was nearly everywhere).

The second thing BlackBerry got right was to focus its sales effort on IT departments rather than on individual consumers. This was a brilliant move because, at the time, all but the most technical users needed IT support to receive push emails from Microsoft Exchange Server. This put IT in control—which is exactly how IT likes it.

More to the point, BlackBerry designed its product to suit its customers' wants and needs—which in the case of IT meant easy integration, broad control capability, and decent security (although there were some security issues). The strategy worked brilliantly. By 2010, BlackBerry boasted 36 million users worldwide. However, many people point to this strategy of selling to IT as the root cause of BlackBerry's subsequent rapid decline.

In 2007, Apple introduced the iPhone, the first of the so-called smartphones. The Android phone quickly followed. Both of these devices (along with others) could also receive push emails from Microsoft Exchange Server. Where they differed was their focus on consumer satisfaction and, in the case of the iPhone, on individual prestige. Even the initial launch of the iPhone, which supported no third-party apps, was touted as a BlackBerry killer. With the release of the iPhone 2 in 2008, and its ability to run third-party applications (along with the unveiling of the App Store), the end was near for BlackBerry.

By this time, it was a relatively simple matter to connect to Microsoft Exchange Server without a lot of help from IT. And while many IT departments had a strong preference for BlackBerry standardization, more and more people began showing up at work with iPhones and Android phones. A small but vocal minority pushed to allow third-party devices. If they were told no, many simply did it anyway via workarounds. As the number of consumer-oriented devices grew, IT was forced to support them.

In the context of this chapter, the real takeaway is that more than any other company, BlackBerry got companies and government organizations accustomed to the idea of employees having mobile devices, giving them near 24/7 access to email no matter where they went. Up to this point, wireless technology had blurred the line between work and not work, but that just meant you could use wireless to connect, shut down and move, and then reconnect. BlackBerry was truly mobile, meaning you could stay connected even as you traveled from place to place. Now, workers could (and did) check and respond to email all the time—at dinner, at their kid's soccer game, and (unfortunately) even while driving. With this newfound connectedness, the line between work and not work was all but erased.

Many critics of BlackBerry point to it as cautionary tale of a company that failed to adapt. But few can deny that BlackBerry changed not only how people work but also the relationship between companies and employees to a degree not seen since the industrial revolution. It also—unintentionally— opened a new front in the battle for IT security.

The Economic Impact of Mobile IP

The economic impact of Mobile IP, the standard that allows IP sessions to be maintained even when switching between different cells or networks, has been nothing short of staggering in terms of both scale and acceleration. As noted, the first smartphones appeared around 2007. Their success quickly led to a proliferation of smart devices. Industry analysts estimate that over 1.4 billion smartphones have been purchased each year for the past 5 years and smartphone users in 2020 is estimated at 3.8 billion.

As remarkable as this growth is, it's dwarfed by the growth in data usage. According to studies, data usage grew by an average of 400 percent per year between 2007 and 2010, and with the emergence of 5G, data usage is expected to grow another 400 percent between 2020 and 2025.

It's instructive to illustrate examples with numbers. **FIGURE 2-8** shows the total mobile data usage on a per-month basis. "Exabytes" is a hard number to understand. For an

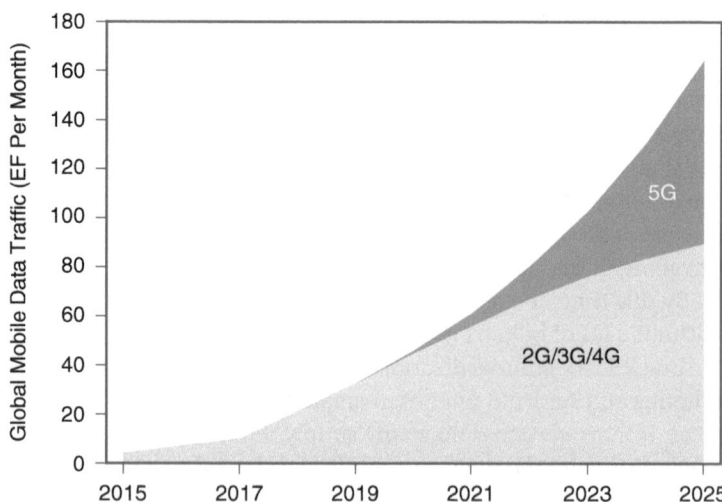

FIGURE 2-8

The incredible growth of mobile data use per leading countries by consumption.

Data from Ericsson Mobility Report, June 2020; https://www.ericsson.com/en/mobility-report/reports

Note: This graph does not include traffic generated by fixed wireless access (FWA) services.

individual, it means that if the average monthly mobile data usage was 20 MB in 2005 (which was a lot back then, and would have been quite expensive), the same average user would consume 20 GB per month in 2010—a mind-bending 1,0000,000 percent increase. It's worth noting that in the first edition of this book published in 2015, this same chart references "petabytes," which is 1,000 times smaller than the updated "exabytes" referenced here.

According to a report by the communication company Ericsson, "The monthly average usage of mobile data in North America is expected to reach 45 GB per smartphone by 2025."

For the carriers, the amount of data consumed over mobile connections far exceeds even the boldest predictions made 10 or even 5 years ago. What's more, the rate of data consumption seems to be accelerating. Mobile providers have been scrambling to keep up with demand, which has boosted subscription rates and driven a great deal of innovation in the areas of compression, streaming, caching, and other data-delivery efficiencies. Interestingly, though, mobile access is beginning to show signs of commoditization, with some providers now giving away data that used to generate lucrative data plans. For example, one major carrier now offers unlimited music streaming outside the data plan. This is a great way to capture a group of users (mostly teens and young adults) who represent potential lifelong customers.

Data providers have also seen incredible growth and are rapidly becoming media creators in addition to hosting media from other sources. With users now expecting high-performance data over mobile connections, data providers have been compelled to build massive, high-performance data centers in many regions to ensure customer satisfaction. This has proven to be an economic boon for switch and equipment providers, as well as to the economies of many small rural markets where the data centers are built. Just 20 years ago, many considered the availability of downloadable music to be just short of a miracle, even though it took 56 minutes per song. Today, kids complain if the high-definition (HD) movie they are watching on their phones from the back seat of a car traveling 70 miles per hour buffers for more than 10 seconds. Clearly, the world has changed.

Unfortunately, all the life-changing benefits of high-speed mobile data come with a significant security risk. As more and more facets of our personal lives have an associated mobile app, more and more personal data will end up on people's phones. This is a gold mine for would-be thieves, who are way ahead of the average unwitting mobile handset user. For cybercriminals who have honed their skills against trained IT adversaries, the average person who may or may not know anything at all about cybersecurity is no match at all. For the IT security specialist, this would be nothing more than a cautionary tale—except for the fact that many of these same unwitting users have access to corporate servers.

Most big-city tourists worry about pickpockets taking their wallet, which might contain some cash, a few credit cards, and a picture ID. These same people, however, often fail to consider that if their phone or device were compromised, they could find all their credit cards run up, their bank accounts cleared, and new credit cards issued in their name and maxed out as well. For good measure, the phone might then be sold to a third party on a **cybercrime** version of eBay (which not only exists, but even has holiday sales) to someone who might then use it to breach the victim's company. This may seem far-fetched, but it's all within the realm of the possible.

The Business Impact of Mobility

It almost goes without saying that the business community has taken great advantage of mobile data—perhaps to an even greater extent than of Wi-Fi. For all of the justifiable security concerns over BYOD, it seems that the boost in productivity is well worth the trouble.

Viewed from a business's perspective, this is easy to understand. For a business, the promise of BYOD is that for the very small cost of a data plan and a phone (about $1,200 per year for a data plan, plus a one-time $100 phone cost), the business claims access to workers for a much greater percentage of their time, including nights, weekends, and vacations. It's rare today for a business to operate only during the 9 a.m. to 5 p.m. shift, but even if you extend the workday by 2 hours (8 a.m. to 6 p.m.), you still only have a 10-hour day. Assuming people sleep 7 hours per night and reserve 2 hours each evening during which they turn the phone off (this may be the biggest assumption yet), that leaves 5 extra hours per day during which employees can make decisions and communicate with colleagues, customers, and partners. In global companies that operate across different time zones, this is especially impactful. Assume as well that BYOD makes available to the company 4 hours per weekend, and that people check their email several times a day while on vacation.

Consider a salaried employee who makes $100,000 per year. Their hourly rate comes to about $37 per hour, assuming they work 60 hours a week for 48 weeks of the year, taking 4 weeks of personal time annually. Applying the preceding assumptions, if this person has a smartphone or device, they will spend 90 extra minutes per weekday, 180 extra minutes per weekend day, and 90 minutes per day of personal time off (in slices) checking, creating, or replying to work emails. Those slices add up to nearly 700 extra hours of work from an employee that, if billed on an hourly rate, would cost more than $25,000. That's a great return on a $1,200 investment!

It's not all positive, of course. There are real security vulnerabilities with which the company must contend, as well as regular occurrences of serious security issues such as attacks or breaches. There is also the argument that smart devices are ready-made time wasters because of their support of apps and social media. Nevertheless, business clearly sees Mobile IP as a big positive, viewing these potential issues as things for the IT security team to worry about. In other words, they assume IT security will fix a problem they probably don't truly understand.

Business Use Cases

The list of actual and potential business cases for mobility is quite long. This section focuses on some general uses as examples.

Any Business Involving the Moving of People or Things

The arrival of Mobile IP brought with it the ability to track people and assets that, until that point, went into a void from the moment they left the physical perimeter of an office or warehouse until the moment they arrived at the destination (if in fact they arrived at all). This greatly improves delivery accuracy and can also be used to calculate the most efficient routing for drivers.

From a security standpoint, this could lead to the false routing of cargo. (Why hijack a truck if you can change a manifest and have it come to you?) There are also complaints from labor unions about privacy invasion—for example, does the company have the right to know exactly where a driver eats lunch?

Delivery (Drop Off) Loss Mitigation

At first glance, this might seem like the same issue as the one discussed in the preceding section, but it's a special case. In the construction business, the theft of supplies is a big problem—one for which material suppliers often bear the burden (and incorrectly so, according to them). The problem is that material suppliers routinely drop off materials—such as lumber, lighting fixtures, plumbing supplies, and even appliances—at job sites that may or may not be secure, and that may or may not have a supervisor present at the time of delivery to take possession of the materials. Even if there is a supervisor present, the site will be unsupervised at some point, and often there is little or no security. Theft of materials is an enormous problem because these materials can easily be sold for cash. Unfortunately for suppliers, they often had to shoulder an unfair portion of this loss because it was not always possible to prove that the right materials were delivered to the right sites, and delivery crews simply could not afford to wait around for signoffs.

In recent years, many suppliers have adopted the practice of using the built-in camera in a smartphone to snap photos of the delivered goods and sending the photos to the customer and back to the supplier. Embedded in each photo is a timestamp and geolocation information—that is, the exact geographic coordinates where the photo was taken. In the event of a loss dispute, this proves without a doubt what was delivered, when it was delivered, and where it was delivered. Here, Mobile IP actually *solves* a security problem.

Information Dissemination

Another big benefit of Mobile IP is the near-instant availability of information to all employees, regardless of where they are. For sales and field personnel, this is a great improvement over the old method, in which they often found themselves with outdated information. When that happened, they were faced with using the old information (which they may have even done unwittingly) or, if the old information was not usable, going without until they could get access to a wired or wireless Internet connection to obtain the right information.

With smart devices, employees can access all manner of corporate information in real time. That means technicians always have the latest manuals, sales and marketing people always have the right information on promotions or competitive intelligence, and support personnel always know the status of orders.

Another great benefit of high-bandwidth mobile connections is live video. Using live video feeds, field personnel can broadcast a problem that is beyond their experience to someone who is more knowledgeable and even get a walkthrough from an expert on the spot. This can work for problems ranging from washing machine repairs to emergency medical situations.

From a security standpoint, the concern here is what has come to be called the **C-I-A triad** where C-I-A refers to Confidentiality, Integrity and Availability. That is, the information in these settings must be private (confidentiality), accurate (integrity), and reliable (availability). The loss of devices also becomes a concern, especially if the devices store or otherwise provide access to private company data.

Enterprise Business Management Applications

Many, if not most, companies that provide enterprise management software now offer mobile applications. This can greatly improve enterprise efficiency, especially for companies with a widely distributed sales force. Applications ranging from **customer resource management (CRM)** systems to corporate expense systems help companies run more smoothly. Having a mobile version of these applications removes the inefficiency that existed in the gap between the actual activity and data entry or retrieval from the system.

With CRM systems in particular, this is a particularly powerful tool, enabling sales personnel to place orders on the spot, ensure the latest/correct pricing, verify order compatibility, and even get discount approval. The security risk is steep, though. If someone gains unauthorized access to the device or app, the intruder could gain deep and closely guarded information about a company. If the company is publicly traded, the risk is even higher because it could result in insider trading or other violations of law.

CHAPTER SUMMARY

Wi-Fi may have untethered computers from network ports, but Mobile IP and smart devices extended the range of connectivity to seemingly every part of the planet. Wi-Fi made the Internet portable from place to place, but mobility connected everyone everywhere in between. For better or worse, people are now connected to the rest of the planet in real time.

It's hard to overestimate the impact of this change. Few technologies have profoundly changed the way humans live and interact with each other to such a degree. More incredible is that the mobile Internet took what was thought to be one of the biggest communication technology breakthroughs of humankind—the Internet—and somehow made it even more impactful. However, from a security standpoint, the mobile Internet took the already difficult task of information security, multiplied it by 1,000, and turned it into a constantly moving target.

KEY CONCEPTS AND TERMS

Advanced Mobile Phone System (AMPS)
Base controller station (BCS)
Base transceiver station (BTS)
BlackBerry Enterprise Server (BES)

Bring Your Own Device (BYOD)
Cell towers
Cellular
C-I-A triad
Code Division Multiple Access (CDMA)

Customer resource management (CRM)
Cybercrime
Diameter
Digital Advanced Mobile Phone System (D-AMPS)

CHAPTER 2 | The Mobile Revolution

Frequency Division Multiple Access (FDMA)
Frequency reuse
General Packet Radio Service (GPRS)
Global System for Mobile (GSM)
Groupe Spécial Mobile (GSM)
Hackers
Handoff
High Speed Downlink Packet Access (HSDPA)
International Mobile Telecommunications-2000 (IMT-2000)
International Telecommunications Union (ITU)
Long-Term Evolution (LTE)
Macrocell
Microcell
Picocell
Short Message Service (SMS)
Signaling System 7 (SS7)
Subscriber identity module (SIM)
T1/E1
Time Division Multiple Access (TDMA)

CHAPTER 2 ASSESSMENT

1. Which of the following are the main design considerations for cellular systems?
   A. iPhones and Androids
   B. Data rates and subscriber plans
   C. Frequency sharing and cell handoffs
   D. Cell handoffs and forward passing
   E. None of the above

2. In FDMA, timing and synchronization are key considerations.
   A. True
   B. False

3. CDMA was predominant in which generation of mobility?
   A. 1G
   B. 2G
   C. 3G
   D. 4G

4. Cell phones in the same cell can communicate directly with each other without going through the base station.
   A. True
   B. False

5. Which of the following describes EDGE and GPRS?
   A. Key 4G technologies
   B. Members of U2
   C. Frequency sharing techniques
   D. Pre 3G data-sharing technologies

6. 4G phones support IPv6 addressing.
   A. True
   B. False

7. The BES server allows which of the following?
   A. Push emails to mobile devices
   B. Netflix on phones
   C. GSM and CDMA compatibility
   D. SMS

8. The C-I-A triad is named that in recognition of the US government having access to all cell phones.
   A. True
   B. False

9. Companies tend to lose money on BYOD.
   A. True
   B. False

10. Which of the following was one thing BlackBerry Limited got right that opened the door for the BYOD phenomenon?
    A. It invented the first smartphone.
    B. It was the first to roll out 3G mobility.
    C. Its phones could run third-party apps.
    D. Its devices could receive push emails from Microsoft Exchange Server.

11. What was the first mobile generation to support Internet access?
    A. 1G
    B. 2G
    C. 2G+
    D. 3G
    E. 4G

Anywhere, Anytime, on Anything: "There's an App for That!"

CHAPTER 3

THE INTRODUCTION AND POPULAR ACCEPTANCE of the smartphone, back in 2007, heralded the mobile apps phenomenon, which transformed how we interact with the world. The variety of mobile apps that became available over such a short period of time was astonishing and defied categorization. Indeed, there appeared to be a mobile app available for anything you could imagine.

With hindsight, we probably shouldn't be so surprised at the huge success and mass proliferation of mobile apps because many factors came together to fuel their massive growth. For starters, the smartphone became not just ubiquitous but essential, such that owners not only always had their phones with them, they also rarely went for more than a few minutes before checking for notices, updates, and likes. These devices not only became part of our lives but in some cases were hard to distinguish from physical appendages.

As prevalent as smartphones are, however, they are simply the tip of the iceberg because the number of connected industrial and smart home devices that make up the Internet of Things (IoT) dwarf the number of smartphones and tablets, and the gulf is widening.

This chapter summarizes the security issues brought on by always on, always connected, and ever-present smartphones, smart home, smart building, and smart city devices that now number in the tens of billions. This chapter also provides an overview of the many wireless protocols that support varying use cases.

Chapter 3 Topics

This chapter covers the following concepts and topics:

- How application growth drove technology development and put further pressure on security
- What Zero Trust security is
- Why Bluetooth is not an ideal option for IoT needs
- What a wireless node network is
- What a piconet is

© Cherezoff/Shutterstock

- Why IoT requires many different connection technologies
- What lingering security issues still exist with WLAN security

Chapter 3 Goals

When you complete this chapter, you will be able to:

- Describe how application vendors circumvented privacy controls
- Understand how to evaluate wireless technology choices based on use cases
- Discuss the difference between service-orientated architecture and native mobile apps
- Describe the differences between the castle-and-moat security model and the Zero Trust security model
- List the advantages of cloud-based architecture
- Understand the basic tenets of mobile cloud computing
- List the key low-power long-range wireless technologies used for IoT connectivity
- Describe a wireless node network
- Describe the trade-offs and considerations made when choosing IoT connection protocols
- Evaluate wireless technology choices based on use cases

Anywhere, Anytime, on Anything

Newly emerging smartphones differed from their predecessors because they bristled with features and functions such as an array of sensors including an accelerometer, a gyroscope, a compass, a proximity sensor, a thermometer, GPS for accurate location, distance, and speed traveled, as well as a flashlight, camera, a microphone, barcode/QR code readers, and a video/voice recorder. Indeed, later models added even more functions such as a pedometer, a barometer, a heart-rate sensor, as well as fingerprint and face recognition functions (and in one vendor's phone, a Geiger counter). This vast array of inbuilt functionality meant that developers had a treasure trove of possibilities when it came to dreaming up potential applications.

A notable feature of mobile apps was their single purpose, as opposed to web-based applications that were designed as Swiss Army knife-style apps. Instead, native mobile apps were designed to do a single thing and to do it easily and to do it well.

The advent of vendors' own App Stores, which authorized, authenticated, and verified apps, made it easy for users to safely download apps with confidence. Also, the App Stores allowed developers to easily publish their mobile apps and earn revenue when users downloaded the apps to their phones. From the developers' perspective, easy access to

software development kits (SDKs) and the low development costs combined with the ease of publishing made mobile app development a very profitable alternative to developing and marketing desktop apps.

Furthermore, the vendor-driven App Stores provided a means of democratization because now any developer, large or small, could create and publish an app that could go viral overnight. Popular apps would always find success through user recommendations and referrals to friends and family.

From the users' perspective, market studies found that consumers loved the convenience and simplicity of mobile apps. However, a recurring theme throughout this text is that where there is a choice between convenience and security, security typically loses out. Seldom has that maxim been truer than in the case of native mobile apps.

Convenience Trumps Security

The huge security problem that was bound to follow the meteoric rise in mobile apps adoption came about through the combination of too much functionality and too little user security and privacy awareness. This was simply because developers could request access programmatically to any function regardless of its relevance to the application. The operating system (OS) would police these actions and seek explicit permission from the user before allowing the app to use the features—for example, the GPS, camera, or microphone. This was to prevent a mobile app directly spying on the user or more surreptitiously accessing their photos, contacts, or tracking the user's location and movements.

The problem was not the technology itself; the issue was that users simply granted permission to the mobile app without any consideration as to why it required that particular functionality. The result was a disaster for user privacy but a field day for developers who harvested user personal data and then sold it to advertisers. Worse, many mobile apps were actively spying on the user or stealing their data and contacts for nefarious purposes. There is no reason that can justify having a flashlight app collect GPS data on your whereabouts other than selling it to researchers, advertisers, or political interests. Most people, if asked "can a random company track your whereabouts at all times, for free and profit by selling it to third parties with no restrictions on their use of that data," would likely give an emphatic "NO." Yet for years, this is exactly what millions of people were allowing.

The threat of malicious mobile apps was countered by the major App Stores because they diligently checked all uploaded mobile apps for unjustified feature requests, and this ensured that most never made it to the market. However, the manufacturers can only do so much and the onus for security and privacy remained with the end user and this is where customer security awareness became so important.

Always Connected, Always On

There once was a time not so long ago when telecommunications were a hit or miss affair. If you wished to talk to someone, you would call their landline number from your desk or home phone or, if out and about, from a nearby payphone. The hope was that they were at their desk or at home close to their telephone at the time. This naturally resulted in communication gaps. Post-1990, the advent of wireless mobile technology filled a lot of the communication gaps because someone could be contacted when they were out of office or not at home so long as you knew their mobile (cell) phone number. Nonetheless, finding someone and successfully communicating with them could still be a problem.

The modern smartphone resolved many legacy communication issues and filled a lot of the gaps because it was a true multichannel device. This was because they could support not just voice calls but SMS messages and email. The advent of Voice over Internet Protocol (VoIP) opened up new wireless channels such as online chat, presence detection, and mobile to Wi-Fi handover. Indeed, the concept of unified communication was built upon the merging of distinct wireless technologies into one omnichannel communications platform. Today, those advances in mobile wireless technologies now allow us to be found and followed wherever we go. Further, by placing this communication platform—a virtual Private Branch Exchange (PBX)—in the cloud, enables communications that are truly always on.

Further, the confluence of mobile (cellular) technology with Wi-Fi VoIP, Bluetooth connectivity, and cloud omnipresence—all features that are common to the smartphone—allows for multichannel communication, anywhere, anytime, and on any connected device.

Unified communication brought the vast potential of communication, collaboration, productivity, and business application integration to the masses, but it came at a price—security. Despite heralding an era of tremendous convenience and efficiency, wireless unified communications came with its own Pandora's box of wireless security issues brought about by this newfound gift of mobility.

The Rise of the Mobile Workforce

As we have seen, a smartphone could provide the user with a multichannelled communication device, which meant they were no longer tethered to the office or home telephone. Instead, they were free to go out and about safe in the knowledge that they would not miss a call. For the consumer, this was all about convenience, but for business, this was a game-changing event that employees were more than happy to embrace. Yes, there were initial concerns regarding work time encroaching into personal time and vice versa, but, generally, most employees were enthusiastic converts. Indeed, such was their enthusiasm they wanted not just access to the company's telephone, messaging, or email systems, but they also wanted to interact with business applications. Now, employees had become fully mobile and were no longer tethered to the office and the working dynamics had changed.

The mobile revolution heralded the era of secure remote computing along with the tremendous advances in mobile carrier broadband and Wi-Fi technologies it rode upon. Now users could safely and securely access company computing resources and servers from anywhere, at any time, and from any connected device. This freedom of access to resources fueled development and innovation. However, it also tore down the concept of traditional network fortification.

Prior to the mobile revolution, networks were designed with traditional layers of fortification in place. The perimeter boundary was designed to be nigh-on impregnable to anything other than authorized traffic. The inner boundaries would build upon this fortification by only allowing traffic filtered by authorized users. Remote-access users were scarce and were accommodated through custom preconfigured virtual private networks (VPNs) that allowed them access to internal network assets. The mobile revolution changed all that because now every user wanted to be mobile and working on their personal devices. Importantly, if everyone wanted access at any time, from anywhere, and on anything, it made the geographical centralization of computer assets behind a fortified barrier pretty much redundant. Clearly, architectural changes were needed.

Mobility brought about a systematic change in how applications were developed and presented to users. In traditional networks, business software tended to be client-server or **service-orientated architecture**. These monolithic applications were unsuitable for mobile users who required either native mobile apps or web-based applications that they could access over the Internet from their device's browsers. The advantage of the latter being that the user's device could then be just about anything that could run a browser. The problem with this method was that the user's device was in effect a slim-client and was totally dependent on the network—in remote scenarios, it would be the Internet—to be able to work.

The alternative was to build the application's user interface (UI) along with some limited capabilities such as storage onto the device itself, such as in a native mobile app. This meant the device and mobile app would only need to connect to the Internet intermittently to synchronize or get updates. Now of course this was not without its own challenges as building, testing, and supporting applications on diverse OS versions such as on Android and iOS is expensive and certainly not trivial. However, if the remote user devices are effectively on the Internet and having to tunnel via a VPN through the corporate perimeter defenses to get to the business application servers, then why not shift the applications to the Internet as well?

By 2010, businesses of all sizes began to see the benefit of cloud computing, especially with regards to mobile and wireless applications. This was because it soon became clear that mobile and wireless technology perfectly complemented each other. Hence, a hybrid style of mobile application appeared that was inherently native with a local UI but utilized **application programming interfaces (APIs)** on cloud-based applications for their processing, intelligence, and functionality such as Google Maps or Yahoo Finance.

Cloud integration with mobile apps came thick and fast because developers could now build mobile mash-ups (collections of external API services linked together with some code to easily build highly functional and often very complex applications) in minutes instead of months. However, the trouble was that rapid development and deployment almost always augurs badly for security, and it was no different this time. The result of the early cloud bonanza in cloud and mobile integration via mash-up applications was unsecured data communications that resulted in loss of data integrity and privacy. Nonetheless, the advantages of mobile and cloud integration were indisputable, particularly for mobile developers and the workers who used these apps.

From a security perspective, it was not so clear cut. There was the mass proliferation of cloud-based applications, storage, and even infrastructure that seemed to pop up overnight. This accelerated the acceptance and expectation of users for mobility in the workplace. Soon, riding alongside was the business agenda for Bring Your Own Device (BYOD) and even **Bring Your Own Cloud (BYOC)**. But, giving users access to company servers and applications from their own consumer devices or, worse, to shadow IT storage in the cloud was a huge step away from the castle-and-moat security blueprint. Indeed, it augured the end for perimeter fortifications as security professionals knew it. But that was not necessarily a bad thing.

From Castle-and-Moat toward Zero Trust

This rapid change in the IT landscape revolutionized the way IT approached security. Gone was the traditional fortress mentality, with its insider and outsider approach, and so too went the VPN. This was because if mobility had blurred the edges between inside and outside, then the cloud had obliterated any distinction for good. In its place came more flexible and granular policy driven systems for identity verification, access authorization, and fine-grained permissions-based actions on profile and context.

The challenges that BYOD had forced upon IT security was that it was near impossible for administrators to deploy and enforce an effective security policy that spans all of those different makes and models of BYOD devices. This brought about a change in mindset whereby the focus on security moved from the user/device to the network itself.

When security operates at the user device level, it tends to assume that some user devices (inside) are trustworthy and others are not to be trusted (outside). However, at the network level, the security model determines that there is no difference between inside and outside because all devices are inherently untrustworthy. Instead of employing the castle-and-moat model used by VPNs built upon clear distinction of inside and outside, the network level never has an assumption of trust. Every request, regardless of origin, to every application is checked, digitally verified, and treated as Zero Trust.

Google first published the Zero Trust model in 2016, which explained how the tech giant "considers both internal networks and external networks to be completely untrusted." This new, decentralized authentication model was quickly adopted throughout enterprise, industry, and commerce as a blueprint for cloud and hybrid architectures where on-prem, cloud, and software as a service (SaaS) infrastructure coexist.

Some advantages to moving to a cloud security Zero Trust solution were clear because now onerous security tasks such as Single-Sign-On, firewall rules, as well as robust **Identity and Access Management (IAM)** could be outsourced to a cloud provider, as could server maintenance and OS administration. Indeed, all that remained that was necessary was to ensure data on the user's device was stored securely and communication was done via **Hypertext Transfer Protocol Secure (HTTPS)**. The remaining security tasks would now be simply administrative such as developing and applying company IAM policies, group permissions, and the like, which would have had to be done anyway. Hence, in only a few years, business networks changed dramatically from being centralized data centers built as castle-and-moat structures to being focused on cloud-based architecture utilizing the Zero Trust security model.

The Mobile Cloud

Over the last decade, cloud computing has emerged as a major disruptor in enterprise security architecture. Businesses of all sizes have shifted operations away from the data center toward the cloud. Cloud architecture provides a convenient, efficient, and often cost-effective way of hosting applications over the Internet. The cloud model also brings on-demand as well as self-service access to virtual and physical computing resources via the Internet, delivering true anytime, anywhere, on anything-style computing.

Cloud-based architecture has many advantages; one of the main advantages of the cloud is to provide the users or clients with easy access to business applications and a huge amount of

Mobile Cloud Computing

When we refer to **mobile cloud computing (MCC)** as opposed to the more generic cloud computing, we are focusing on the distinct combination of cloud computing, mobile computing, and the wireless network. All of these functions are required in order to bring the vast amount of cloud-based computational and storage resources to mobile users. The basic concept here is to make it possible for cloud-based mobile applications to be executed on a vast array of diverse mobile devices.

Cloud Apps versus Native Mobile Apps

When designing mobile apps, a key design criterion will be whether to deploy the app as a mobile cloud app or as a native mobile app. Therefore, we need to consider the differences between the two.

On one hand, we have native mobile apps and these are designed specifically for the type of user device such as the OS version because it will be installed and run directly on the device. As a native mobile app is device/OS specific then a different version will need to be created for each of the mobile devices/OSs supported. A native mobile app is typically downloaded from an App Store.

Mobile cloud apps, on the other hand, are more like web-based applications because the application runs on external servers and not on the device itself. This means the app is device/OS agnostic because it only needs a browser on the mobile device as the UI. Some differences that are worth considering are:

- Native apps have direct access to the device's sensors and features like GPS, camera, location, and microphone. Cloud apps, however, do have access to these resources but are limited to the availability of APIs that access the device itself.
- Native apps are a lot faster than the cloud apps because they access the UI directly, whereas cloud apps need Internet connectivity via a browser.
- Native app development takes longer and is typically more expensive because separate apps need to be built for each of the development platforms like iOS and Android. Mobile cloud apps are device and OS agnostic because they are written in web-based languages such as HTML5, CSS3, or JavaScript, or in server-side web application frameworks giving mobile cloud apps cross-platform and device/OS compatibility.
- Security is a concern for both types of mobile apps. The concerns are data protection and privacy. These concerns relate to how data is protected on the device, in flight, and in cloud storage. Each method has its own concerns and controls.

Deploying Wireless: Different Strokes for Different Folks

Wireless technology has developed over a long period but really has transformed over the last couple of decades as new use cases have emerged. Back in 2000, hardly anyone in business used Wi-Fi (802.11b) other than as a stop gap for a lack of wired Ethernet. However, success in the consumer market came quickly and Wi-Fi was on its way by 2005 to usurping cabling and the connection of choice in the home and the office. Nonetheless, around that time, if you had a remote office, a pool of vending machines, or ATMs scattered around town, the choice of wireless connectivity was sparse. You had the choice of broadband wireless point-to-point radios or GPRS as a 2G+ data protocol. However, what if you had a field of tiny smart sensors on a farm that collect environmental data? These devices are too primitive to be connected to Wi-Fi, broadband point-to-point radios, or GPRS. So what did you do then?

The huge spike of interest in IoT in consumer and industrial services brought the need for wireless technologies and protocols that addresses specific use cases. These new wireless technologies would have to work with resource-constrained devices and/or be able to share resources across an ad hoc mesh network. Among other things, these IoT use cases needed to work over low power, have much shorter or longer range, and support low data throughput. They also had to cater for a vast range of potential use cases in industry, health care, commerce, and even in oil and gas.

Another key criterion for successful cloud hosting was of course having sufficient bandwidth on the Internet pipe connecting the local premises with the cloud organization. It's okay running all your processor intensive apps in the cloud, but for adequate performance you will need very high throughput and low latency Internet connections. Therefore, high bandwidth links and low latency protocols will become essential in any cloud deployment. This is clearly demonstrated when we look to IoT and how businesses manage thousands of remote devices via the cloud.

The Industrial Internet of Things

A notable lasting effect of the mobile revolution was the emergence of new wireless technologies that met the requirements of diverse industrial and consumer use cases. We can be forgiven for thinking that wireless technologies boil down to mobile carrier broadband, Wi-Fi, and Bluetooth because these are the most commonly deployed in IT organizations and the ones supported on our smartphones so we are most familiar with them. Nonetheless, over the last couple of decades, many innovative wireless technologies have come to the fore to meet the specific requirements of, amongst many others, IoT, **machine-to-machine (M2M) communications**, global communication systems, and even ultralong-distance space travel. These can all be considered to be applications of Industrial IoT and the one thing they have in common is that they have specialized operational profiles.

To meet the specific challenges thrown up by the emergence of IoT in consumer, commercial, and industrial scenarios has required a complete reimagining of wireless technologies and protocols. After all, mobile carrier solutions such as 3G, 4G, 5G, Wi-Fi (802.11), and Bluetooth may well be the answer for IT enterprise use cases but fall well short when a selection of low cost, low power/high efficiency, long range, and longevity become the determining criteria. As such, a vast array of specialized wireless technologies and protocols have emerged that can operate on resource-constrained devices in order to meet the challenges that these new use cases present.

IoT Wireless Technologies

The choice of communication technology in the proximity network, which is where we connect all the diverse types of sensors and devices, will directly affect the cost, reliability, and performance of the IoT system. However, because system designers can deploy wireless devices in so many ways, no one technology or protocol is suited to all scenarios—there is no one size fits all.

As an example, if we look at a smart building design, there could be thousands of sensors required to monitor the environmental conditions, such as temperature, humidity, smoke, sound, or pressure. Physically hard wiring all these devices or sensors for power and communications, typically in awkward locations, would be expensive and time-consuming. Hence, deploying a wireless sensor network for power and communications would be the preferred design. This is the basis of the **wireless sensor network (WSN)**.

A WSN consists of many wireless nodes connected as a local mesh or star configuration that cover a local area such as a smart building's floor space. Each WSN node relays data until the data reaches the edge node, which is typically a hub controller device. Return traffic from the backend application, which is typically a control function, travels to the WSN nodes in a similar fashion.

Another characteristic of a WSN node is that they may be resource-constrained and use low power. Therefore, they can run off a battery or can harvest energy through solar, kinetic, wind, or electromagnetic radiation. WSN nodes often use very basic communication interfaces and so they need to connect to an IP-aware edge device able to communicate with the rest of the IoT system.

Low-Power Technologies

The latest research and development is focused on low-cost, low-power wireless connectivity solutions. There are, however, serious trade-offs that need to be taken into consideration. For example, if the goal is higher data rates then this requires higher power levels to enable the radios to transmit at these higher frequencies. As a result, designers tend to opt for lower frequencies to conserve and optimize battery life. However, lower frequencies mean slower data rates and longer transmission—for the same amount of data. This means that the radio will stay on for longer, thereby consuming more power and potentially draining the battery. Therefore, designers have to carefully consider several key factors:

- The frequency (data rate)
- The size of the message being sent
- The technologies and protocols available
- The radio power/transmit time
- Preservation of battery life
- How to secure data transmissions

Designing Low-Power Device Networks

The problem is that not all technologies and protocols fit every scenario and wireless technology must meet a required standard for each specific deployment. After all, it is of little use producing a start-up product that uses a nonstandard interface and this is where standards become so important.

In this text, we have already covered how computers communicate in IT architectures and the protocols that they adopt such as Ethernet at the Physical and Data Layers and IP protocol at the Network Layer, TCP/IP for transport, and HTTP at the Application Layer. This is all well and good for PCs, laptops, and tablets, but when we consider resource-constrained IoT devices then there is a clear distinction. Actuators and sensors are typically resource-constrained, that is, they have little processing power, local storage, memory, or communications capability. Therefore, they need their own set of specific standards.

A well-established standard for resource-constrained-type devices is the IEEE 1451 or smart transducer interface standards. This family of standards describes a set of open, common, network-independent communication interfaces for connecting transducers (sensors or actuators) to microprocessors, instrumentation systems, and control/field networks.

IEEE 802.15.4

One of the early IoT enablers is the low-power IEEE 802.15.4 standard. It was first introduced back in 2003, and updated up to and including 2012, to set the standards for low-power commercial radios in wireless personal networks. The 802.15.4 standard can be used with IPv6 over Low-power Wireless Personal Area Network (6LoWPAN) and standard IPs to build a wireless-embedded Internet.

The introduction of the 802.15.4 standard addressed the Physical Layer. However, change was required not just at the Physical Layer but at the higher layers, so we also needed a different protocol structure that is dependent on vastly more efficient protocols.

Wireless Communication Technologies

The advent of a plethora of wireless communications over the last two decades has been one of the drivers for adoption of IoT both by the consumer and in industry. The array of wireless technologies has arisen because they serve different purposes. Some technologies, such as Wi-Fi (802.11), are designed for high bandwidth and throughput where power was not an issue, for example for web browsing and downloading from the Internet to a computer. Bluetooth, on the other hand, was designed to provide wireless connectivity in a **personal area network (PAN)**. It became hugely popular by connecting the devices that you wear or carry, such as a smartphone, or a watch, to a headset. Zigbee found a niche in the smart home market by proving to be adept at interconnecting devices from different vendors around the home. As a result, Zigbee has a range and capabilities that fall somewhere between Wi-Fi and Bluetooth.

Today, wireless communication technologies and protocols are constantly evolving or being repurposed to match new use cases. Zigbee has, for example, evolved to cover neighbor area networks (NAN) in industrial IoT scenarios and Bluetooth's range and capabilities have also improved beyond the scopes of the PAN.

Whichever technology we choose to adopt will always have constraints, such as range, throughput, power, physical size, and costs. Therefore, we have to take a diligent approach when evaluating the wireless technology best suited for the purpose. In the following section, we will review some of the more common technologies and protocols out there.

Bluetooth Low Energy

Bluetooth Low Energy—which is also known as Bluetooth 4.0 or Bluetooth Smart—is the version of the Bluetooth technology specifically designed for IoT. As its name suggests, this is a power and resource-friendly version of the technology, which is designed to run on low-power devices that typically run for low periods either harvesting energy or powered from a coin-sized battery. However, what makes Bluetooth suitable for WSN nodes is its ability to form an ad hoc network called a **piconet**.

The definition of a piconet is a grouping of paired Bluetooth devices that form dynamically as devices enter or leave radio proximity. The piconet consists of 2 to 8 concurrently active devices that communicate over short distances—remember Bluetooth's original purpose as a PAN to wirelessly connect headphones, smartphones, and other personal electronic devices. Additionally, Bluetooth's ubiquity, miniature format, low cost, and ease of use makes it a very attractive short-range wireless technology.

Zigbee IP

Zigbee is an open global wireless technology, which is specifically designed for use in consumer, commercial, and industrial areas. Zigbee operates in three license-free bands at 2.4 GHz, 915 MHz for North America, and 868 MHz for Europe. What makes Zigbee different from other short-range low-power wireless technologies is that it is a superset of IEEE 802.15.4. What this means is that it provides its own higher layers of application and security support, which provide compatibility with products from different manufacturers.

Zigbee excels in supporting control and monitoring applications that typically have low data throughput, short range, and low-power requirements. Zigbee's range is about 70 meters, but it supports three different network topologies, namely the star, mesh, and cluster tree or hybrid networks, so relaying communications from one Zigbee node to another in a network vastly extends the range.

The original Zigbee standard did not support IP by default, which made it difficult to interface with consumer applications. Zigbee IP takes advantage of its higher autonomous application and security layers to add support for 6LoWPAN and RPL (Routing Protocol for Low-Power and Lossy Networks) that optimize meshing and IP routing for wireless sensor networks. This blend of technologies results in a solution that is well suited to extend IP networks to IEEE 802.15.4-based MAC/PHY technologies.

Z-Wave

Z-Wave is a low-power radio-frequency (RF) communications technology that is optimized for reliable and low-latency communication of small data packets. The primary use case for Z-Wave is in home automation scenarios for products such as smart lighting and thermostat controllers and sensors. What makes Z-Wave great around the home is that it is impervious to interference from Wi-Fi and other wireless technologies in the 2.4-GHz range such as Wi-Fi, Bluetooth, or Zigbee.

Z-Wave uses a simpler protocol than some other RF techniques, which enables faster and simpler development. In addition to being quick to market, it is very scalable because it supports full mesh networks that can control up to 232 devices.

RFID

Another extremely popular wireless technology used in retail, commercial, and Industrial IoT is the radio-frequency identification (RFID) system. RFID uses tiny tags to store electronic information that can be communicated wirelessly via electromagnetic fields. RFID technology is pervasive across many industries, typically for the purpose of identification and tracking of inventory or assets.

NFC

Near-field communication (NFC) evolved from RFID technology and is commonly used today for short-range wireless communication. A popular implementation is in contactless payment systems. The advantage of NFC is that no device pairing is necessary and the range is limited to only a few centimeters to prevent interference or hijacking. As a result, NFC has become popular with smartphone application developers in Android and iOS phones to create payment systems or methods to transfer contacts between phones simply by placing them close together.

Thread

In 2014, a new IP-based IPv6 networking protocol launched called Thread, which is aimed at the home automation environment. Based on 6LoWPAN, Thread, from an application point of view, is designed as a complement to Wi-Fi's limitations in a home automation environment.

Thread is designed to work on existing IEEE802.15.4 devices and supports a mesh network capable of handling up to 250 nodes with high levels of authentication and encryption.

6LoWPAN

The problem that 6LoWPAN addresses is that IP protocols have far too much packet overhead for most embedded sensors. These low-power devices typically need to transmit only tiny amounts of data. This means that a far more efficient communication protocol is required that provides encapsulation and header compression suitable for very short transmission times—that protocol is 6LoWPAN.

In addition, 6LoWPAN is configurable as a robust, scalable, and self-healing mesh network. Mesh devices can route data destined for other devices, while hosts are able to sleep for long periods of time to conserve battery life.

Cloud VPNs, WANs, and Interconnects

When you deploy in the cloud, there will be the obvious decision as to how you will securely connect to the cloud infrastructure from your wireless devices, networks, or environment. The common choices are to use the cloud service provider's VPN or interconnect configurations. These may be direct or supplied via a third-party carrier but either way, in the case of interconnects, the data must be encrypted. This is because many third parties just provide a communications channel; it's up to the business to encrypt the data as they see fit.

Other types of WAN links are surfacing such as WWAN (wireless wide area network) and SD-WAN (software-defined wide area network). Typically, these will start off as wireless solutions such as carrier (4G) or Wi-Fi backhaul before being aggregated at a fixed-fiber media convertor. In metro areas or networks, WWANs are typically just 802.11 unlicensed spectrum broadband point-to-point radios or they may be used as back-up redundant links for wireless free space optical links.

Free Space Optics

Free space optics (FSO) are laser wireless broadband WAN links that have a range of 3–5 kilometers line-of-sight and a throughput of 10 Gbps. These laser beam point-to-point links are used for data backhaul from remote offices in enterprise or campus networks or, more recently, within remote IoT sensor networks. FSO and Wi-Fi complement each other very well as if the optical network links fail due to sand/dust storms, fog, or heavy rain then the broadband Wi-Fi radios can kick in to restore the connection, albeit at a lower throughput. This is because broadband 802.11 wireless operates at frequencies unaffected by sand/dust storms, fog, or heavy rain. By combining FSO as the primary link, there is the default condition of high speed, high throughput over 3–5 kilometers with a fail-over medium speed and medium throughput over the same distance as an emergency backup link. However, if the primary or backup links sometimes need considerably greater reach, such as hundreds of kilometers, then you might look to another wireless technology—WiMAX.

WiMAX

WiMAX is an alternative last-mile solution to carrier network's Long-Term Evolution (LTE) and digital subscriber line (DSL) networks because it is a high-speed RF microwave technology that provides up to 1 Gbps bandwidth. WiMAX can work over 100 kilometers or more at a low bit rate, which makes it ideal for many rural IoT applications. If, however, you need higher bit rates over longer distances, perhaps thousands of miles, then you need to look to the heavens.

vSAT

A very-small-aperture terminal (vSAT) is a satellite communications technology used to deliver broadband Internet and private network communications over a vast coverage footprint. When you need to connect to the Internet when you are hundreds or even thousands of miles away from a base station or you need to broadcast to clients over an entire continent then satellite wireless communications is the solution. Satellite communications allows for high bandwidth, high speed, and secure communications, but it does have high latency due to the distances the data must travel. Nonetheless, if the application is not low-latency dependent then satellite communications is a viable wireless technology. As such, satellite communication is the norm for shipping, aircraft, oil and gas platforms, trucking, and many forms of asset tracking in logistics and science.

SD-WAN

As we have seen, there are many diverse methods for wireless WAN communication and each has its own properties and challenges. This is one reason why SD-WAN tends to aggregate several diverse wireless technologies into the mix to ensure a robust network configuration regardless of the weather, geographic, environmental, or atmospheric conditions.

By extending SD-WAN into the wireless network, it enhances network performance while improving security for all types of on-premise and cloud-based applications. This is because pushing the security and network monitoring out to the cloud provider ensures that changes in IAM policies or new rules and updates will be automatically pushed out to the SD-WAN and Wi-Fi gateways. Then, the SD-WAN and wireless LAN systems can enforce the business policies and control the traffic flow and thereby improve the security profile.

WAN Technologies for IoT

IoT encompasses such a broad spectrum of industries and use cases that no one low-power WAN technology will fulfill every use case requirement. Therefore, several technologies have emerged that offer different choices, and sometimes these are trade-offs between range and throughput or battery life and operating frequency. However, each technology has its own set of attributes so it should be possible to find one that matches requirements.

There has been a lot of research and development into producing low-power IoT wireless technologies suitable for remote resource-constrained devices. The results have been a proliferation of new technologies and protocols that are constantly evolving. This section discusses the current main technology players in the low-power WAN field.

Sigfox

Sigfox low-power WAN is an end-to-end system consisting of the presence of a certified modem at one end and a web-based application at the other end. Developers must acquire a modem from a certified manufacturer to integrate into their IoT end node device. Alternatively, there are third-party service providers who make Sigfox-compatible access point networks available to handle traffic between the end nodes and Sigfox servers. The Sigfox servers manage the end-node devices in the network, collect their data traffic, and then make the data and other information available to the user through a web-based API.

To provide security, the Sigfox system uses frequency hopping, which mitigates the risks of message interception and channel blocking. Furthermore, Sigfox provide antireplay mechanisms in their servers to avoid replay attacks. The content and format of data sent in the transmission is user-defined so only the user knows how to interpret their device data.

LoRaWAN

LoRaWAN (long-range wide area network) architecture is a "star of stars" topology whereby end node devices connect through gateways to network servers. The LoRaWAN protocol is based upon a chirp, which is a spectrum-spread wireless hop between the end nodes and the gateway. The end nodes communicate by chirping the gateway when they have data to transmit. LoRaWAN allows for a trade-off between payload and range, which is dependent on local radio conditions. Security for LoRaWAN is provided for by using unique network and device encryption keys.

Low-Power Wi-Fi (HaLow)

Wi-Fi 802.11ah, or HaLow as it's more commonly known, operates in the ISM bands (industrial, scientific, and medical radio bands) just below 1 GHz, because these lower frequencies enable battery-operated equipment because they use lower power. While most Wi-Fi 802.11 have maximum ranges around 100 meters, HaLow can reach up to a kilometer radius with the right antenna.

The design goal of 802.11ah is to deliver a trade-off between low power and long-range communication. This typically means that user stations will have a sleep mode to conserve battery charge and only communicate when they must and by using short data packets. An important network design factor is ensuring optimized distancing between stations because optimizing the contention access procedures will minimize the transmit time and the power usage.

In order to extend the communication range, 802.11ah uses a relay agent, which is a special station type used to pass messages over longer distances at low power. The network itself can support around 8,000 stations, which vastly increases the network footprint.

Millimeter Radio

Recently, a lot of interest has been shown in the unlicensed 60 GHz wave band in the millimeter range. Nonetheless, one fact is immutable—you can have high bandwidth or you can have long range, but you can't have both.

The problem is simple: high data rates and throughput is directly related to high frequency and long range is directly correlated to low frequencies. At the 60GHz band, millimeter radio undoubtedly has the potential to send high throughput, but it will be over short distances, less than 1 kilometer, and it will need to be with a clear line of sight.

However, there is another operational problem, for there is another immutable law of radio—the higher the frequency, the more susceptible it is to rain fade and atmospheric conditions. Therefore, millimeter radio is highly susceptible to not just objects in its line-of-sight path but also atmospheric conditions because it is susceptible to even light rain. In adverse conditions, millimeter radio requires a high-power boost to overcome the rain fade margin and that would rule it out for most IoT purposes.

Today, when we think about wireless networks, we most likely think about the public carrier mobile networks or private Wi-Fi WLANs (802.11x) because these have a role in most consumer homes, businesses, and enterprise architectures. However, in some circumstances, Wi-Fi and public LTE technologies and protocols are not ideal. There will be many use cases where an alternative solution is required that falls somewhere between the two. This is where private LTE comes into the wireless solutions mix.

Private LTE Networks

There is a definite requirement today for large-scale wireless LANs. Nonetheless, the present capability of Wi-Fi to meet that organizational need is very limited. As a result, many organizations are moving away from Wi-Fi toward adopting private LTE.

Private LTE has emerged in recent years as a solution to fill the gap between high-end private Wi-Fi networks and public LTE carrier networks. The advantage of a private LTE network is that it can deliver dedicated, fixed-cost connectivity and better information security for all IT assets and IoT devices.

Private LTE, just like carrier LTE deployments, uses licensed, shared, and unlicensed RF spectrums to reduce costs, congestion, enhance traffic flow, and ultimately to improve information security. The private LTE network replicates, on a smaller scale, carrier LTE networks and is built upon micro towers and very small cells, making it similar to a WLAN with its access points.

The way that it works is down to how the RF spectrum is managed. In some cases, a mobile carrier may use their own licensed LTE spectrum to build a private LTE network for a third party. Alternatively, an organization may build their private LTE network using the lightly licensed spectrum such as CBRS (Citizen Band Radio Spectrum) on 3.5GHz (this model is termed shared spectrum). Finally, an organization may choose to build their private LTE network on the existing unlicensed spectrum but avoiding the congested ISM bands. Either way, private LTE can deliver specific benefits such as lower data costs, improved network performance, and tighter information security.

The way that private LTE provides these benefits is that it works over different frequency bands from the saturated 802.11 Wi-Fi networks. This is hugely beneficial in industrial use cases where keeping the competition for available frequency and potential RF interference to a minimum is essential to the running of critical autonomous devices such as robots.

However, the benefits of private LTE are not just about frequency separation and isolation; it also comes about through its ability to handle priority, pre-emption, and its inherent capability to scale. This means that private LTE networks can handle vast amounts of locally attached devices and their traffic without having to go out onto expensive pay-per-bit public carrier LTE networks.

Finally, private LTE is considered by many organizations to be the ideal stepping-stone to 5G. Hence, these organizations see that deploying unlicensed private LTE is a roadmap for their future 5G rollouts.

Wireless Network Security

Historically, WLAN technology has been viewed by security professionals as convenient but an inherently unsecure medium. This is because early WLAN implementations, often found in **small office/home office (SOHO)** environments, were typically installed by technicians who were not knowledgeable about radio technology or security. The result was unauthorized users poaching the network—that is, using its bandwidth without permission. Worse, some users would gain unauthorized access to the data on the network, even injecting or forging data.

Of particular concern was the use of omnidirectional antennas with wireless networks. With such an antenna, the wireless access point broadcasts to anyone in the vicinity who has a receiver, whether they are authorized to access the network or not. As shown in **FIGURE 3-1**, this area can and often does extend beyond the physical walls of an office or building. This puts the network at risk. Another risk with wireless networks is intruders using directional antennas to eavesdrop on the network, even from quite some distance away.

With these problems in mind, the IEEE set out to develop a security standard as part of the 802.11 standard. The goal was for WLANs to be considered as secure as their wired

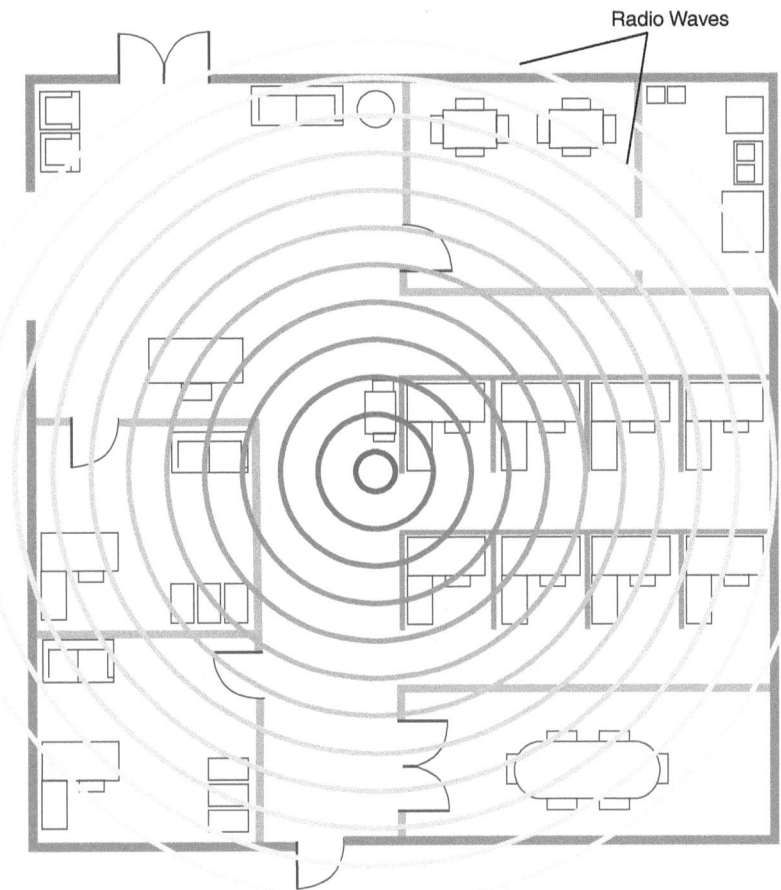

FIGURE 3-1

Omnidirectional coverage areas can extend beyond an office's walls.

equivalents, allowing confidentiality, integrity, and availability. Unfortunately, early attempts to secure WLANs proved to be flawed.

The first attempt called **Wired Equivalent Privacy (WEP)** used a challenge/response mechanism. Unfortunately, an eavesdropper could intercept the challenge and response. Although the response was encrypted, the challenge was not. The attacker could simply capture and replay the response to successfully respond to a challenge and in doing so could gain access to the network.

In response to WEP's inherent weaknesses, the Wi-Fi Alliance introduced **Wi-Fi Protected Access (WPA)**. It too was short lived because it proved to be only more difficult to break. However, its successor **Wi-Fi Protected Access 2 (WPA2)** did become the standard for many years. WPA3 was introduced in 2019, which has more advanced features to accommodate today's business and industrial security requirements.

 NOTE

Encryption and strict access control can secure most networks from unauthorized use. If an authorized device falls into the hands of an attacker—for example, if a laptop is lost or stolen—then the attacker has all the information and configuration needed to beat the encryption and even the strictest access control.

Lingering Security Issues

Although there have been considerable advances in WLAN security, network and security managers still cast a suspicious eye over anything wireless. And to be fair, wireless networks do give cybercriminals additional opportunities.

Mobile IP Security

Until the latter part of the noughties, mobile phones were predominantly voice only. The advent of 3G networks made high-quality Internet access from a mobile device a reality. The demand for data access soon became a torrent around 2010, with mobile data exceeding voice traffic across mobile carrier networks.

Unfortunately, cybercriminals were not far behind. This was mostly because phone manufacturers made instant and easy access a higher priority than basic security. As such, smartphones and tablets were enabled for Bluetooth discovery out-of-the-box. The criminal element rejoiced, because they now had direct access to these devices, which they could surreptitiously use to make voice calls, send data, listen to or transfer calls, gain Internet access, and even transfer money.

Today, mobile phones are no longer shipped with Bluetooth enabled in discovery mode. In addition, security has been hardened to prevent unauthorized connections and remote access to the phone's features. Confidence is such that smartphones are trusted for use in eBanking, eCommerce, and email. However, despite these improvements in securing wireless mobile devices and the underlying radio networks, there is no room for complacency. Cybercriminals, who have become adept at intercepting signals over unencrypted wireless networks, are never far behind.

CHAPTER SUMMARY

The availability of applications drove the growth of smartphone subscribers, which in turn encouraged more subscribers in a virtuous (or viscous) cycle that resulted in the creation of more mobile devices than people on this planet. This growth of course attracted a market for people keen on taking advantage of all the personal and private data collected on the app, creating yet more attack vectors for hackers and data thieves and miners to exploit.

Just as developers, and manufacturers, were getting their collective arms around this, the IoT revolution began, and now the number of connected devices is multiples of the world's population. With each new use case, new attack vectors, exploits, and fixes are discovered and deployed.

Such is the nature of wireless and mobile security. The game of cat and mouse has no end and even if some theoretical limit of technology was reached, the cleverness of the bad guys is usually no match for the carelessness of users who want things to be easy and intuitive and unobtrusive and secure. When push comes to shove, they freely sacrifice the security requirement. While baffling, it certainly keeps many people actively employed.

KEY CONCEPTS AND TERMS

Application programming interface (API)
Bring Your Own Cloud (BYOC)
Free space optics (FSO)
Hypertext Transfer Protocol Secure (HTTPS)
Identity and access management (IAM)
Machine-to-machine (M2M) communications
Mobile cloud computing (MCC)
Near-field communication (NFC)
Personal area network (PAN)
Piconet
Service-orientated architecture
Small office/home office (SOHO)
Software development kit (SDK)
Wired Equivalent Privacy (WEP)
Wi-Fi Protected Access (WPA)
Wi-Fi Protected Access 2 (WPA2)
Wireless sensor network (WSN)

CHAPTER 3 ASSESSMENT

1. Application vendors gained access to user information mostly by
   A. Hacking their way into operating systems
   B. Lying to phone manufacturers
   C. Users simply allowing it
   D. Buying passwords on the black market

2. The Zero Trust security model assumes there is no difference between inside and outside because all devices are inherently untrustworthy.
   A. True
   B. False

3. One of the main cloud-based architecture advantages is
   A. It provides the users or clients with easy access to business applications from anywhere that has an Internet connection
   B. It's more secure than native applications because all clouds are secure
   C. It comes free with Gmail
   D. It solves data privacy issues

4. Which of the following is NOT true about native apps?
   A. Native apps have direct access to the device's sensors and features like GPS, camera, location, and microphone
   B. Native apps are a lot faster than cloud apps because they access the UI directly
   C. Native app development takes longer and is typically more expensive
   D. Security is not a concern for native apps

5. Nearly all IoT use cases can be covered with the two main IoT wireless protocols.
   A. True
   B. False

6. Which of the following is not a determining criterion for industrial use cases of IoT?
   A. Cost
   B. Low power/high efficiency
   C. Range
   D. Battery life
   E. Speech quality

7. Which of the following is true about IoT wireless trade-offs?
   A. Higher data rates require the same power as lower data rates
   B. Lower frequencies mean slower data rates and longer transmission
   C. Data rates do not impact battery life
   D. There is no correlation between data rates, frequency, and power

8. Bluetooth Low Energy was specifically designed for IoT.
   A. True
   B. False

9. A piconet
   A. Is an ad hoc grouping of paired Bluetooth devices
   B. Consists of two to eight devices that communicate over short distances
   C. Is an attractive and affordable short-range wireless technology
   D. All of the above

10. Wired Equivalent Privacy (WEP) replaced Wi-Fi Protected Access (WPA) as an unbreakable wireless security protocol.
   A. True
   B. False

11. More mobile-capable tablets are sold than PCs.
   A. True
   B. False

CHAPTER 4

Security Threats Overview: Wired, Wireless, and Mobile

THIS CHAPTER provides a general overview of network security threats and considerations, with a particular focus on wireless devices—including Internet of Things (IoT)—and mobile devices (where it applies). In an ideal setting, security professionals could easily lock down any network or network-enabled device to mitigate the risk of data loss or unauthorized access. Unfortunately, this practice must be applied in the real world, where the business's needs, third-party requirements, and employee satisfaction must be balanced with the security team's priorities. This chapter aims to give you an understanding of the threats as well as an appreciation for the many factors that go into implementing a security strategy. A pragmatic approach that provides the right balance of security and access is often the best path to success.

Chapter 4 Topics

This chapter covers the following concepts and topics:

- What you should protect
- What the general threat categories are
- What the threats to wireless networks and mobile devices are
- How to mitigate risk
- What authorization and access control are
- What information security standards are
- What regulatory compliance entails

Chapter 4 Goals

When you complete this chapter, you will be able to:

- Describe the general threat categories in wireless and mobile environments
- Describe the five aspects of threat identification

- Describe the security vulnerabilities of Wi-Fi, Wi-Fi-connected IoT devices and mobile devices
- Describe defense in depth and how it is used
- Understand the purpose and methods of authentication, authorization, and accountability
- Describe examples of information security standards
- Understand the benefits and consequences of regulatory compliance
- Understand how to reduce risk using risk-mitigation techniques
- Describe the threat-mitigation techniques for Wi-Fi and mobile devices

What to Protect?

In essence, **information security** refers to the processes and practices that must be implemented to secure the digital assets you wish to protect from various threats. The first step to formulating any security plan is to ask questions, such as the following:

- What are you trying to protect? Is it corporate data, intellectual property, customer data, financial assets, or remote control of a physical device?
- Why are you trying to protect it? Is protection mandated by a government or industry agency, or is it an internal best practice?
- What is the value of the asset? Have the assets been quantified? Has the cost of a data breach been estimated?
- What are you protecting it from? Are the threats internal or external? Are they aimed at data theft, device control, or system access? Are the threats environmental or human in origin?
- What constraints prevent you from protecting the asset? Is broad access required? Does the data change or move around?

The answers to these basic questions will help you identify the assets you must secure, as well as the priority and value of each individual asset. This is vitally important. Security must be cost-effective. An organization should spend enough on security to meet the risk-reduction objectives and no more.

Securing an organizational network is a lot more complex than most people think. Even securing a small business is not easy; it requires standards and best practices that meet the business's needs and, in many cases, the industry's requirements. Consequently, you need flexible guidelines but must also adhere to mandatory regulations.

In addition, the motivation for securing a small company may not be the same as that for securing a publicly listed company or a large government agency. Similarly, the head of sales or product development in a publicly held company may have a vastly different opinion than the chief security officer regarding the balance between ease of authorized access and robust external security.

General Threat Categories

What are the different threats to an organization's assets? To determine this, you must first consider who or what are you protecting the assets from. Is it external attackers or employees? Some major business failings in recent times have come about through internal manipulation of company financial data or from malicious insiders—the lesson for companies being that they should treat both employees and external intruders with caution. However, many have been the result of unwitting employees being taken advantage of by hackers who are as adept at social engineering as they are at cracking security measures. It's much more fruitful to have someone let you in than to try to overcome security measures head on. Most companies employ the practice of **least privilege**, whereby personnel are given the rights and permissions to perform their jobs and nothing more. This does not solve all problems, but it does greatly reduce overall exposure to risk.

You must also consider which assets you wish to protect, as well as the constraints you face, both physical and from a business perspective. Not all assets can or even should be rigidly secured. Some assets, such as a web server or an automatic teller machine (ATM), will need to have public exposure. Others, such as information on an internal database server, can be more robustly secured, because access requirements are more limited.

Your security procedures should align with the business objectives and the assets' functions. For example, you cannot reasonably secure an e-commerce server by disconnecting it from the Internet. Yes, doing so would make it secure from threats of Internet attacks, such as denial of service attacks, sync attacks, or hacking attempts. But doing so would also make it useless for its purpose.

In short, the security measures should be proportional to the value of the assets, and should not create an impediment to the assets' purpose and function. This is a key point that many security professionals miss—information security should not become an impediment to authorized users performing authorized business functions. That said, having reasonable but visible security practices ensures that security is top of mind for employees who use or have access to protected data or systems.

The key is to ensure that information security processes, practices, and techniques are aligned with the business's plans, goals, objectives, and functions. This requires user education, because many problems are the result of a lack of awareness or understanding. An information security policy is also key. Indeed, an information security policy is one of the

---

A Cultural Change

The information security policy should match the business requirements, objectives, and culture of the company it serves. This is especially true in these days of mobility and wireless technology, where employee productivity and mobility trump rigid access controls. Nowadays, businesses encourage employees to use their own devices and applications and provide them with freedom of access to information from any location and any device. Therefore, information security must adapt to support these policies, initiatives, and practices, while at the same time securing and protecting the company's assets. This is a significant cultural change from the norms of a decade ago.

most important security documents in any business. It is the foundation and the rationale behind all security measures implemented by security officers.

Initiatives and practices may change over time as businesses adapt to new technologies. However, the core principles of information security and the C-I-A triad of confidentiality, integrity, and availability will remain. They have, however, been expanded to cater to the specific requirements of new technologies and business models. These expanded principles of information security are as follows:

- Confidentiality—Preventing unauthorized disclosure of information
- Integrity—Preventing unauthorized modification of information
- Availability—Preventing unauthorized withholding of resources or services
- Accountability—Making users accountable for their actions
- Nonrepudiation—Preventing the denial that an action has been taken

Confidentiality

Traditionally, confidentiality has been about securing access to information on storage devices and servers via access policies and rights. Administrators may have encrypted data or hidden it from view to protect it from unauthorized access by applying strict access policies on directories and folders. Similarly, administrators would protect data by encryption and authentication mechanisms, if deemed necessary, when it was traversing the local area network (LAN), wide area network (WAN), or Internet.

In today's modern networks, when it comes to confidentiality, you must consider both the privacy of information (protecting data from being seen) and its secrecy (hiding knowledge of data's existence or whereabouts).

Additionally, you must consider the confidentiality of data stored on laptops, smartphones, and tablets. These devices are of particular concern because any time that information leaves the organization's property, it becomes vulnerable. This is also an issue because the decreased size of these devices, along with the increased amount of time these devices are with people outside the office, has greatly increased the likelihood that the physical device will be lost. For example, it's doubtful that a decade ago, many people would have taken their work laptop to a coffee shop and left it behind. The smaller size of smart devices, however, makes them easier to leave behind. In addition, the fact that their owners tend to carry them everywhere significantly increases the likelihood of their being lost. As a result, device loss, and the remote locking or wiping of a lost device, have become major challenges for information security.

Another form of confidentiality concern stems from IoT devices. A great deal of information can be gleaned about people and organizations by tracking devices usage, on and off times, and settings.

Integrity

In a security sense, integrity refers to the assurance that information is genuine, that it remains true to its original form, and that it has not been manipulated or tampered with. Many applications use mathematical algorithms called message-digests or hashes to ensure the integrity of a document or file—that is, to ensure that the document or file has not been altered either by accidental corruption or deliberate manipulation. These algorithms produce a digest, or a **hash**, of the file or data.

This approach works because an identical file, when passed through the algorithm, will always produce the same hash number or numerical result. However, if you change just one single bit within the file, the algorithm will produce a different result. If you compare the hash number on each file and they match, you can be confident that both files are identical.

Hash algorithms provide assurance of data integrity when data is passing over unsecured networks such as the Internet. They also provide a method for mitigating man-in-the-middle (MITM) attacks, whereby an attacker intercepts files or messages and alters them before passing them on to the intended recipient. In a wireless, Wi-Fi, or mobile environment, session hijacking is a particular concern.

Availability

With regard to information security, availability is concerned with ensuring that systems and services are available to users when they need them, and are not withheld by unauthorized means. This is quite different from availability from a system or network perspective, which is more a measure of reliability and is ensured through machine clusters and fail-over master/slave configurations.

In security, availability is largely about preventing denial of service (DoS) attacks, in which an attacker tries to prevent legitimate access to a system or service by making it unavailable. Attackers commonly launch DoS attacks through synchronization (SYN) flooding of Transmission Control Protocol/Internet Protocol (TCP/IP) devices on the Internet to disrupt the three-way handshake used by TCP to establish a session. This handshake involves the following steps:

1. The requester sends a synchronization request, called a SYN.
2. The server responds with an acknowledgment to the request, called a SYN-ACK.
3. The requester responds to the acknowledgment with a message of its own, called an ACK.

After this occurs, the client and server can communicate.

With a SYN flood attack, the attacker blasts the server with fake SYN requests. The server is obligated to respond, but the requester never replies with an ACK to establish a session. This leaves the host with incomplete connections. When this process is replicated thousands of times, the host's finite capacity to open TCP connections will be consumed, and no more connections will be possible. This is a crude but highly effective attack on a system's availability.

On a smaller scale, jamming techniques can prevent availability on Wi-Fi and mobile systems. In some cases, the mobile system itself is targeted. This may send users into a near panic, because they have come to expect uninterrupted connectivity.

With regards to IoT, any interruption to the availability to access devices can prevent control of lighting, HVAC, physical security systems, and even network access to controlling physical access via security doors.

Accountability

Accountability is important from a security standpoint. Despite diligent efforts to ensure confidentiality, integrity, and availability, a system's access controls can still be bypassed either accidentally or deliberately. Serious security breaches sometimes occur from within the network, originated by authorized users rather than unauthorized attackers outside the firewalls. Therefore, there must be mechanisms in place for the accountability of internal users. These normally include audit trails and logs that authenticate users and record their actions.

The key point is authentication. A user cannot be accountable if the system has not authenticated them. Authentication on its own, however, is of limited use because it simply verifies that the user is who they claim to be and places no restrictions on access to resources. In order to restrict access on a per user basis, we need authorization. The standard method for authentication is through network logons and combinations of usernames and passwords. Authorization, on the other hand, is typically applied though the granting of permissions based on the user's functional requirements or their role; which, again, raises the lost-device issue. If a lost device falls into the hands of a knowledgeable attacker, devastating damage can result. Timely reporting of the loss of devices and access tools (cards, fobs, and so on) is critical in these cases to limit the threat and mitigate damage.

Nonrepudiation

Nonrepudiation addresses when someone denies they took a certain action. That is, nonrepudiation provides undeniable evidence that the action was taken and by whom. Nonrepudiation is important in e-commerce and financial transactions such as online trading.

A simple application of nonrepudiation is when sending and receiving documents. The sender may attach a read receipt, which the recipient signs and returns. That would appear to meet the requirements. Unfortunately, however, there is no guarantee that the person who read and signed was indeed the intended recipient. Therefore, the recipient can repudiate, or deny, the action.

With nonrepudiation, digital signatures are used. If the sender signs the document with their own digital signature, that is considered nonrepudiation of origin. Similarly, when the recipient receives the document, that person can use their digital signature to sign the receipt. That is considered nonrepudiation of delivery.

At issue here is users connecting over unsecured Wi-Fi links, where passwords can be stolen or where sessions can easily be hijacked. This makes the validity of nonrepudiation difficult to ascertain.

Threats to Wireless and Mobile Devices

While the expanded principles of information security are good in a general sense, they are not very instructive with regard to the day-to-day protection of mobile and wireless clients, devices, and networks. This section looks at some specific threats, which are broken into three categories:

- Data theft
- Device control
- System access

These categories will have some overlap, but they cover the full breadth of threats.

Before looking at specific mobile and Wi-Fi threats, it's important to remember that virtually every threat that exists on wired networks also exists on wireless and mobile networks.

In addition to these threats, some threats are specific to wireless and mobile or now have wireless and mobile variants. This section will focus on these threats.

In a wireless or mobile environment, you can also deny service by creating radio interference that prevents clients from communicating with access points. Another method is by physically taking out a mobile backhaul connection to deny network access to smartphone subscribers. This is where the complexity of wireless and mobile security comes into play. On the surface, they are merely new means of accessing a network. But they represent entirely new dimensions of threats and vulnerabilities. That means managing mobile and Wi-Fi–enabled devices requires a broader approach to security management.

Data Theft Threats

Regardless of the method used to compromise or gain access to a mobile device, the hacker or cybercriminal typically has specific goals in mind, many of which pertain to data theft. Often, these hackers go after a target of opportunity rather than instigating a targeted attack on a specific company. In such a case, they will look for **personally identifiable information (PII)**, which is information that can be used to identify, contact, or locate a single person, or to identify an individual in context, with any business-related data considered a bonus. Typically, the areas of interest for cybercriminals are:

- Credentials for personal or business accounts
- Credentials for business or personal information
- Credentials for remote access software for business networks
- Access to data and phone services

With the huge proliferation of mobile devices that support LTE and broadband wireless data over the last few years, there has been an increase in mobile-specific attacks designed to steal data. The most common threats are as follows:

- **Sniffing (also called snooping)**—The most obvious vulnerability of any radio-based communication is that signals can be easily intercepted through a process called sniffing or snooping. No physical access to the medium is needed. Fortunately, this is easily circumvented with encryption. Although not all encryption methods are created equal, even so-called "weak" encryption is better than nothing. Yes, some methods of encryption can be cracked. But the reality is that most sniffing and snooping hacks are simply targets of opportunity. As such, having at least basic security will prevent most of these attacks. Basic or weak encryption will not stop a determined and talented hacker from breaching a specific target, but any company likely to be targeted would be expected to have robust data protection. For the vast majority, simple steps go a long way.
- **Malicious applications (malware)**—While malicious applications, or malware, have existed for many years, the 10,000-fold increase in the number of applications now available on smart devices (as compared to software downloaded on PCs) has opened many more attack vectors. These applications range from malware that can be autoinstalled on phones to **spyware** that can copy emails, texts, and contacts. There is also a significant privacy issue due to the ability to track the location of a mobile device, not to mention the significant amount of PII that these devices contain. This can result in some new, formerly

unheard-of, implications. For example, a malicious application capable of logging Global Positioning System (GPS) locations and text messages on the phone of an executive—who, say, schedules a hotel rendezvous with a person who is not their spouse or performs some other potentially damaging deed—could be used to blackmail that executive for insider information or access. It may sound far-fetched, but this sort of thing actually happens, and is the reason strict restrictions on downloading apps must be put in place.

- **Browser exploits**—Specifically targeting mobile users, these exploits take advantage of vulnerabilities on mobile web browsers. This a big consideration for Bring Your Own Device (BYOD) environments because unlike company-owned assets, on which updates can be mandated and managed, personal devices can and often are several updates behind. That means that simply by visiting an unsafe webpage, a user can trigger a browser exploit that installs malware or performs other actions on their device. This is another reason that **mobile device management (MDM)** is a critical tool in a BYOD environment.

- **Wireless phishing**—**Phishing** involves sending fake emails or Short Message Service (SMS) messages to a target in an attempt to get the victim to click a link that will take them to a fraudulent website. Once there, the user will be prompted to enter account credentials or other confidential information, which can then be used by the cyberthief to access the real account. Phishing is exacerbated by smart devices with small viewing screens, which can make it difficult to notice some of the telltale signs of phishing. Hackers can also take advantage of users who connect to rogue access points (and variants called **evil twins**).

- **Lost or stolen devices**—This is obvious and low tech, yet it remains one of the most common threats for mobile devices. Not only do lost or stolen devices result in data loss they also can result in unauthorized system access. When a device is lost or stolen, the data on it is also lost. In addition, the device is compromised, especially if it has remote access software configured for access to the corporate network. Often, these secure virtual private network (VPN) connections are set up to autoconnect for convenience using weak passwords. Passcodes can prevent some issues—at least with the casually curious person who finds or takes a device—but once in the hands of a skilled hacker, these passcodes are easily circumvented. Timely notification and blocking or wiping the device are the best options if it is lost or stolen.

- **System or device takeover**—With IoT, specifically with wireless-controlled IoT devices, attackers can block physical or remote access. This can be used to physically damage assets (for example, remotely turning off the AC in a server room causing servers to overheat). It could also just allow someone with an anarchist bent to wreak havoc with lighting, door access, and so on.

 **NOTE**
Data theft can occur without depriving the owner of access. Any data or information that can be seen can be replicated and used elsewhere.

Device Control Threats

In addition to obtaining data resident on a device, a hacker often aims to control the device itself. With control of the device, the hacker not only has ongoing access to data but can also use the device to launch other attacks or leverage the permissions on the device to gain access to higher-valued targets such as internal servers. This occurs through a process of lateral movement across the network called *lily padding*, or *island hopping*, in which the hacker "hops" from one device to another, with each hop getting the hacker closer to the target. Examples of device control threats include the following:

- **Unauthorized and modified clients**—Users sometimes create vulnerabilities when they try to circumvent policies or device configurations. This can open backdoors and other vulnerabilities. Examples of this include user hacks found on the Internet to alter a smart device (referred to as **jailbreaking**) and opening a smartphone hotspot (without security). This is a problem because the corporate network views these devices as authorized clients, which may be used by hackers to access systems or data. Again, this is a new issue brought about by BYOD, where some people see no issue with using their device however they see fit.
- **Ad hoc connections and software-based access points**—Ad hoc networks have been possible for many years, but setting one up formerly required a high level of technical skill. With new smart devices, however, it's quite easy to make these connections, which are easily exploited.
- **Endpoint attacks**—Several tools now exist that can attack wireless clients directly. An automated tool called Metasploit, for example, can be used to probe Wi-Fi clients for thousands of known vulnerabilities. Once exposed and exploited, the Wi-Fi client can be controlled and/or monitored.
- **Bluetooth Wi-Fi hacks**—Traditionally, vulnerabilities in Bluetooth protocols have enabled hackers to gain access to and control of mobile devices. However, this is no longer as easy as it used to be because Bluetooth is now switched off and set to non-discovery mode by default. If a user changes this setting, however, hackers can easily take control of a Bluetooth-enabled mobile device—a real concern for BYOD.
- **Near field communication and proximity hacking**—One technology that allows an ad hoc wireless connection between two devices that are within a few feet of each other is near field communication (NFC). Unlike Bluetooth, the pairing process is automatic. Already used extensively for social media and to exchange contact information, the future of NFC includes the ability to autopay via credit card at point-of-sale (PoS) terminals and will likely become a prime target for hackers.

> **NOTE**
>
> Jailbreaking modifies Apple's iOS to allow unsigned code to run on Apple devices such as iPhones and iPads. This allows the user to download and install third-party applications from sources other than Apple's App Store. **Rooting** is a similar process, but it only applies to Android devices. Rooting grants the user access to the root account of Linux.

System Access Threats

As noted, hackers are often interested in deeper access into a network. For them, device control is simply a means to an end. However, there are cases in which hackers are more interested in breaking the network or disrupting network access for political or financial gain or, in some cases, to exact revenge for a real or perceived insult or injury. Examples of these types of system access threats include the following:

- **DoS attacks**—As discussed, wireless local area networks (WLANs) and mobile networks are vulnerable to both network-based DoS attacks and those created specifically to attack the inherent weaknesses of radio-based systems. In the case of Wi-Fi, using the less-crowded 5 GHz band reduces the chance of an accidental DoS but does not help with targeted attacks. Modulation techniques that spread communication over multiple frequencies and channels help a great deal, but there are also sophisticated jamming techniques that hackers can bring to bear.

- **Evil twin access points**—An access point can easily be set to the same network name—**service set identifier (SSID)**—as a legitimate WLAN or hotspot, fooling unsuspecting users into connecting. This is not a new problem, but there are new hacker tools that can listen to clients to see what SSIDs they are looking for and then configure themselves to look like one of those networks. The client will then connect without the user having done a thing. Once connected, the client is subject to a full host of network attacks.
- **Rogue access points**—Unauthorized or rogue access points have been a problem for as long as Wi-Fi has been commercially available. Today, the appearance of rogue access points is usually due to poor site planning, which results in wireless dead zones. Out of frustration, an employee may set up a rogue access point to gain access to the network. But if a hacker gains entry to a building, they can easily set one up as well. Unless regular site survey sweeps are conducted, rogue access points may go unnoticed by IT for some time, resulting in a lingering vulnerability.

Risk Mitigation

Several risk-mitigation methods can be applied to mobile devices and Wi-Fi clients to address the wide variety of threats discussed in the previous section. Typically, the worst-case scenario is a lost or stolen device that is completely vulnerable. The practices for protecting data on corporate-liable devices involve encryption, wiping lost devices, and remote locking. These practices are also necessary when employee-liable devices are allowed to access and download company data. Therefore, they should be considered mandatory when developing a BYOD or **corporate owned personally enabled (COPE)** strategy.

The key methods to consider are as follows:

- **Mobile device screen locks and password protection**—These are the first line of defense against unauthorized access to business data and accounts residing on a mobile phone or tablet.
- **Remote locks and data wipes for mobile devices**—Passcode locks are usually sufficient to prevent a casually curious person from gaining access to a smart device, but a knowledgeable hacker can easily circumvent them. If a device is lost or stolen, a remote lock can temporarily secure a device. If the device is not recovered, a remote data wipe or swipe will prevent all future access to business data and accounts stored on the phone.
- **Mobile GPS location and tracking**—During the lifetime of any mobile phone device, it will collect and transmit a whole host of details regarding the phone's environment and location. This is very useful data if people or assets need to be tracked. It's also useful in locating lost or stolen devices.
- **Stored data encryption**—In most cases, device locks and data wipes are sufficient to mitigate the risk of data theft, data loss, and data leakage. But as an added security measure, executives and other employees with access to sensitive information should encrypt

data on their personal devices. (There is little need for most employees to have data encrypted on their personal devices, however.)

These are four basic precepts of mobile device security practice. However, the best practice is to establish controls that ensure **compliance** with a business plan and policy. One of the key policies must be immediate notification when a device is lost. Once notified, IT can remotely lock the device and perform a GPS location trace. If the device can be retrieved, the remote lock can be removed. If not, the device is wiped. The key to risk mitigation is to minimize the window of vulnerability via timely notification. It's worth noting, however, that thieves who target mobile devices know to put stolen devices in metal boxes so that they can't receive radio signals (for example, the signal to lock the device). They then bring it to a room protected from radio signals, where they can pull data from it.

It's also important to balance user satisfaction with risk mitigation. To maximize employee satisfaction (and policy acceptance), it's essential to realize that some level of freedom and autonomy in the use of employees' own devices makes for a happier workplace.

Mitigating the Risk of BYOD

The proliferation of wireless mobility throughout the general population has caused the BYOD trend to become a norm in business. It started when companies began issuing BlackBerry mobile devices to enable employees to access email remotely. Soon, employees began requesting that their personal BlackBerry devices be configured for use with the BlackBerry server. Today, some form of BYOD is the norm for most companies and the BYOD market is greater than $300 billion at the time of this writing, up from $94 billion back in 2014 (according to Global Market Insights).

It wasn't long before BlackBerry found itself usurped by Apple, Android, and even Windows Mobile. It became common practice for employees, including CEOs and CTOs, to bring their personal devices to work and to use them to gain access. This allowed employees to consolidate all their communications, whether business or personal. Business policymakers, who were less risk-averse than IT, soon saw the potential benefits of having employees use their own devices. These benefits, which included increased productivity, creativity, collaboration, and mobility, were too good an opportunity to pass up, despite the obvious risks. Therefore, it was not a case of IT allowing BYOD; it was a matter of securing it.

To address the problems associated with having employee-owned devices active on a secure company network, it was imperative to develop access and acceptable use policies. After all, a business may be deemed culpable and responsible for a device using unlicensed or pirated applications on its network. Furthermore, it might find itself deep in litigation if an employee were to use a private mobile device with illegal content, were to be the source of a hate crime, or were to engage in pornography or something similar, which the authorities traced back to the company network. This is why MDM and **mobile application management (MAM)** are so important to a company's security policy.

Mobile Device Management

MDM is a technology that has emerged to enable network security administrators to manage mobile devices. MDM typically sends over-the-air signals to mobile devices to distribute applications and configuration settings for all makes of mobile phones and other mobile devices. The intent is to provide a central point of control and policy from which to enhance the functionality and efficiency of mobile communications while reducing costs and risk.

MDM architecture consists of a server element, which is the central management system, and a client element, which resides on the mobile device. The MDM server provides automatic identification and configuration of any new mobile device that attempts to join the network for the first time. Subsequently, it maintains a history of all configurations and updates sent to the devices on the network and can send new updates over the air (OTA) when required. However, the job of MDM is not just to keep mobile devices up to date. The application typically provides other essential services such as remote locking, location tracking, and wiping of the device in case of loss or theft.

The server can initiate MDM security features, which send the commands OTA to the remote mobile device. This is the case when activating remote locking and device wiping, which are crucial security measures when dealing with mobile devices that may contain sensitive or valuable company data. Device tracking, however, relies on the GPS functionality of the mobile device to report its current GPS location. With both company and employee-owned devices, these measures—especially mobile device tracking—are sensitive subjects. While employees may allow the remote locking and wiping of data from their phones under certain circumstances, they are less likely to be thrilled by the possibility of their movements being logged, especially during non-office hours. For this reason, GPS tracking is usually activated only after a device has been reported lost or stolen.

MDM provides a valuable management facility for administering and securing mobile devices on large networks. MDM can be expensive for smaller companies, but that capital expense can be avoided through the use of a Software as a Service (SaaS) cloud-based MDM solution. No matter which MDM solution is implemented, it will facilitate the ability to manage and administer a potentially large number and wide variety of mobile devices.

Mobile Application Management

MDM provides a method for configuration and policy management, but what about application management and control? How do you securely manage the provisioning and delivery of mobile applications? The inherent risk of BYOD, which makes it so unpalatable to IT, is not so much about the physical devices. Rather, it is about the data they contain and the access they enable. Of particular concern are applications of unknown origin or quality residing on the employee's device. After all, the employee is likely to download all sorts of recreational applications, some of which open backdoors to the device. To mitigate this risk, the following steps must be taken:

- **Secure applications**—Mobile applications have few, if any, access restrictions, and can open up all sorts of phone features without the user's knowledge or explicit consent. These include location tracking and camera operation. There may or may not be some indication of the **End User License Agreement (EULA)**, the legal agreement between a software developer and users.
- **Secure network access**—Because a mobile device could well be shared by many authorized users, it is imperative to authenticate not just the user but also the device. You must also authenticate both the user and the device with regard to context—that is, time, location, and destination.

- **Encryption**—You must ensure that data is encrypted when in transport and, ideally, in local storage, too. Also, it's a good idea to compartmentalize data storage, separating company files from personal data such as the employee's vacation photos.

MAM is responsible for administering and managing applications on mobile devices. MAM software controls the provisioning and distribution of in-house mobile applications and, in some cases, commercially available applications through an enterprise application store. With MAM, the IT department can verify and authorize the download of in-house and commercial applications from the central store. This goes a long way toward establishing a secure application management system.

When you combine MAM with MDM, you typically have the features required to authenticate users, provide and deliver applications, handle application revision management, handle updates, produce performance and status reports, and control access to users and groups. With these technologies in place, you can:

- **Empower people**—You can let people use the device of their choice. This improves collaboration, mobility, and, the ultimate goal, productivity.
- **Protect sensitive information**—You can restrict access to and the downloading of sensitive information.
- **Protect devices and data from unauthorized access**—By using techniques such as password blocking, remote locking, and device wiping, you can secure data on a remote device, even if it is stolen.

Other Risks with BYOD

While there may be positive claims with regard to BYOD, reduced costs for IT management and security are not among them. Logic would dictate that the cost of securing a remote mobile device is higher than the cost of securing a desktop computer just based on the number and variety of smart devices and software versions in service. In addition, BYOD and, to some extent, COPE suffer from some inherent vulnerabilities that must be addressed. These include the following:

- **Legal separation between personal and business use**—How can IT reasonably insist that users comply with its version of "acceptable use" on a device that is not company owned? After all, it is not the role of IT to apply its view of ethics to an employee outside business hours. But adhering to acceptable use policy is something that BYOD participants must agree to and accept under a BYOD policy.
- **Leakage of company data into the wild**—The biggest danger that arises with regard to data security is the downloading of information onto a device that is, strictly speaking, out of the company's control. To address this, there must be an MDM application that controls and manages employee devices. Employees must accept this, even though the device belongs to them. Employees must also accept and sign a BYOD policy that enables the company to lock or remotely wipe a device if it is lost or stolen.
- **Enforcement of policy and governance**—Applying policy and governance on fixed desktop systems is straightforward, but it is not so easy to apply the same rules and policy to remote, employee-owned devices.

- **Threat of loss and theft**—Once a device is in the hands of an unauthorized user, half the security measures have already been circumvented. All that's left is to break the username and password combination. This may seem like an acceptable risk until you realize that for many sites, users employ the same or similar login credentials. It is especially troublesome when employees use several types of mobile devices and have access to many different layers of security.

BYOD for Small-to-Medium Businesses

If you are a **small-to-medium business (SMB)** that cannot afford MDM or MAM, what are your alternatives? Implementing BYOD or COPE in an SMB environment is actually quite simple, and is an approach that businesses of all sizes should consider. Not only is it secure, but it is also very cost-effective. One solution is **desktop virtualization**.

Desktop virtualization is simply the reproduction of the user's desktop on an Internet-accessible server. By connecting to this virtual representation of their own company desktop, users can circumvent many of the security problems related to remote access. In this scenario, all execution and read/write operations are performed on the company server. Users may execute, write, read, and edit files, but the files remain on a company server, just as they would be if the users were to perform the same actions at their desks. There are no downloads to the remote device because only keystrokes or screen changes traverse the network. This ensures that there is no data leakage. Furthermore, by using a virtual desktop, remote users have a common desktop and set of applications. There is no longer the problem of compatibility between applications on home, personal, and company devices.

Critics of this approach note that many virtualized applications suffer performance limitations on touchscreen devices, resulting in lower productivity. As always, a reasonable balance must be maintained between security and user satisfaction.

Defense in Depth

To defend against many different kinds of attacks, security professionals must put various types of controls in place. They include the following:

- **Physical controls**—These are the physical security measures that safeguard the environment, such as doors, locks, cameras, security gates, and fences. Physical and environmental protection is included as one of the 18 control families in NIST SP 800-53 (discussed in the upcoming section "NIST SP 800-53").
- **Logical/technical controls**—These include the more obvious hardware and software devices and appliances that protect the network, such as antivirus software, firewalls, host intrusion protection, and network intrusion protection. Wireless mobile devices such as smartphones rarely have these features installed by default.
- **Administrative controls**—These include security policies, processes, and procedures.

MDM and MAM can provide configuration tools and can be used to enforce company policy, download antimalware software, and sandbox business and personal data (that is, run it in isolation) before a device joins the network. For wireless networks, this is an essential part of building a secure multilayer defense policy that adheres to the principles of **defense in depth**, or the thoughtful and strategic application of all these components in concert.

With defense in depth, security controls are applied to networks and systems in layers, the rationale being that should an attacker breach the perimeter, there will be layers of security devices to protect assets located deeper inside the network. By building security from the inside out, you can robustly defend the inner, high-value assets, with each subsequent layer radiating outward toward the perimeter having a reduced level of access control (see FIGURE 4-1).

This model allows for lower security requirements at the perimeter, where web servers and intranet services will be located. This is preferable to a heavily defended perimeter model, sometimes called M&M security, based on an ad campaign for the popular candy that was touted to be crunchy (hard) on the outside and chewy (soft) in the middle. This latter model is effective only if no web services are offered to public or remote users, which is rare. Besides, if you place all the defenses at the perimeter, it might keep intruders out, but if one should break through, there is nothing to prevent them from gaining access deep in the network.

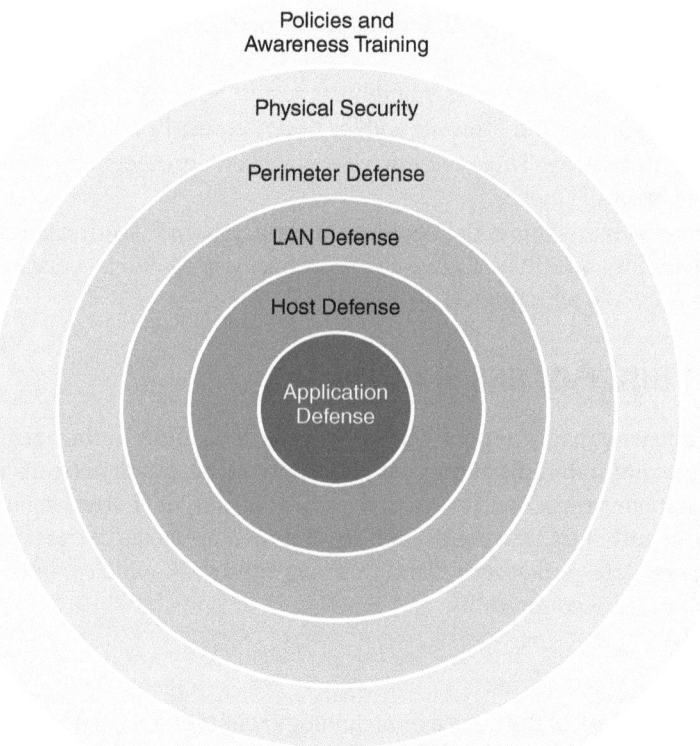

FIGURE 4-1

Defense in depth is the practice of deploying multiple forms of security to reduce the risk of deep penetration from unauthorized users. Unauthorized users would have to breach several forms of security to reach an intended target.

To secure a multiaccess (wired, wireless, mobile) network, best practices call for the use of multiple layers of defense. These include the following:

- **External network layer**—This is the perimeter layer where web servers and services are exposed to the Internet. Typically, layered firewalls provide a secure **demilitarized zone (DMZ)**, which exists between the untrusted Internet and the trusted (secure) internal network. The DMZ is used for web services and secure VPN access for remote users. Other controls may include host intrusion detection, logging, and vulnerability/penetration testing. If Internet/intranet access is being provided via Wi-Fi, then the access point should be attached to one of the less-secure outer-firewall interfaces. This will ensure that the firewall permits guest devices to access the Internet but denies them access to the company network.
- **Perimeter network layer**—This layer uses an inner firewall to segregate the external network from internal resources. This layer normally hosts more secure and restricted web services—typically for external front-end email servers and intranet web servers for employees, partners, and vendors. Internet proxies and external **Domain Name System (DNS)** servers, which resolve IP addresses to URLs, may reside here. These may be protected by **intrusion prevention systems (IPSes)** (which prevent authorized access), stateful packet inspection firewalls (which monitor the state of the network while blocking traffic), and **deep packet inspection** firewalls (which look at the payload as well as the headers for signs of malicious traffic). This is similar to a DMZ but is not intended for external anonymous access. Rather, it is a layer that requires secure access via login. It is not a fully internal system.
- **Internal network**—This is where the user and hosts reside. Security administrators protect this layer using IPS appliances and firewalls and/or router access lists between subnets.
- **Application server network**—This is an inner security zone protected by another layer of highly restrictive firewall rules.
- **Database server network**—This is the heart of the data network. It often features very tight security policies and multitiered access restrictions, segregated by very high security configuration and firewalls with limited open ports.

Authorization and Access Control

The key to applying security in any network, large or small, is to allow authorized users to go about their business unhindered, yet prevent unauthorized access to network resources. To achieve this, mechanisms must be in place to control access to the network. Furthermore, there must be a way to authenticate a user's identity and then apply authorization rules and access controls to users, data, and other assets. This is termed **AAA**, which stands for authentication, authorization, and accountability.

AAA

Access to a network regardless of the device or technology, whether a fixed desktop or mobile wireless device, requires authentication, authorization, and accountability, defined as follows:

- **Authentication**—The process of validating a claimed identity, whether a user, device, or application

- **Authorization**—A process that works in conjunction with authentication to grant access rights to a user, group, system, or application
- **Accountability**—A chronological record of system activity that can be forensically examined to reconstruct a sequence of system events

Having robust AAA measures is important to a network's general security. Securing assets from unauthorized access via the network is as important as preventing unauthorized access to the network by intruders. This is especially true today, where there has been shift from the traditional model of users being internal to the network toward one in which many users are external and beyond the firewalls. Additionally, access control must cater to the different security and access controls necessary to support employees' remote access.

Only a few years ago, remote-access users had company laptops with secure VPN connections that were superficially protected by a username and password challenge. The actual security, however, was much deeper, with the device also being authenticated. In other words, there was a two-part measure of authentication: the laptop and the user. This level of security was sufficient because an unauthorized user could not simply copy the VPN client application or transfer it to another machine. A user could download the client and install it from the Internet, but the authentication would fail because the VPN server application had to create and publish the security keys to the VPN client. The devices used these keys to authenticate one another and to encrypt the traffic between the pair. Therefore, the VPN client server model was very secure.

Today, it's common for employees to use their own devices for work, whether they be laptops, smartphones, tablets, or home PCs. Therefore, they require the same level of access and authorization on their own devices as they have on their company-assigned desktop PC. This new way of accessing network resources requires a new security approach. Now, security must take into account not just the authorization of the user and their device, but also the context.

Context-aware security devices came about as a method to provide greater granularity in applying access controls. Security policy dictates the rules that apply to the authentication process. These rules take into consideration not just the user's details but which device that person is using and even the location and time. By considering these extra criteria, security administrators can apply different access rights and authorization to different contexts. For example, a context-aware security device identifies the user (who), the application or website that the user is attempting to access (what), the time of the access attempt (when), the location or origin of the request (where), and the device on which the request has been made (how).

Context-aware firewalls operate up to and including Layer 7 of the OSI model. This brings a level of granularity not possible with traditional or even second-generation firewalls. This is due to their flexible, application-based rule structures. Non-context firewalls, which rely on IP addresses and port numbers when constructing access rules, lack the flexibility required in today's business environments. For example, a context-based firewall will have business rules such as "deny or allow Skype" or "allow Yahoo! Messenger but not file sharing." While a typical firewall cannot handle these common security requirements well, context firewalls can and do. In addition, they can identify and make decisions about the required attributes—including user, application, location, time, and device—to build simple but effective layered security policies for handling remote access from BYOD phones and tablets.

However, access and authentication are merely the gatekeeper functions of security. Once the security device has authenticated and authorized the user, you must still police activity and protect company assets, including preventing information from leaving the network. The big difference between mobile devices and company desktops is that when a user downloads a file to a mobile device, the data leaves the company network to reside on said device, outside the company's physical control. This is potentially a big problem and is one of the many security challenges brought about by the proliferation of smartphones and tablets as network access devices.

Information Security Standards

How can IT manage all these wireless devices and enforce a common policy? To assist with this and with the broader aspects of IT security, standards have been developed. Two such standards arose from joint subcommittees of the International Organization for Standardization (ISO) and the International Electrotechnical Commission (IEC), called the ISO/IEC. These voluntary standards, ISO/IEC 27001:2013 and ISO/IEC 27002:2013, address different aspects of and approaches to IT security. The National Institute of Standards and Technology (NIST) developed another standard, the NIST SP 800-53. Mandatory for all unclassified U.S. government-run networks, the NIST SP 800-53 has also become a de facto standard for many foreign governments and for private organizations and businesses throughout the world.

ISO/IEC 27001:2013

The purpose of the ISO/IEC 27001:2013 standard is to "provide requirements for establishing, implementing, maintaining and continuously improving an Information Security Management System (ISMS)." This model of establishing and improving was heavily reliant on the **PDCA cycle**. As shown in **FIGURE 4-2**, PDCA stands for plan, do, check, act.

Later revisions of the standard moved the focus to evaluating and measuring the performance of IT security management. Its purpose, however, remained the same: to assist in the design and implementation of an organization's security system. ISO/IEC 27001:2013, although still focused on PDCA, addresses establishing, reviewing, and improving the information security management system and its inherent processes.

FIGURE 4-2

The PDCA cycle.

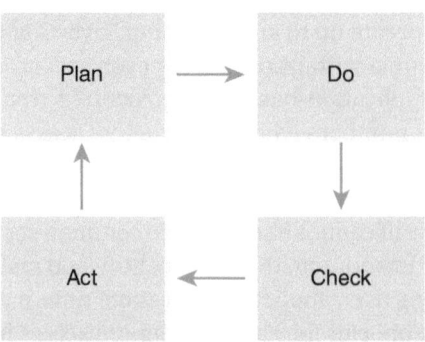

ISO/IEC 27002:2013

The ISO/IEC 27002:2013 standard consists of guidelines, techniques, and general principles for initiating, implementing, managing, and improving information security within an organization. The ISO/IEC 27002:2013 standard addresses 14 general groups, each with its own objectives. It also suggests 114 controls that are seen as best practices for achieving those objectives. The 14 groups are as follows:

- Information security policies
- Organization of information security
- Human resource security
- Asset management
- Access control
- Cryptography
- Physical and environmental security
- Operations security
- Communications security
- System acquisition, development, and maintenance
- Supplier relationships
- Information security incident management
- Information security aspects of business continuity management
- Compliance

Data from ISO/IEC 27002:2013 Information Technology—Security Techniques—Code of Practice for Information Security Controls." International Organization for Standardization. Accessed August 30, 2014.

This wide breadth of concern and responsibility for each of the 14 groups falls under the purview of information security within a business environment. However, it isn't that simple, because these security standards are not mandatory. Rather, they are suggestions and best practices for addressing areas of concern. They may be highly recommended techniques and practices, but the standard remains flexible enough to allow security teams to tailor it to any situation. The ISO/IEC designed these standards to be comprehensive enough to cover even the largest of organizations.

NIST SP 800-53

The NIST SP 800-53 standard, titled "Security and Privacy Controls for Federal Information Systems and Organizations," outlines a risk-management framework that addresses security controls for federal information. The specification covers 17 areas, including access control, incident response, business continuity, and disaster recovery.

To ensure that these safeguards are properly implemented, the NIST requires vendors seeking selection and implementation on federal networks to undergo a certification and accreditation process. These controls include management, operational, and technical safeguards, as well as countermeasures aimed at ensuring the confidentiality, integrity, and availability of systems and information.

Regulatory Compliance

In 2002, after the failure of such businesses as WorldCom and Enron, the U.S. government introduced new regulations aimed at preventing companies from intentionally or unintentionally losing, masking, or altering securities-related information. The collapse of these companies made it clear that ISO/IEC standards were not sufficient to protect investors and employees. Executive officers sometimes felt they were above the rules, and security officers did not question them. It simply did not occur to these security officers that financial data would need protection from modification by a company's executive officers!

The failure of self-regulation brought about new, widespread regulations. These regulations dictated that the integrity of financial reporting was now the responsibility of auditors, executives, and members of the board. This spawned a wave of both government and industry regulations, all of which must be complied with. This has prompted the addition of the word compliance to the language of IT; in this case, meaning obedience to the entire scope of security and business laws and regulations.

The Sarbanes–Oxley Act

The **Sarbanes–Oxley Act (SOX)**, also known as SarbOx, was enacted to address investor confidence and corporate financial fraud through reporting standards for public companies. It ensured that public accounting firms would be liable for failures in financial reporting. Interestingly, SOX did not address information security directly. In fact, it barely got a mention. What it did address, however, was corporate governance and the integrity of financial reporting. However, because financial auditors were one of the principal targets of SOX—and they had experience implementing and auditing information security systems—the emphasis was on ensuring the integrity of financial data, which is an information security role.

The impact of this regulation was felt by all public companies, which were obliged by law to adhere to it. Of course, there were additional audits and regulations, but it also had a huge impact on information security—even though its actual aim was to regulate financial reporting and governance. Specifically, it required that "each annual report ... contain an internal control report ... [that] contains an assessment of ... the effectiveness of the internal control structures and procedures of the issuer for financial reporting."

This is of course a direct and unavoidable reference to securing the integrity of financial data. But for executives (and auditors), it was also a clause about information security. In one regard, that was a good thing, because it brought security to the forefront of corporate strategy (and increased security budgets along with it). Unfortunately, it also brought a shift in emphasis, because information-security efforts now had to comply with SOX. Consequently, companies targeted their information-security initiatives accordingly.

The Gramm–Leach–Bliley Act

The purpose of the **Gramm–Leach–Bliley Act (GLBA)**, originally enacted in 1999, is to secure and protect personally identifiable information held by financial institutions. The legislation explicitly states that institutions must protect the confidentiality and integrity of the financial information stored on their systems. GLBA concerns itself with confidentiality, integrity, and availability. The requirements are as follows: "Each bank shall implement a comprehensive written information security program [policy] that includes administrative, technical and physical safeguards."

The Health Insurance Portability and Accountability Act and the Health Information Technology for Economic and Clinical Health Act

Although the **Health Insurance Portability and Accountability Act (HIPAA)** focuses on privacy and security for patients receiving health care, it has a direct impact on IT with regard to how electronic information is stored and transferred. HIPAA is concerned with C-I-A. The requirements are to "[i]mplement reasonable and appropriate policies and procedures to comply with the standards, implementation specifications, or other requirements of this subpart."

Supplementing HIPAA is the **Health Information Technology for Economic and Clinical Health (HITECH) Act**, which was enacted as part of the American Recovery and Reinvestment Act of 2009. The HITECH Act addresses privacy and security concerns associated with the electronic transmission of health information and supplements and strengthens the enforcement of HIPAA rules.

The Payment Card Industry Data Security Standard

Arguably, the most pervasive industry regulation is the **Payment Card Industry Data Security Standard (PCI DSS)**, a comprehensive industry standard aimed at ensuring the safe and secure handling of credit cardholder information at all steps of the payment process. This mandatory industry regulation, which began as a series of separate programs at each of the major credit card companies, was developed in 2004. It now covers credit, debit, and ATM cards, as well as other forms of electronic payment.

PCI DSS specifies 12 requirements, which are organized into six control objectives:

- Build and maintain a secure network
- Protect cardholders
- Maintain a vulnerability management program
- Implement strong access control measures
- Regularly monitor and test networks
- Maintain an information security policy

At all levels of the payment process are checks, audits, and annual assessments that, depending on the size of the retailer (or processing company), can range from self-reported checklists to in-depth onsite visits and inspections by a team of auditors.

Given the huge financial gains for cybercriminals and the far-reaching implications of a breach of credit card information (not to mention the headlines and devastating consumer reaction), the PCI Standards Council has become one of the most powerful nongovernment regulatory bodies in the world.

GDPR & CCPA

Prior to the introduction of the European Union's (EU's) General Data Protection Regulation (GDPR), data security and regulations were primarily focused upon the C-I-A of the data. However, when the EU's GDPR came into force in May 2018, privacy and more specifically the user's personal data became the focal point of IT security. The privacy protection regulation was designed to modernize laws aimed at protecting the personal information of individuals. It also provides the controls and enforces the rights of individuals and gives them more control over their personal information. Even though the GDPR is an EU regulation, it has extraterritorial reach in that it applies to anyone, anywhere, who does business with an EU citizen.

The GDPR has seven key principles:

- **Lawfulness, fairness, and transparency**—User data collected should be on a legal basis, be fair—clear, open, and honest—as well as transparent about how you will use the data—and not be misleading or unexpected to the user.
- **Purpose limitation**—A user's personal data can only be collected for a specific and agreed purpose.
- **Data minimization**—User data collected should be only the minimum required to fulfill the agreed stated purpose.
- **Accuracy**—User data must be accurate, and the user has the right to view and edit their personal data.
- **Storage limitation**—User data can only be stored for a specific and stated time period.
- **Integrity and confidentiality (security)**—User data must be adequately protected.
- **Accountability**—The business must be responsible for the way they handle user personal data and how they comply with the other six principles.

The California Customer Privacy Act (CCPA) is the State of California's version of the GDPR and will come into force in 2020. The CCPA relates to anyone who does business with a Californian citizen. Similar to the GDPR, its aims are to protect the privacy of Californians' personal data from misuse and monetization by big tech businesses.

Both the GDPR and the CCPA will have profound effects on mobile application development, because developers were previously harvesting vast amounts of user data without any permission or product relative purpose. For example, wearable health monitors became popular around 2015, but it came as a surprise to many users just how much personal data was being harvested that went well beyond user expectations and was then transmitted and stored remotely. Now, with these regulations in force, businesses in the data harvesting market must be very careful that they do not fall foul of the regulators by misusing their customers' personal data.

Detrimental Effects of Regulations

For all their good intentions, there are negative sides to regulations—and some would argue they are big negatives.

One downside is that many regulations, such as SOX and HIPAA, do not actually address security standards, techniques, or practices. Instead, they address inflexible business requirements. Consequently, there is confusion within the security profession and in business in general about the difference between being secure and being compliant. Unfortunately, some fail to understand that regulatory compliancy is simply a subset of security, not the overriding goal.

This is a critical point, and one worth stating explicitly: Compliance and security are not the same thing. In fact, several companies that have suffered significant security breaches over the past few years—including retailers such as Target, Neiman Marcus, and Michaels— were certified as being compliant at the time of the security incident. This is not to claim that regulations such as SOX, HIPAA, GLBA, and PCI DSS do not make companies more secure or that the practices they outline are not valid. It just means that, on their own, they do not provide the strength and depth to secure an organization. In a nutshell, the following equation applies:

$$\text{Security Best Practices} + \text{Regulatory Compliance} = \text{Corporate Security}$$

Another downside of regulation is the bandwagon effect that has occurred in recent years. Worried about being left on the sidelines or motivated by the opportunity to introduce new legislation, virtually every industry, country, state, and, in some cases, city has created a data privacy regulation of its own, forcing companies that do business in their area of control to comply. In most cases, these regulations are redundant. Even so, companies bear the burden of proving compliance. Herein lies the problem. Security teams spend a significant portion of their time on compliance—or more specifically, on compliance paperwork. This takes them away from actually securing networks. Ironically, by focusing corporate governance and compliance on regulations, the industry may have made companies less secure.

Proponents of regulation argue that regulation is necessary. They believe that even more specific laws should be enacted to make industries compliant with best practices, the rationale being that companies will enforce security only if they are forced to do so. On the other end of the spectrum, opponents of regulation claim that regulations stifle innovation and prevent companies from using initiative and creativity to mitigate threats in a more economical fashion. They believe that the measure of compliance is through the validation of documented practices, procedures, and work orders. However, that means all these documents must be in order if the company is to pass an audit. The time and expense required to produce and maintain this documentation, not to mention train IT to follow the regulations and to police others in the organization, are prohibitive, especially given the scale of documentation required.

This is the problem with enforcing regulations (and indeed, any standards): An auditor can only audit what is documented. If the company keeps procedures and work practices to a minimum, then it is easier to pass the audit. The more documentation and procedures, the harder it is to comply. After all, the purpose of the audit is to check that the company's employees do what the procedures say they will do through reasonable documented procedures, processes, and work practices. Therefore, as long as the company takes reasonable and documented measures to secure the company's assets, that is sufficient to pass an audit.

CHAPTER SUMMARY

The existence of wireless and mobile smart devices on corporate networks presents some unique challenges to IT security professionals. When those devices are user-owned, the challenges are multiplied. Security teams must not only account for the inherent vulnerabilities of wired and mobile technologies, but must also deal with user behavior, which can be unpredictable. In addition, the increased risk associated with lost or misplaced devices far eclipses the risk associated with PCs. In response to this, IT security teams can employ processes and controls to reduce risk.

Sound planning backed by adherence to best practices must be balanced with the inherent need for legitimate access in support of the business. Likewise, compliance with government and industry regulations helps ensure that basic best practices are applied. However, compliance is merely a component of a comprehensive security practice. It should not become a means unto itself.

Understanding the specific threats that target Wi-Fi and mobile devices and employing tools such as MDM and MAM are the keys to bringing risk down to acceptable levels without getting in the way of the business or being at odds with user norms.

KEY CONCEPTS AND TERMS

- Authentication, authorization, accountability (AAA)
- Compliance
- Corporate owned personally enabled (COPE)
- Deep packet inspection
- Defense in depth
- Demilitarized zone (DMZ)
- Desktop virtualization
- Domain Name System (DNS)
- End User License Agreement (EULA)
- Evil twin
- Gramm–Leach–Bliley Act (GLBA)
- Hash
- Health Insurance Portability and Accountability Act (HIPAA)
- Health Information Technology for Economic and Clinical Health (HITECH) Act
- Information security
- Intrusion prevention systems (IPSes)
- Jailbreaking
- Least privilege
- Mobile application management (MAM)
- Mobile device management (MDM)
- Nonrepudiation
- Payment Card Industry Data Security Standard (PCI DSS)
- PDCA cycle
- Personally identifiable information (PII)
- Phishing
- Rooting
- Sarbanes–Oxley Act (SOX)
- Service set identifier (SSID)
- Small-to-medium business (SMB)
- Spyware

CHAPTER 4 ASSESSMENT

1. What is the purpose of ISO/IEC 27002:2013?
   A. To provide rules and methods for wireless security
   B. To provide a standard for cross-vendor solution compatibility
   C. To provide requirements for establishing, implementing, maintaining, and improving an information security management system
   D. To give regulators something to do

2. PDCA stands for which of the following?
   A. Plan, document, check, audit
   B. Plan, do, check, act
   C. People, documents, computers, access
   D. None of the above

3. Compliance with the ISO/IEC 27002:2013 standard is mandatory.
   A. True
   B. False

4. Compliance with government and industry regulations is the best way to ensure network security.
   A. True
   B. False

5. The concept of least privilege does which of the following?
   A. It identifies the haves and the have-nots
   B. It gives access to critical systems with minimal approval
   C. It limits access approval to one system per day
   D. It blocks access from all systems by default, giving access on an as-needed basis

6. Mobility and Wi-Fi have made AAA-based access much easier.
   A. True
   B. False

7. Which of the following describes the strategy and practice of implementing multiple layers of security?
   A. Defense in depth
   B. Perimeter security
   C. Least privilege
   D. Trust but verify

8. Which of the following is a common security threat for mobile and Wi-Fi–enabled devices?
   A. Physical loss and theft
   B. Malicious applications
   C. Phishing
   D. Unsecure or rogue wireless access points
   E. All of the above

9. Which of the following describes MDM?
   A. It is an important security certification
   B. It is a technology that enables network security administrators to manage mobile devices
   C. It is a data privacy regulation that pertains to the health care industry
   D. It is a technology that enables network security administrators to manage applications on mobile devices

10. MDM helps to do which of the following?
    A. Empower employees by letting them choose their own smart devices for use at work
    B. Protect sensitive information
    C. Protect devices and data from unauthorized access
    D. All of the above

PART II

WLAN Security

CHAPTER 5 How Do WLANs Work? **93**

CHAPTER 6 WLAN and IP Networking Threat and Vulnerability Analysis **125**

CHAPTER 7 Basic WLAN Security Measures 153

CHAPTER 8 Advanced WLAN Security Measures 175

CHAPTER 9 WLAN Auditing Tools 199

CHAPTER 10 WLAN and IP Network Risk Assessment 221

How Do WLANs Work?

CHAPTER 5

BEFORE LOOKING AT THE VULNERABILITIES of and security measures for wireless local area networks (WLANs), it's important to gain a fundamental understanding of how WLANs work. This chapter provides a basic overview of WLAN design and discusses the operation and behavior of wireless in general and 802.11 WLANs in particular. This overview includes not only WLAN topologies and Institute of Electrical and Electronics Engineers (IEEE) standards but also a primer on radio frequency (RF) behaviors as well as antennas and site-survey considerations. Armed with this knowledge, the information security specialist is in a much better position to recognize, mitigate, and prevent security threats targeted at or enabled by wireless networks.

Chapter 5 Topics

This chapter covers the following concepts and topics:

- What wireless LAN topologies exist
- What the 802.11 standards are
- How 802.11 unlicensed bands operate
- How wireless access points work
- How wireless bridges work
- How wireless antennas work
- What a site survey is and why it's important

Chapter 5 Goals

When you complete this chapter, you will be able to:

- Describe the common WLAN topologies
- Identify the differences between the common 802.11 standards

© Cherezoff/Shutterstock

- Describe the common methods of frequency spreading
- Describe the basic functionality of access points
- Understand the difference between an access point and a wireless bridge
- Describe and understand the different types of antennas, along with their uses
- Understand the reasons for and steps behind site surveys

WLAN Topologies

To understand how WLAN systems work, you must consider the basic components that make up a WLAN topology. The fundamental requirement of any wireless communication is that devices communicate over a common radio frequency. WLAN technology is no different because it requires each device in the network to be tuned to a specific radio frequency within the **ISM Unlicensed Spectrum.**

ISM Unlicensed Spectrum

ISM is short for the industrial, scientific, and medical band of the unlicensed wireless spectrum. In addition to these specific purposes, the ISM band is also available for use with short-range wireless communication techniques such as in personal area networks (PAN), WLAN, and wireless wide area networks (WWANs). A significant objective of the ISM is to provide easier access to the wireless spectrum for innovators and product developers through access to unlicensed RF bands. Nonetheless, there are still strict regulations that apply and devices must comply with. Therefore, if you are developing a product that is going to use the ISM band, you will certainly need to get prior approval from the government. This is required to ensure that the device adheres to the local rules, the regional policies of the government, as well as the international policies of the International Telecommunication Union (ITU).

The ITU created the standards that specify the frequency band, the spectrum, the transmission power regulations, and various other characteristics. However, not all of the ISM frequency bands are available to use worldwide—some only have regional availability. In the subsequent table, we can see all of the currently available ISM unlicensed bands that have worldwide or only local availability.

CENTER FREQUENCY	BANDWIDTH	AVAILABILITY
6.780 MHz	30 kHz	Local Acceptance
13.560 MHz	14 kHz	Worldwide
27.120 MHz	326 kHz	Worldwide

CHAPTER 5 How Do WLANs Work?

40.680 MHz	40 kHz	Worldwide
433.920 MHz	1.74 MHz	Local Acceptance
915 MHz	26 MHz	Local Acceptance
2.450 GHz	100 MHz	Worldwide
5.80 GHz	150 MHz	Worldwide
6.00 GHz	1200 MHz	Local Acceptance
24.125 GHz	250 MHz	Worldwide
61.250 GHz	500 MHz	Local Acceptance
122.500 GHz	1 GHz	Local Acceptance
245 GHz	2 GHz	Local Acceptance

Most wireless devices operating today work in the 2.4 gigahertz (GHz) band. This is the most popular ISM band because Wi-Fi, Bluetooth, radio-frequency identification (RFID), Zigbee, 6LoWPAN, and near-field communication (NFC) technologies all use this spectrum. From a radio performance perspective, the 2.4 GHz range has great signal propagation characteristics because signals will travel through walls and other solid objects very well without any shadowing (or changes in power level). In addition, the wide bandwidth of 100 megahertz (MHz) is sufficient to support broadband data rates for multiple users simultaneously.

The other very popular worldwide unlicensed band is the 5 GHz range, which has a bandwidth of 150 MHz, which makes it suitable for higher data rate communication. There are other high frequency ISM bands available with very high bandwidths of up to 2 gigabytes for example at the 244 GHz range, which is significantly wide. However, for the rest of this chapter, we will concentrate on the ISM bands used by the most common wireless protocols such as Wi-Fi (802.11) and Bluetooth. Both of these wireless technologies have become hugely popular in PAN and WLAN networks due to their inherent support in smartphones, wearable technologies, and in the case of Wi-Fi in providing the network backbone to connect all these diverse wireless devices together.

In February 2020, a new frequency band at 6 GHz was approved in the United States by the Federal Communications Commission (FCC) and is awaiting approval in the European Union late in 2020 for a new ISM frequency band at 6 GHz. This new available unlicensed spectrum for Wi-Fi is considered to be a huge advance in wireless communications because it will allow routers to broadcast over a new uncluttered bandwidth. This is going to go a long way to alleviating many current Wi-Fi issues such as spectrum congestion, which is a major bugbear in Wi-Fi communications. However, the additional swath of available bandwidth should not only improve connection reliability, speed, and reduce interference, it will also improve spectrum efficiency. This is because the new 6 GHz band with 1200 MHz available spectrum can support up to seven Wi-Fi 6+ channels simultaneously without any bandwidth overlap or contention issues. The introduction of the 6 GHz band effectively quadruples the available spectrum to Wi-Fi and this is why it is considered to be one of the most significant advances in Wi-Fi for the last 20 years.

WLAN Anatomy

The main component of any WLAN is a radio card, which 802.11 standards refer to as a **station (STA)**. A STA can be any 802.11-capable device, including a laptop, smartphone, tablet, or plug-in card. If the device is an endpoint, it is referred to as a *client station*. An access point, which is also a STA, serves as the central hub that communicates with all the client stations within range. Typically, an access point is wired to a fixed-line switch, which acts as the gateway through which client stations access the Ethernet network. However, an access point may also act as a bridge, connecting to another access point (which then connects to the switch).

Wireless Client Devices

Any device that meets the following criteria can act as a wireless client:

- It contains a radio card or integrated transmit (TX) and receive (RX), noted as TX/RX
- It contains an antenna (many of which are internal to the device)
- It operates under 802.11 protocol standards

As far as their radio abilities go, there is no difference between a smartphone, tablet, or laptop as compared to, say, a wireless thermostat. If a device supports the 802.11 standard, it can act as a client station. This also means that the device is constrained by the half-duplex mode of communications and must contest with other clients for the RF medium.

A client station is configured to associate with an access point by creating a Layer 2 connection. If the device is to be mobile, however, then the client must be able to hand over to another, stronger signal when appropriate. In most cases, the client's radio card will attempt to connect and associate with whichever access point has the strongest signal. This is how a client manages to roam around large areas with multiple access points while maintaining a session. Every access point is identified by a **service set identifier (SSID)**, which is a configurable name or alphanumeric code. If multiple access points have been configured with the same SSID, and if the correct security credentials have been supplied (if required), then the client will be seamlessly handed over to the strongest signal (as long as the access points use different channels to avoid interference).

Of course, before a client can connect to an access point, it must detect one's presence. A client does this by one of two methods:

- **Passive scanning**—With **passive scanning**, the client listens for a beacon, which an access point continually emits. When the client "hears" a beacon advertising an SSID for which it has been preconfigured, it will select that access point. If it hears more than one beacon, it will select the access point with the strongest signal.
- **Active scanning**—With **active scanning**, the client proactively scans the network by sending out probe pulse requests. These requests can contain the SSID of a specific preconfigured network, but they can also "discover" new networks by leaving the SSID field blank. A request probe set with an SSID is called a *directed probe*. A request probe with a blank SSID variable is called a *null probe request*. When a client transmits a directed probe request, all access points configured with that SSID respond with a probe response. The contents of a probe response are the same as the information held in a beacon. This gives the client all the configuration information that it needs to join the **basic service set (BSS)**,

which is discussed in the next section. Similarly, when a client sends out a null probe request, all available access points respond with a probe response.

With passive scanning, the client listens for beacons from access points advertising their SSIDs. With active scanning, a client sends out probe requests and then listens for responses. Active scanning has several advantages:

- A client device can find and connect to the access point with the best radio signal.
- The client device can build a map of available access points and their relative signal strengths, which can assist in handover speed and efficiency.
- The client can occasionally go off network to check for stronger signals on other radio channels. In this way, the client can actively strive to maintain the highest quality signal.

However, active scanning can pose a security risk. For example, a client that uses active scanning could fall prey to a rogue access point (a user- or hacker-installed access point) that is set up with a powerful signal to attract clients. The rogue access point will then have access to all traffic passing through it, and can either copy, capture, or redirect the traffic.

802.11 Service Sets

The 802.11 standards define four topologies, called *service sets*, for how Wi-Fi devices connect with each other. They are as follows:

- Basic service set (BSS)
- **Extended service set (ESS)**
- **Independent basic service set (IBSS)**
- **Mesh basic service set (MBSS)**

These topologies enable radio cards to communicate with each other. To understand how they communicate, however, you must first understand their abilities and constraints. By nature, 802.11 radio communication is **half duplex**. That is, both devices are capable of transmitting and receiving, but only one can transmit at a time. This is in contrast to **simplex communication**, used in broadcast radio (FM), in which one main station transmits, while all other devices receive only. A third option is **full duplex** communication, used in Ethernet connections. With this type of communication, devices can both transmit and receive simultaneously. For 802.11 radios to work in full-duplex mode, there must be two separate channels: one to transmit and one to receive. This is why Ethernet performs so well compared with 802.11 wireless radio communications.

A client station can operate in one of two configurable modes:

- **Infrastructure mode**—The most common WLAN topology is **infrastructure mode**. It uses an access point as a central connection point and portal to a distribution system. Infrastructure mode enables clients to communicate via a BSS or an ESS.
- **Ad hoc mode**—A second mode, **ad hoc mode**, enables wireless clients to communicate directly.

Basic Service Set

The BSS is the cornerstone of wireless networks. It defines a common topology, or arrangement, in which a single access point connects and associates with several client stations. In practice, the access point is typically also connected to a distribution network such as an Ethernet LAN, although that is not required. With the BSS, all communication goes through the access point. That is, in a BSS, client stations cannot communicate with each other directly. Rather, they must pass through the access point. The range or coverage of the wireless network defines the basic service area. The size and shape of the service area depend on several factors, including radio power, antenna gain, and the surrounding environment. Because environmental factors can affect radio conditions, the basic service area varies.

BSS is typically used in homes or small office/home office (SOHO) environments. For larger areas with many clients, several BSSs can be combined to provide the required coverage and support.

Extended Service Set

The ESS is used in larger networks to connect several access points to an Ethernet LAN. An ESS is a combination of two or more BSSs connected via a distribution system medium such as an Ethernet network. The design of an ESS may require **seamless roaming**, where a client station can roam from the service area of one access point into the service area of another access point without disrupting the connection. For seamless roaming to work, a 15- to 20-percent overlap in coverage areas is required.

The alternative is **nomadic roaming**, in which there is no overlap, with coverage areas remaining autonomous. With nomadic roaming, a client station moving from one access point's area of coverage to another's will lose the connection until it enters the new coverage area and associates with the new access point.

A third type of ESS deployment is the **collocation model**. With this model, there is 100-percent overlap between the two access point service areas. This duplication of coverage is used to address client capacity issues for large departments with many clients or in areas such as conference rooms, which may see spikes in the number of clients seeking a connection.

Independent Basic Service Set

In an IBSS configuration, no access point is used. Instead, client stations form peer-to-peer relationships with other client stations. These peer-to-peer relationships are set up in ad hoc mode on an as-needed basis.

Mesh Basic Service Set

With an MBSS, clients, access points, and gateways are all meshed together, enabling client-to-client and access point-to-access point communication. Only STAs that are assigned to the meshed network can communicate directly. Communications to non-meshed STAs are routed through a gateway.

The 802.11 Standards

The IEEE 802.11 standards outline the protocols, methods, and controls for building, operating, and communicating over WLANs. Originally drafted in 1997, the 802.11 standard has seen multiple revisions, each with its own letter qualifier (for example, 802.11a, 802.11n, and so on). Several of these revisions specify new frequencies, channels, or modulation techniques, which is not all that uncommon with IEEE standards. However, because Wi-Fi routers have become a mass-market consumer category, many manufacturers use the letter designation as part of their product marketing and naming to signify differentiation (in the case of having the latest technology) and to ensure that consumers purchase compatible devices. For example, many laptops and wireless cards are compatible with 802.11a and 802.11b, but not 802.11n. By specifying that a Wi-Fi access point is 802.11n, it avoids consumer confusion.

New Wi-Fi Alliance Naming System

The Wi-Fi Alliance has previously followed IEEE's 802.11 naming system based upon the 802.11 framework. However, due to the increasing confusion among customers who have to try and make sense of the alphabet soup, the Wi-Fi Alliance is finally introducing its own simplified naming scheme that does away with the letter notation. Instead, the Wi-Fi Alliance has decided to categorize each standard as a generation. Thus, the latest version 802.11ax will be known as Wi-Fi 6. Its predecessor 802.11ac will now be known as Wi-Fi 5.

The new Wi-Fi Alliance naming system will run concurrently with the current IEEE 802.11 naming system. Theoretically, this should make the process of buying suitable and compatible new devices with better connectivity easier.

Here's how the Wi-Fi Alliance naming standards correlate with the IEEE standards:

- Wi-Fi 6—802.11ax (2019)
- Wi-Fi 5—802.11ac (2014)
- Wi-Fi 4—802.11n (2009)
- Wi-Fi 3—802.11g (2003)
- Wi-Fi 2—802.11a (1999)
- Wi-Fi 1—802.11b (1999)

The 802.11 standards are now discussed in detail.

Wi-Fi 1 (802.11b)

This standard uses the unregulated 2.4 GHz band, with a throughput of 11 Mbps. Because other unregulated devices such as microwave ovens and cordless phones commonly used this frequency, 802.11b access points and clients were subject to interference. However, thanks to their lower price point and greater range, which enabled them to cover most homes with a single access point, 802.11b became the preferred technology among consumers, despite their lower throughput.

Wi-Fi 2 (802.11a)

This standard operates at 5 GHz and provides data rates between 1.5 Mbps to 54 Mbps. Originally, 802.11a access points were more common in office environments and in residential settings, but this is no longer the case. While 802.11a does provide faster data rates than Wi-Fi 1 (802.11b) (which was developed at the same time), the higher frequency used results in a shorter range. 802.11a also suffers more attenuation from walls, doors, and other surfaces. Although 802.11a access points historically cost more, many businesses preferred them because of their higher throughput, which helped to lessen the gap between wired Ethernet and wireless access.

Wi-Fi 3 (802.11g)

Released in 2003, 802.11g offered the best of breed of both 802.11a and 802.11b. Supporting bandwidths of up to 54 Mbps, 802.11g utilized the 2.4 GHz frequency band, allowing greater range than 802.11a. Because 802.11g was backward-compatible with 802.11b, it saw rapid adoption in the consumer market; there was no longer any reason to invest in 802.11b access points.

Wi-Fi 4 (802.11n)

Ratified in 2009, 802.11n offered a significant breakthrough in data rates, with speeds up to 600 Mbps. Operating on both the 2.4 GHz and 5 GHz bands and using **multiple input/ multiple output (MIMO)** antennas, 802.11n also provided greater range and experienced less interference. The 802.11n standard improved the network throughput using MIMO technology, which uses multiple input, multiple output antennas to send and receive, hence the extended data services at higher data rates. Because it supports both the ISM bands at the 2.4 GHz and 5 GHz range, 802.11n was suitable for dual band operation.

Wi-Fi 5 (802.11ac)

This standard was released in December 2013 as an amendment built upon 802.11n. The objective was to focus on even higher throughput through higher data rates. The new features that Wi-Fi 5 introduced included support for a wider channel in the 5 GHz band. In 802.11n, the bandwidth was 40 MHz, but with 802.11ac, the bandwidth available went up to 80 or 160 MHz, which enabled higher data rates. This was accomplished by using **multi-user MIMO (MU-MIMO)** technology along with high-density modulation, hence the data rate could be very high going up to 866.7 Mbps using the 160 MHz bandwidth.

802.11ad was released by the Wi-Fi Alliance in March 2013. This uses the 60 GHz ISM band to support data rates up to 7 Gbps using the 60 GHz band's high frequency to get a wider bandwidth and that enables very high data rates to be exchanged. 802.11ad was around 10 times faster than 802.11n.

802.11ah was released in December of 2016 to support emerging Internet of Things (IoT) technologies and devices. Because IoT devices typically do not require high throughput then a narrow bandwidth is sufficient. 802.11ah operates in the 900 MHz band. However, a critical design criterion was that low power consumption needed to be included such as a wake and doze period protocol to save energy when the device is not needing to use the channel access. The 802.11ah protocol provides connectivity to thousands of IoT devices using a single access point within a 1 kilometer by 1 kilometer square area. 802.11ah supports machine-to-machine (M2M) communication, smart metering, and smart home IoT devices. With a data rate of up to 347 Mbps, this is sufficient for simultaneous support for multiple IoT applications.

Wi-Fi 6 (802.11ax)

Despite Wi-Fi 5 (802.11ac) offering an impressive theoretical maximum rate of 1.3 Gbps, which was sufficient for most business-orientated usage, it performed poorly under high network concentration conditions. The IEEE addressed this problem through the introduction of the 802.11ax, or High-Efficiency Wireless standard, which promises a fourfold increase in average throughput per user.

Consequently, the IEEE came up with the 802.11ax (Wi-Fi 6) standard and it is designed specifically for high-density public environments, like stadiums, shopping malls, and airports. But it also will be useful in IoT use cases and in businesses that use bandwidth-intensive applications like video conferencing.

Wi-Fi 6 (802.11ax) was released in 2019 and is the next evolution from 802.11ac. The 802.11ax standard offers theoretical network speeds of up to 10 Gbps, and up to 12 Gbps over very short distances, which is around a 30–40 percent improvement over the 802.11ac standard. However, the new 802.11ax standard also means MU-MIMO will be available in all Wi-Fi routers. This is important because MU-MIMO provides a means of supplying constant data streams to multiple users concurrently in both uplink and downlink, which was something only available previously on top-end wireless routers. Moreover, 802.11ax uses **orthogonal frequency-division multiple access (OFDMA)**, which increases the data rate by dividing the available spectrum into smaller units, which makes better use of the available transmission frequencies within the 2.4 GHz and 5 GHz bands.

Previously with the 802.11ac standard, the dual-band router would broadcast on both the 2.4 GHz and 5 GHz spectrums. However, with 802.11ac to make better use of spectrum efficiency, those bands were divided into 64 20 MHz-wide channels to form 160 MHz blocks. 802.11ax, however, further subdivides the current 64 20 MHz channels into 256 individual subchannels. This means that 802.11ax can talk simultaneously with up to 9 devices per channel, or 74 devices over a 160 MHz block. Also, Wi-Fi 6 updates Wi-Fi 5's 256 Quadrature Amplitude Modulation (QAM) up to 1024 QAM, allowing broadcasts of up to eight simultaneous streams.

To better understand why 802.11ax came about, we have to understand the problems that mobile operators had backhauling vast amounts of carrier data across their hugely expensive RF spectrum. 802.11ax is designed for that purpose to facilitate easy cellular data offloading. In this scenario, the cellular network offloads wireless traffic to an 802.11ax Wi-Fi network to free up their own expensive and limited bandwidth.

The fundamental problems with the older versions of Wi-Fi, up to and including Wi-Fi 5, are that bandwidth is shared among endpoint devices and only one device can communicate at a time. To compound the problem, there are overlapping RF coverage areas, especially in dense deployments, and end users roam between access points.

Why is this a problem? Well, Wi-Fi 5 is still based on an old technology called Carrier Sense Multiple Access with Collision Avoidance (CSMA/CA), which requires devices to listen for an all-clear signal before they start transmitting. In the event of network congestion, the device will initiate a back-off procedure, whereby it waits for the all-clear before it transmits. In most WLANs, this is typically transparent to the user because they are more concerned with reading, updating, or browsing rather than transmitting data, so raw data speed is rarely an issue. However, it is not just about raw data rates or high performance because, as we have seen, most business users would not benefit from these gigabit speeds. The problem is network capacity that reveals itself when users try to use Wi-Fi in a crowded stadium, an airport lounge, or in a shopping mall where thousands of end users are attempting to access the Internet at the same time. With legacy Wi-Fi standards, the system can only handle a single user at a time and performance suffers.

However, the introduction of Wi-Fi 6 promises vastly improved performance, extended coverage, and longer battery life. For example, Wi-Fi 6 delivers a nearly 40 percent increase in pure throughput thanks to higher order QAM modulation, which allows for more data to be transmitted per packet. It also achieves more efficient spectrum utilization. For example, 802.11ax creates broader channels and splits those channels into narrower subchannels. This increases the total number of available channels, making it easier for endpoints to find a clear path to the access point. Not only can an 802.11ax device deliver a single stream at 3.5 Gbps, but with MU-MIMO technology, 802.11ax-enabled devices can also deliver eight simultaneous streams to a user device. The improvement is that with 802.11ac, MU-MIMO is limited to downlink transmissions only; however, 802.11ax creates MU-MIMO connections that are bidirectional.

What is more, MU-MIMO deploys a technique called beam focusing, which means it can target its transmission toward a specific receiving device rather than broadcast the transmission in all directions. This means that MU-MIMO with beam focusing can support many users simultaneously, which is a big improvement on the early Wi-Fi standards because they only permitted one transmission at a time per access point. Significantly, 802.11ax piggybacks on MU-MIMO with Long-Term Evolution (LTE) cellular base station technology OFDMA. This modulation protocol allows each MU-MIMO stream to be split into four additional streams, boosting the effective bandwidth per user by four times.

Another difference between the 802.11ac and 802.11ax standards are that the former operates only in the 5 GHz range, while 802.11ax is dual channel and operates in both the 2.4 GHz and 5GHz bands. For example, 802.11ax will support 12 channels—8 in the 5 GHz and 4 in the 2.4 GHz range—providing a greater spread of bandwidth.

Wi-Fi 6E

As we have learned, Wi-Fi works by broadcasting over the unlicensed RF bands: 2.4GHz and 5GHz. In 2020, the FCC added a third band at 6 GHz, which effectively quadruples the total space available to traditional Wi-Fi. However, to make use of this new swath of airwaves, equipment will need to support a new Wi-Fi 6E (E for enhanced) chipset. This is because Wi-Fi 6E is an extension to the existing Wi-Fi 6 standard and not a new Wi-Fi standard in its own right. Consequently, Wi-Fi 6E has the same theoretical speeds as the existing Wi-Fi 6 standards, working over the 5 GHz band, but where it gains is that it will have more available spectrum of 1200 MHz in the 6 GHz band to work over so will be able to support more fully utilized channels without having to share an available spectrum. This should mean that Wi-Fi 6E devices will be able to perform close to mobile carrier 5G millimeter speeds when connecting to 6E enabled smartphones. The potential for Wi-Fi 6E is tremendous but it will require buying new Wi-Fi routers and devices such as smartphones and smart TVs to be able to benefit from this advancement.

802.11 Unlicensed Bands

The use of radio waves (specifically, the transmission of radio signals) falls under the supervision of the FCC, which controls their usage through licensing. Most of the RF spectrum is licensed, with different frequency bands used for specific functions (radio, television, etc.). The licensed use of a frequency allows a government or commercial entity to employ its designated frequency without fear of interference.

There are also unlicensed bands on which users can operate without an FCC license. To do so, however, they must use certified radio equipment and must comply with certain technical requirements, including power limits. Those who use these unlicensed bands do not have exclusive use of the bands and are subject to interference.

> **NOTE**
> When configuring an access point to use either 2.4 GHz or 5 GHz, it is important to make sure the client device can support the 5 GHz spectrum. Many phones and tablets do not.

Wi-Fi technology that adheres to the 802.11 standard operates on multiple unlicensed frequency bands. The most common of these is the 2.4 GHz spectrum. The 5 GHz spectrum is also becoming very popular, and there are now a number of dual-band radios that support both 2.4 GHz and 5 GHz.

A third spectrum, operating at 60 GHz, has been proposed, and is covered by the draft 802.11ad amendment. This high-frequency band will provide **very high throughput (VHT)** of up to 7 Gbps over short distances. However, because ultra-high frequencies do not travel through obstacles such as doors and walls, the indoor range will be restricted to line-of-sight implementations. There are also working groups in the process of defining and ratifying tri-band standards for 2.4 GHz, 5 GHz, and 60 GHz, which will allow for handovers if the user passes from the short coverage areas that use the 60 GHz band into wider coverage areas, which use either the 2.4 GHz or 5 GHz spectrum.

Narrowband and Spread Spectrum

There are two techniques for radio transmission over a given spectrum:

- **Narrowband**—As its name suggests, **narrowband** uses very little bandwidth by transmitting over a narrow beam of frequency (for example, a 2 MHz-wide channel at 80 watts). Because its tight frequency bands are easy to target, narrowband is susceptible to frequency jamming. Narrowband utilizes much greater power to transmit over its narrow range of frequencies. As a result, there is a real possibility of interference if two narrowband stations are operating in the same proximity. Because of this, narrowband transmitters are licensed and regulated to ensure that nearby stations do not compete by cranking up the power of transmission.
- **Spread spectrum**—With **spread spectrum**, the transmission is spread across the entire frequency space available (for example, over a 22 MHz band at 100 megawatts, which is 11 times the channel width but a nearly 1,000 times lower power). Because of its wide frequency band, spread spectrum is much harder to block (jam). Spread spectrum signals are transmitted using milliwatts of power, so the risk of interference is not so great. As a result, spread spectrum transmitters do not require a license or fall under the control of a regulatory body.

Multipath

An inherent issue with radio waves is that they reflect off certain materials and surfaces. This results in multiple versions of the same radio waves bouncing around, known as **multipath**. If these radio waves reach the intended antenna, the reflected signals are often delayed and often out of phase, having taken a less direct path. In some cases, depending on the amount of delay and phase shift, the two signals (the original and the reflected) can actually attenuate or nullify each other (see **FIGURE 5-1**).

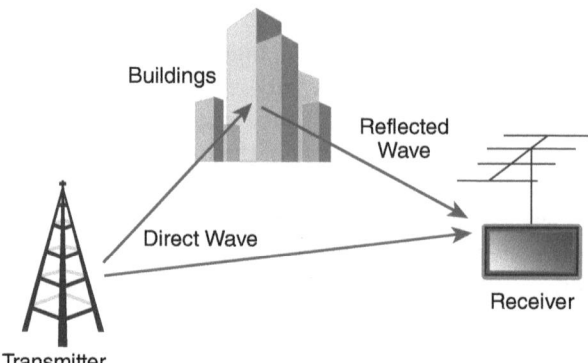

FIGURE 5-1

Multipath occurs when multiple versions of the same radio waves bounce off certain materials and surfaces.

Spread spectrum systems are less susceptible to the effects of multipath because the broad frequency range used mitigates them on any given frequency. MIMO antennas also help to mitigate multipath, in their case by using it to facilitate higher throughput and data rates.

Frequency Hopping Spread Spectrum

Frequency hopping spread spectrum (FHSS) works by transmitting data using a small carrier space in short bursts and then continuously changing, or "hopping," to another frequency during transmission. The length of time the transmitter stays on a given frequency, called the *dwell time*, is set for a predetermined period. The hopping sequence is not random, but follows a pattern. Once the pattern is complete, the cycle begins.

The 802.11 standard limits the hop steps to 1 MHz in size. There are typically between 75 and 79 hops per sequence, but this can vary from country to country. For FHSS to work, the transmitter and receiver must be on the same frequency, so they must follow the same hop sequence. The data needed for this synchronization is stored in the beacon management frame.

One obvious problem with FHSS is that the transmitter and receiver can only communicate during dwell times. Because no communication can take place during the hops, throughput is lowered. A shorter dwell time results in lower throughput, and a longer dwell time results in greater throughput. However, the reverse is true with regard to the effect of interference. A shorter dwell time allows for greater protection from interference and jamming. Because of these tradeoffs, FHSS is rarely used for 802.11 (Wi-Fi), although most Bluetooth devices still use FHSS.

Direct Sequence Spread Spectrum

Like FHSS, **direct sequence spread spectrum (DSSS)** uses hopping. However, unlike FHSS, which hops between frequencies, DSSS stays on a fixed channel, hopping within the frequency space on that channel. It does this using a technique called *data encoding*.

Data encoding encodes each bit of data and then transmits it as multiple bits of data. The idea behind data encoding is that each bit of data is converted into a series of bits called *chips*. Using a logic process, the data is converted to a binary sequence, which looks like noise. The receiving device receives the chips sequence and converts it back to a data bit.

This is where the advantage of DSSS becomes evident. If some of the chips are not received correctly or if the signal is corrupted, the receiving card can work out what the correct signal was based on chips that were successfully received. This technique is so robust that as many as 9 out of 11 chips can be lost in transmission without degrading communication.

Wireless Access Points

A **wireless access point (WAP)** is simply a half-duplex switch that contains a radio card and an antenna that can be tuned to one or more unlicensed radio frequencies (specifically, the 2.4 GHz and/or the 5 GHz bands). The radio card is under the same constraints as a client station in that it can only transmit if no other station is already doing so. A WAP, however, does a little bit more than a hub, because it actually works at the control and data-functional layers. In other words, it has some switch-like intelligence. For example, a WAP must make a connection and then maintain that connection over the air. Therefore, wireless protocols must support a greater array of messages than a protocol for a wired connection. Wireless protocols such as 802.11 include messages to associate, authenticate, and disassociate, because there is no real physical connection that can be monitored. In contrast, Ethernet needs none of these messages. Simply plugging in or unplugging the cable does the trick.

That said, WAPs are still reasonably simple. Different WAPs can operate on several channels within each frequency. Unfortunately, however, these unlicensed channels are very heavily populated, especially the 2.4 GHz band. This can result in conflicting radio signals from neighboring WAPs, which degrades the signal.

A common misconception is that tuning multiple WAPs to the same frequency and channel—for example, 2.4 GHz and channel 6—will boost the network signal. Unfortunately, the opposite is true. If neighboring networks share the same frequency/channel combinations, it actually attenuates the overall radio signal quite significantly. (Interestingly, from a security perspective, this can be a simple and inexpensive way to jam or deny service to a WAP.) Network designers configure WAPs to ensure that non-overlapping channel plans are used to avoid interference. **FIGURE 5-2** illustrates what non-overlapping channels look like for the 2.4 GHz band.

How Does a WAP Work?

A WAP works by acting as a portal for another network, typically on another physical medium such as wired Ethernet. Its job is to connect wireless clients over the air and then either redirect the traffic to another wireless client or, more often than not, to a wired network. Wireless 802.11 can be described as an access network, although it can also have a role in backhaul. Its main role in most network architectures is to connect local wireless devices and to act as a portal to another physical network, called a *distribution network*.

A WAP works by using RF signals as a carrier band frequency. The frequencies assigned to 802.11 are the unlicensed bands at 2.4 GHz and 5 GHz. WAP radio cards are configured to work on one of the two unlicensed bands and on several channels. However, because the channels must be spaced five channels apart to prevent overlap and interference, there are only three possible selections for multichannel WAPs: channel 1, channel 6, and channel 11. A modern WAP typically supports both frequencies and has several radio cards and antennas.

FIGURE 5-2

Channels are designed to not overlap to prevent co-channel interference.

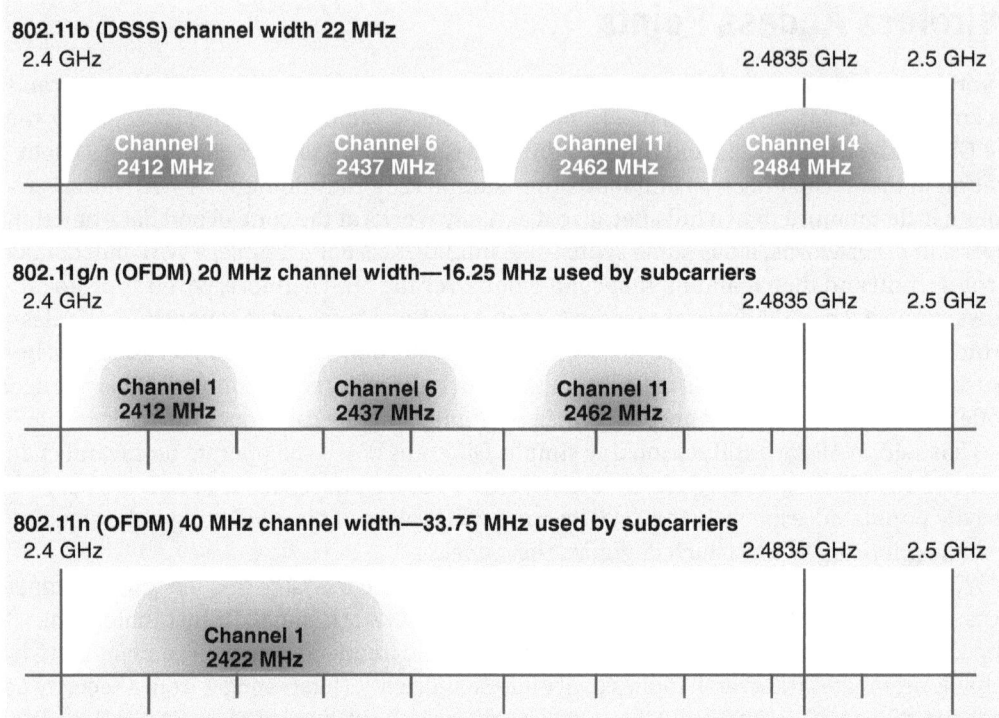

A WAP continuously transmits beacons on the selected RF band and channel to advertise its presence and configuration to any client station in broadcast range. A beacon is transmitted as a broadcast, with the destination media access control (MAC) address being set to all ones (which is the address used for broadcast messages that all devices will receive) and the source address being the MAC address of the WAP. Listening client stations receive and process all broadcasts on the same radio channel.

The beacon transmitted by the WAP contains timing and synchronization information needed for connecting. These beacons are the heartbeat of the wireless network. Much depends on precise timing between wireless clients and the WAP. In BSS and ESS configurations, only the WAP transmits beacons. However, in ad hoc mode, where there is no WAP, all client stations transmit their own beacons. Information transmitted by a beacon includes the following:

- A timestamp for synchronization timing
- The channel
- The data rate

- Spread spectrum parameters
- The SSID
- **Quality of service (QoS)** parameters

Client stations listen for WAP beacons through passive or active scanning. To join a WLAN and transmit data to and through the WAP (to other client stations or to servers on the distribution system beyond), a client station must authenticate and associate. Therefore, when a client discovers a WAP with a matching SSID, it will go through an authentication process.

The 802.11 association process should not be confused with network authentication—that is, entering a username and password. Authentication at the 802.11 Layer 2 level is similar to what happens when a cable is connected to a wired network: A connection is formed between the client and the network. Because there is no physical connection with wireless, there must be some other way to form the connection. Wi-Fi Protected Access 2 (WPA2) uses a four-way handshake for association. This process not only provides a mechanism for association but it also produces an encryption key used to read broadcast messages (which are encrypted in WPA2). The process, illustrated in **FIGURE 5-3**, is as follows:

1. The WAP sends a **nonce** (a single-use random number), called *ANonce*. The client uses this value to create a key.
2. The client sends its own nonce (called *SNonce*) to the WAP along with an integrity and authentication message, called the *message integrity code (MIC)*. The WAP uses the SNonce to create the key needed to decrypt messages from the client.
3. The WAP replies with the key used to decrypt broadcast messages, called the *group temporal key (GTK)*, along with a MIC.
4. The client sends a confirmation to the WAP, which completes the association.

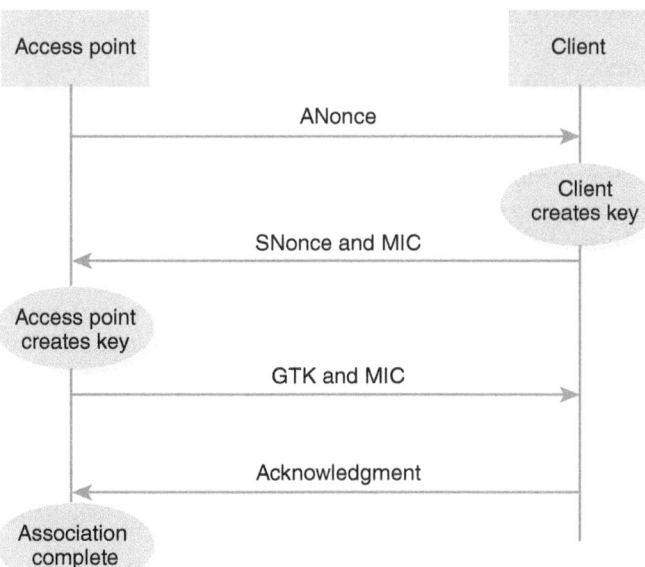

FIGURE 5-3
The four-way handshake.

When the association procedures are complete, the client sends a Dynamic Host Configuration Protocol (DHCP) request to obtain an Internet Protocol (IP) address and Domain Name System (DNS) server settings. The WAP responds with the IP address and other network configuration information, which allows the client to communicate with the network beyond the WAP.

WAP Architecture

There are two types of WAPs:

- **Autonomous access points**—Able to operate at the control and data layers, **autonomous access points** have switch-like intelligence.
- **Thin access points**—In **thin access points**, the switch-like intelligence is stripped out and relocated in a WLAN controller device. The WLAN controller acts as a central administrator and controller for several thin access points. Using thin access clients considerably eases the administrative burden of managing many WAPs on a WLAN.

To understand what this switch-like intelligence is and how it works, you must consider the typical application of a WAP when deployed in a network. A WAP is typically a portal or gateway that enables wireless client stations to connect to resources on another network medium. This is practically always a wired Ethernet backbone network. There are times, however, when the WAP will not be connecting to a fixed-wire Ethernet network but rather to another WAP acting as a bridge or an extender. However the connection is made, there is a wireless 802.11 network, which is where the WAP and the clients receive and transmit, as well as a backhaul backbone network (whether that be Ethernet 802.3 or a radio bridge). Therefore, the WAP must be able to recognize, reframe, address, and deliver packets between those two interfaces or mediums. This added intelligence is referred to as **integration service (IS)** and **distribution service (DS)**.

Integration Service

In a wired network, a switch redirects packets to the correct port based on the destination MAC address contained in the Layer 2 frame header. Similarly, a WAP redirects packets to the network backbone or to a wireless medium, based on the same Layer 2 frame header information. In the majority of cases, the WAP will switch packets from the 802.11 wireless network to an Ethernet wired backbone. To do this, it must reframe the packets to adhere to the Ethernet framing structure. Similarly, when packets are sent to a client station from the Ethernet network, the WAP must reframe them to 802.11 standard frames. This switch-like intelligence is called integration service (IS). A simple way to think of IS is as a frame translation method between 802.11 frames and other mediums such as Ethernet frames.

Distribution Service

By design, WAPs are portal devices that pass traffic across the wireless network or through the IS to a connected non-802.11 medium. This occurs through a distribution service (DS). A DS consists of two components:

- **Distribution medium**—The **distribution medium** is the physical medium to which the WAP's ports connect, typically an Ethernet LAN.

- **Distribution system service (DSS)**—The **distribution system service (DSS)** is internal software that controls the switch-like intelligence and manages client station associations and disassociations.

The control of switching between mediums is determined by DSS. However, DSS does not define the medium to which a WAP translates and forwards data. In the majority of cases, it is onto an Ethernet network, but forwarding can sometimes occur over another wireless medium. Therefore, you can define a WAP as "as a wireless hub with switch-like intelligence that also acts as a translational bridge between two network mediums."

Wireless Bridges

The previous section touched on DSS, which translates traffic between an 802.11 device and the distribution medium used for backhaul. Sometimes, the backhaul medium is wireless. In such cases, a WAP can also serve as a wireless bridge. The 802.11 standard describes a mechanism called **wireless distribution system (WDS)** whereby the frame format can handle four MAC addresses. The real-world deployments for WDSs are in bridges, repeaters, and mesh networks.

WDSs can be used to connect WAPs to form a wireless backbone. This can be particularly useful when WAPs are used to provide both coverage and backhaul simultaneously. However, to do this successfully, the WAP needs to have two separate transmitter/receiver channels. (It is possible to use a single radio device, but performance would be poor because the WAP would have to constantly shift between communicating with the local end points and the backhaul device.) Typically, a wireless bridge employs one radio device operating on the 2.4 GHz channel to communicate with WLAN clients and another radio device configured to operate on the 5 GHz channel to serve the backhaul.

802.11 bridges are not an ideal solution for backhaul, but they can serve a purpose when no Ethernet backhaul is available. Network designers who want to extend the wireless reach (again, where no Ethernet drop is available) often deploy bridges and repeaters. WDSs must be deployed in these circumstances. This requires the use of four MAC addresses: a source and destination MAC address and a transmitter and receiver MAC address. Although repeaters can extend the reach of the WLAN, there are drawbacks with regard to reduced throughput and increased latency (speed of transmission) due to the packets having to be transmitted twice.

Wireless Workgroup Bridges

A wireless workgroup bridge is often used when there are several non-wireless devices, such as Ethernet networked PCs, in a workgroup or in an office that require backup wireless connectivity to the network. This is needed to provide resiliency in the event of a failed wired network or a loss of service due to a router or switch failure. In this case, the wired PCs continue to use their network cables and a switch simply fails over to the workgroup bridge. When the workgroup bridge associates with the nearest WAP, the connected wired PCs will be able to communicate over the wireless network through the workgroup bridge.

Workgroup bridges are very handy for scenarios such as these, when non-wireless equipment needs to connect to a wireless domain. However, there are some drawbacks. Because the workgroup bridge is itself a wireless client, it can communicate only with the WAP rather than with the other client. The wired Ethernet connection on the workgroup bridge is typically a half-duplex, 10 Mbps connection, which is often shared between many clients. As a result, performance can be quite bad. On the other hand, without the use of a wireless workgroup bridge, some devices would have no connectivity at all.

Residential Gateways

A wireless residential gateway is a home wireless router that acts as a gateway to the Internet through a DSL or cable broadband connection. A wireless residential gateway typically has a built-in hub supporting four Ethernet connections and a built-in WAP to create a WLAN. Therefore, a residential gateway is simply two devices in one—an 802.11 WAP and an Ethernet switch. Some Internet service providers (ISPs) even offer a single device that serves as a WLAN router, an Ethernet switch, and a wireless WAP.

A residential gateway's job is to provide Internet connectivity to the various devices that you connect via the wired and wireless WLAN. To assist with this, it has many built-in capabilities, such as automatic IP address assignments via DHCP as well as network address translation (NAT) and integrated firewalls and access lists.

Typically, the administration of the residential gateway occurs via a built-in web browser, accessible on a default address. This requires, however, that the user (homeowner) plug in a network cable to the physical WAP and follow the configuration guide. This is relatively simple to do, but many people simply want to take the WAP out of the box, plug it in, and use it, and manufacturers have complied. This may seem great to consumers, but it's horrible from a security standpoint, especially given that setting up a relatively secure basic wireless network is not all that difficult.

There are typically several sections in the configuration interface. These include the following:

- Local Layer 2 settings (including options for things such as MAC filtering and peripherals such as printers)
- Layer 3 settings (including options for connecting to the WAN, Internet, or a VPN back to a corporate site)
- Wi-Fi network configuration settings (including basic security options such as encryption and authentication)
- Optional firewall and advanced security settings

Of these, the most important settings are the Wi-Fi network configuration settings. Because the Layer 2 and Wi-Fi network are preconfigured, they work out of the box. However, this plug-and-play ability comes at the price of having no security whatsoever. Users who never change their settings are at serious risk of a security breach.

In the past, this was not much of a concern for corporate IT security specialists. But with many people now working at least some time from home and the increased adoption of Bring Your Own Device (BYOD), many company assets are now exposed on home networks. In light of this, IT security leaders should insist as part of a BYOD policy that some simple but effective steps be taken by employees who work from home or who have BYOD devices. These steps include the following:

- **Implementing WPA2 encryption**—This is critical because it protects data, including private login and password information that is often in the clear and easily snooped. WPA2 also prevents unauthorized users from accessing the network.
- **Setting (limiting) radio power**—This is especially important for employees who live in an apartment, but should also be done by employees who live in a house. There's usually no need to connect out on the street.
- **Selecting access controls**—For example, employees should specify allowed devices by indicating their MAC addresses. MAC addresses can be spoofed, but this is good security for most home users.

These steps are very easy for most people. For those who find these steps confusing, it may be time well spent for IT departments to assist them, given the high cost of a security breach.

Enterprise Gateways

The difference between a residential and an enterprise wireless gateway is simply a matter of capabilities. An enterprise gateway will typically have a WLAN and a LAN interface, which enables it to act as a translational bridge between the two mediums. Typically, an enterprise gateway is deployed as a guest point of access to the Internet, with no direct access to the corporate network. An enterprise gateway has several integrated capabilities and functions, such as an Ethernet switch, 802.11 WLAN, DSL router, firewall, intrusion prevention device, and logical filtering via access lists. This makes it more robust and more expensive than a residential gateway, even though both perform a similar duty.

The main difference between enterprise gateways and their residential counterparts is their ability to handle large numbers of connections and to scale to size. For example, an average residential gateway might struggle to support more than 10 simultaneous connections. An enterprise gateway, on the other hand, would be expected to support more than 100 devices connecting concurrently, with no noticeable performance drop.

Wireless Antennas

Central to the performance of any wireless device is the antenna. There are three main categories of antenna, which are classified by their mode of operation and are used in different circumstances to address specific coverage issues. The three categories are as follows:

- **Omnidirectional**—These are the general-purpose antennas typically installed by default on WAPs. They radiate radio waves equally in all directions, theoretically providing 360-degree coverage.
- **Semi-directional**—These antennas are used when coverage is required in a specific direction. Their RF radiance tends to be limited to 180 degrees. These antennas are typically used to radiate RF in one direction.
- **Highly directional**—These antennas are very specific. They are comparable to a flashlight with a very focused beam. They are typically used in point-to-point situations, where a very exact, high-gain beam is required.

Omnidirectional Antennas

Radio waves are transmitted in both horizontal and vertical patterns. An **omnidirectional antenna** radiates 360 degrees on the horizontal plane and between 7 and 80 degrees on the vertical plane (assuming the antenna is positioned straight up). Omnidirectional antennas are typically used in point-to-multipoint configurations.

Increasing the antenna's gain, or power, flattens the radiation pattern in such a way that it stretches the horizontal coverage while restricting the vertical. Therefore, the higher the antenna's gain, the lower the vertical coverage.

This is an important consideration when designing site coverage. There is a trade-off between greater horizontal range and greater vertical range. For example, if you install an access point with a high-gain omnidirectional antenna in middle of a building's first floor, you will have excellent coverage throughout that floor. However, the signal is unlikely to be sufficient on the second floor because the vertical range has been curtailed. This might or might not be exactly what you want. What is important is that you understand the effect of increasing the antenna's gain on its horizontal and vertical coverage.

You must also consider that the range of an omnidirectional antenna can easily extend beyond the physical limits of a building. This could make it easy for a hacker to sit safely in a parking lot or in an adjacent building while attempting to probe for security weaknesses. Remember that more power is not always better.

Semi-Directional Antennas

Semi-directional antennas restrict unnecessary radiation, directing the focus in one direction. Semi-directional antennas are often used in short-to-middle-range point-to-point configurations—for example, when connecting campus buildings. Another use for semi-directional antennas is to extend RF coverage down a long corridor with offices on either side. After mounting a WAP with a semi-directional antenna at one end of a corridor, you can transmit only one direction, offering very little coverage on the back side of the WAP. This can be very useful when filling gaps in coverage or providing backhaul links. Semi-directional antennas are particularly useful when connecting buildings of varying heights because they can be tilted downward or upward.

There are several types of semi-directional antennas, the most common being Yagi or planar antennas. Yagi antennas (see **FIGURE 5-4**), which have a range of up to two miles, are more often used in outdoor point-to-point configurations. Planar antennas (see **FIGURE 5-5**) are often used in warehouses and retail stores.

Semi-directional antennas are very flexible and come in a huge range of coverage patterns. As a result, they are the type of antenna most frequently used when addressing coverage issues.

One type of semi-directional antenna, commonly used for mobile large-scale 802.11 networks due to its flexibility and capacity, is the sectional antenna. Sectional antennas demonstrate how flexible and efficient semi-directional antennas can be. A sectional antenna has a pie-sliced radiation pattern. This makes it very efficient for covering corridors or paths, because the RF coverage is well defined and constrained at the edges.

Several pie-sliced sectional antennas can work together to cover 90, 180, or even 360 degrees. Because there is no backward transmission, there is little interference between the antennas.

FIGURE 5-4

Yagi antenna.

© Malekas85/iStock/Getty Images Plus/Getty Images

They can be mounted on the same towers, in a ring configuration. Each antenna covers a specific slice of the 360-degree pie. Sectional antennas can be tilted vertically to alter their coverage area independently of the others. This is a common sight on cellular towers, where a four-antenna configuration with independent antenna tilt is used to cover a 360-degree range.

Because they operate independently, sectional antennas are connected to different radios. This makes them far more efficient than if they used a single radio, as with an omnidirectional antenna.

Highly Directional Antennas

Highly directional antennas are tightly focused. Featuring high gain and a narrow beam, they are used in point-to-point connections—for example, in backhaul links.

There are two types of highly directional antennas:

- **Grid antenna**—This antenna uses a wire grill. The spacing between the wires on the grill is determined by the frequency on which the antenna is designed to transmit and receive. **FIGURE 5-6** shows a grid antenna.

FIGURE 5-5

Planar antenna.
Courtesy of Circular Wireless.

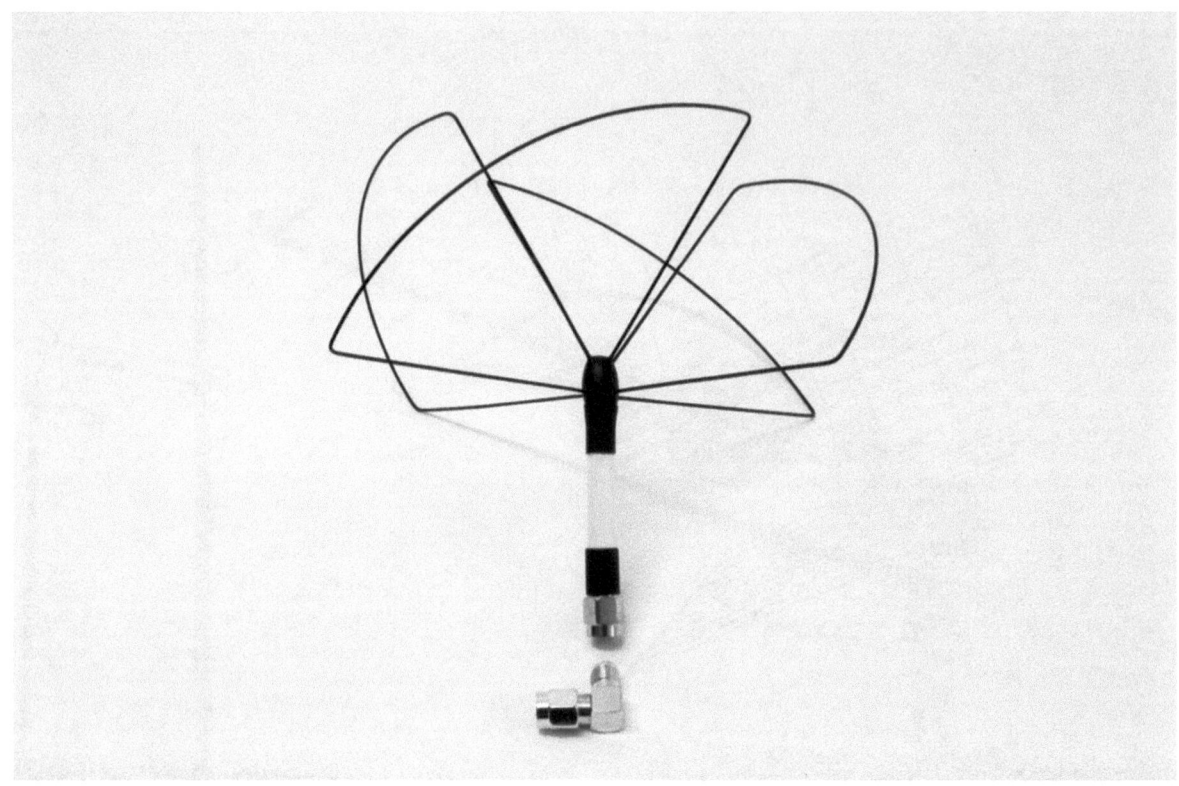

- **Parabolic antenna**—A parabolic antenna resembles a small satellite dish. **FIGURE 5-7** shows a parabolic antenna.

Highly directional antennas are suitable for long-distance, point-to-point connections—up to 35 miles. However, long-distance point-to-point links have several design issues and are subject to wind and atmospheric conditions. They also require high towers for line-of-sight connectivity.

MIMO Antennas

MIMO antennas are a newer technology. The architecture of these antennas allows multiple antennas to transmit and receive concurrently. With MIMO technology, complex signal-processing techniques provide enhanced range, throughput, and reliability. These are key components in mobile 3G and 4G handsets but are also included in the 802.11n standard.

With MIMO, antenna signals must travel a different path for the receiver to be able to distinguish the multipath signals. Therefore, when installing MIMO devices, the groups of

FIGURE 5-6
Grid antenna.
© Luoman/iStock/Getty Images Plus/Getty Images

antennas should be tilted away from each other to introduce a measure of delay between the individual antenna radio signals. MIMO offers a security advantage in that it is hard to jam because it uses multiple radio frequencies and channels. **FIGURE 5-8** shows a MIMO antenna.

Due to the desire for increased throughput, MIMO antennas are becoming the norm for deployment in congested areas such as airports and shopping malls. With 802.11n indoor WAPs, MIMO antennas are usually integrated into the chassis of the device, and no alteration is required or indeed possible. Other vendors supply external antennas in the form of three omnidirectional antennas. These are sometimes removable and can therefore be replaced with other higher-gain antennas. With indoor use, however, that is not usually a concern, because the 802.11n range should be sufficient.

SU-MIMO

Original implementations of multiple radio antennas were introduced in the 3G and LTE mobile carrier networks in an effort to boost data rates and throughput. MIMO technology was created to make use of multiple antennas on a wireless router so that they could be used for both receiving and transmitting, thus improving overall network capacity for wireless

FIGURE 5-7

Parabolic antenna.

© Denyshutter/iStock/Getty Images Plus/Getty Images.

connections. The subsequent success of MIMO in mobile networks meant it was quickly implemented into the Wi-Fi 802.11n specification. The type of MIMO implemented could be either 2x2 MIMO or 4x4 MIMO, which had 2 or 4 antennas, respectively, and could thus support 2 or 4 data streams simultaneously. With early deployments of MIMO, it was typically **single-user MIMO (SU-MIMO)**. What this meant was that the entire bandwidth of the access point was dedicated to a single user/device for the entire timeslice. If the user device could support SU-MIMO then the access point was able to use multiple spatial streams to send a large amount of data on separate streams to each antenna. Many modern user devices support two or three streams, allowing for high speed connections. SU-MIMO is the technology traditionally used in 802.11n and 802.1ac Wave 1 networks.

FIGURE 5-8
A Wi-Fi router with MIMO antennas.

Telecom CMD MIMO antenna designed by Laird.

MU-MIMO

MU-MIMO is an improved wireless technology designed to overcome the single user limitation of SU-MIMO. The design goal was to find a way for access points to support multiple MIMO-enabled devices concurrently. Consequently, MU-MIMO can be considered to be the next evolution up from SU-MIMO. Nonetheless, when the standard 802.11ac was announced, only routers and access points supported the technology. However, compatible devices soon emerged and now there are many MU-MIMO customer endpoint devices. Indeed, any customer device that supports 802.11ac (Wi-Fi 5) will support MU-MIMO; unfortunately, devices such as 802.11n or earlier will not be compatible.

How MU-MIMO Works with Wireless Devices

MU-MIMO was a very important breakthrough in wireless communications because it was an elegant work around the inherent radio problem of only one client being able to communicate over the bandwidth spectrum at a given time. The problem is evident when multiple users are trying to access the wireless network at the same time and from the same access point. The nature of the 802.11 protocol is that users are served on a first-come, first-served basis, but this creates congestion and ultimately constrains the number of devices that can effectively connect to a base station. MU-MIMO alleviates this congestion by allowing multiple users to access the access point. The way this works is that MU-MIMO slices the available bandwidth into separate, individual streams that will then share the connection equally. A MU-MIMO router can come in 2x2, 3x3, 4x4, or 8x8 variations, which refer to the number of streams (two, three, four, or eight) that are supported by the access point. As a result, the access point now offers customer devices from 2 to 8 additional access antennas so the queuing should be effectively reduced and any congestion mitigated.

The 802.11ax standard uses MU-MIMO to increase network capacity in order to better serve dense environments such as airports, stadiums, and shopping malls.

Determining Coverage Area

When dealing with antennas, it is important to understand the antenna coverage charts (often called planar charts or radiation envelopes) provided by vendors. Each antenna type has its own specific characteristics. These are shown in coverage charts from two perspectives: a heads-down azimuth or horizontal aspect, and a side-on elevation or vertical aspect. These charts give a good visual impression of how the RF will propagate throughout the area, showing the typical footprint of RF coverage under ideal conditions. They give you a good idea as to the coverage area you can expect the antenna to deliver (see **FIGURES 5-9** and **5-10**).

> **FYI**
>
> The coverage maps in Figures 5-9 and 5-10 can be very confusing at first because they are represented in a logarithmic format rather than a linear one. An easy way to remember the scale is that in a linear chart, the next line is the same value increase as the one before it, whereas in a logarithmic chart, the next line is a multiple of the value before it. For example, if a chart's first value is 10, the value of the next line in a linear chart would 20. But in a logarithmic chart, the value of the next line would be 100.

Site Surveys

A site survey is used to determine how RF radiation will behave within a facility, taking into account the coverage, interference, and gaps based on antenna placement and power considerations. The process of carrying out a site survey can vary greatly in its complexity, from the deployment of a single access point in a SOHO to a large enterprise covering many buildings and floors. In the enterprise case, site surveys are critical, complex, and time-consuming. The importance of carrying out a site survey, even for the simplest of implementations, cannot be stressed enough.

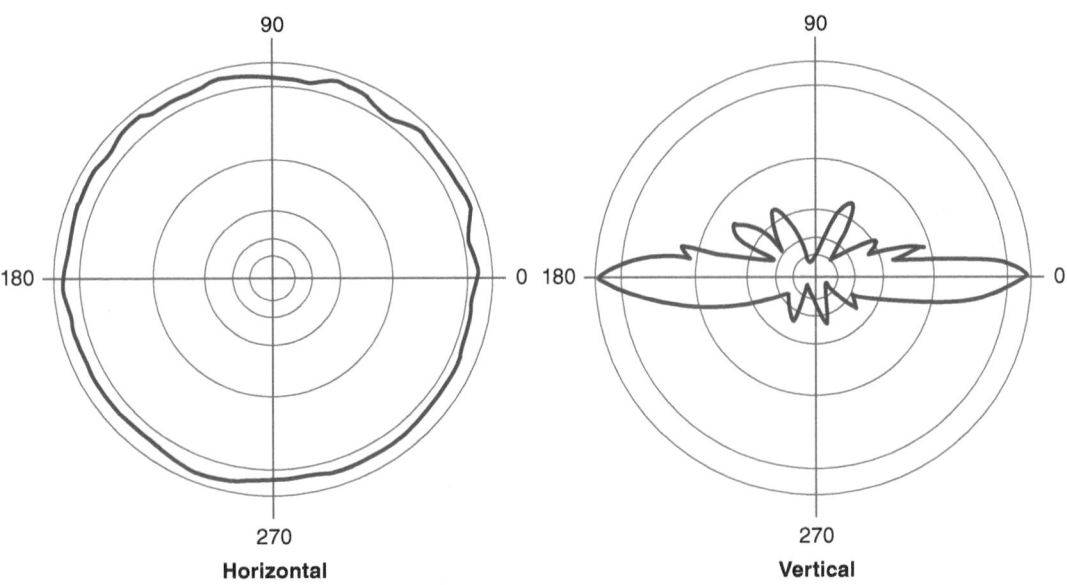

FIGURE 5-9
Omnidirectional coverage map.

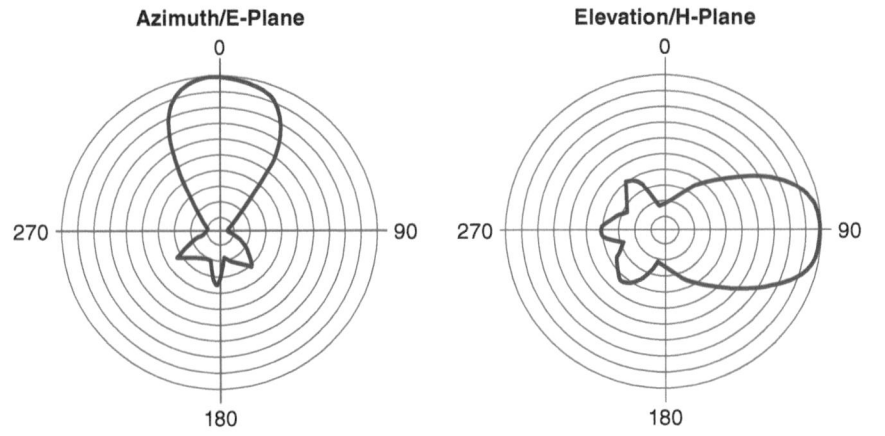

FIGURE 5-10
Semi-directional coverage map.

A site survey should not just be an exercise in RF coverage. Capacity planning must also occur when you consider denser areas of population that will require higher throughput, such as a conference room. There is also the need to identify possible areas of interference. Even as recently as 10 years ago, this was not that critical. Given the proliferation of wireless networks and other devices, however, ensuring that neighboring devices do not cause interference is now very important. Therefore, performing a protocol and spectrum analysis is the first step in any RF site survey.

Spectrum and Protocol Analysis

A spectrum analysis will help detect any type of RF interference that might conflict with your WLAN. When performing a spectrum analysis, a tool called a spectrum analyzer is used. A spectrum analyzer is a test device that measures the amplitude and frequency of electromagnetic signals (radio waves).

Traditionally, spectrum analysis was outside the budget for most site surveys due to the cost of spectrum analyzers. (Rental rates were quite high as well.) In recent years, however, several PC-based versions have become available for 802.11 site surveys. These make it easy to measure the unlicensed frequency spectrum at 2.4 GHz and 5 GHz where WLANs (and many other wireless devices) operate. These solutions detect all radio activity within the site, whether it is 802.11-based or interference in the same spectrum from, for example, a microwave oven.

Performing a spectrum analysis enables you to visualize and map what RF activity already exists within the 2.4 and 5.0 GHz spectrums. If the background noise (that is, any unwanted radiation on those channels) exceeds 85 decibel-milliwatts (dBm), then you can expect the wireless network to suffer performance problems, because the signal-to-noise ratio will result in a degraded signal. This same analysis can be used to ensure that Wi-Fi signals do not reach outside the range specified by network designers. While you might not specify that the signal be limited to indoors, because some companies may allow workers to connect from outdoor break areas, teams should ensure that coverage meets the specific needs and no more.

Being able to visualize signal degradation in RF environments is very important. You must have some way of knowing the RF characteristics of the site where you plan to build the 802.11 WLAN. A noisy environment can have a detrimental effect on a WLAN, causing (among other things) interference, corruption of data transmissions, or a general lack of availability. Transmission failures and high error rates are expected and addressed under the 802.11 specifications, and many applications can survive as much as a 10-percent retransmission rate. Others, however, are not nearly as forgiving. For example, VoIP suffers noticeable performance drops at just a 2-percent retransmit level. High levels of interference and background noise can also prevent a client station from transmitting, because it is always detecting activity on the air (and therefore will wait for a transmission window that never opens). Therefore, it is imperative that you discover what background noise is present.

Sources of Noise and Interference

Noise and RF interference come from many sources. These include the following:

- Microwave ovens
- Cordless phones
- Fluorescent bulbs
- Elevator motors
- Bluetooth radios
- Other 802.11 wireless networks
- Malicious transmitters (jammers)

In most office environments, the main concerns are microwave ovens and other neighboring 802.11 networks. Most microwaves today are well insulated, but it doesn't take much of a leak to interfere with an 802.11 −40 dBm signal. To ensure the least amount of interference, it's a good idea to perform a site survey to determine the presence of interfering signals. If there is a noisy transmitter nearby on the 2.4 GHz band, you may want to choose the 5.0 GHz band. In some cases, a business may be forced to support both based on the types of devices used.

Coverage Analysis

Coverage analysis is what most people think of when they first consider performing an RF site survey. The first step in coverage analysis is to conduct site coverage and capacity-planning interviews. These are information-gathering exercises to establish the capacity and coverage requirements based on the population density in certain areas, which may determine the need for smaller or larger cells or even collocation. These interviews should take into consideration not only the density of client stations but also the applications used and whether support for roaming is required.

No special tools are required for coverage analysis other than the RF signal strength indicator on a laptop or other 802.11-enabled device. Note that it is important to turn down the power on any access points in the area. Many vendors ship access points set to the default power level of 100 megawatts, which is too high for most networks and will likely cause interference with neighboring access points.

NOTE

It can't be stressed enough that more power is not better. In fact, it can cause performance issues and increase vulnerability from a security standpoint.

Access Point Placement

The site survey will dictate the location and the boundaries of the access points. Once determined, the position of each access point must be checked to ensure it is within reach of a wiring closet, because Ethernet has a distance restriction of 100 meters with Cat 5 cabling.

A mixture of omnidirectional and semi-directional antennas should be used because it is often difficult to obtain adequate coverage using only omnidirectional antennas. The goal is to cover only the planned site and to avoid transmitting outside that boundary. Semi-directional antennas can provide that clear definition because they are designed to transmit in only one direction. With the clever application of both semi-directional and omnidirectional antennas, it is usually possible to generate RF coverage that fulfills the requirements while restricting unnecessary propagation or pollution of the RF signal.

Coverage Assessment

When the site coverage and capacity planning interviews are complete, and the access points are placed and enabled, it's time to assess the coverage. You can do this manually or through the use of predictive analytics.

Manual coverage assessment is pretty straightforward. You can do it in one of two modes:

- **Passive mode**—In a passive manual survey, a client card is used to collect the RF measurement of receive signal levels, signal noise, and signal-to-noise ratios. The client card is not associated with the access point, however. It only works at Layer 1 and Layer 2. This will provide the raw numbers, but will not yield any insights into the quality of the connection.
- **Active mode**—In an active manual survey, the client station authenticates and associates with the various access points working at Layer 1, Layer 2, and Layer 3. The advantage of an active survey is that upper-layer performance measurements such as packet loss, **latency**, and **jitter** are available. This gives an indication of connection quality in addition to the raw numbers.

Predictive coverage assessment relies on predictive modeling applications and simulation software to create visual models of RF cell coverage. Using the modeling application eliminates the need to take manual cell measurements because all the cell sizes are predicted using an algorithm. Although predictive coverage assessment might sound like the more expensive option, the opposite is often true—especially for large sites. Engineers tend to use predictive software more often in a bid to reduce the cost and time of performing manual site surveys.

Self-Organizing WLANs

New controller-based access points, or self-organizing WLANs, provide the ability to dynamically reconfigure lightweight thin access points based on changing RF conditions. This dynamic configuration means the network can adjust signals, channels, power levels, and patterns to maintain optimal operating characteristics and network performance. Recall that so-called thin clients have no switch-like intelligence; they simply monitor the network conditions and report back to their respective controllers. The controllers, which control multiple access points, can in turn analyze the data and take corrective action. Specifically, they can send orders to the thin access points to reconfigure the settings or change channels to maintain optimal performance.

CHAPTER SUMMARY

The 802.11 standards that govern the design, installation, and use of wireless networks are but a small fraction of the overall Layer 1 landscape. Moreover, Layer 1 has often been a little discussed or considered aspect of the overall OSI Reference Model. Despite its obscure location in the networking constellation, however, WLANs and the RF considerations inherent in them have had a disproportionate impact on networking, productivity, and security.

For the information security specialist, this weighs heavily. As a result of wireless, access to the network, whether authorized or not, is no longer bound by a physical connection. This makes security far more complicated and far more difficult to manage. Gaining an understanding of this highly impactful medium is a critical aspect of the successful IT security specialist's charter.

CHAPTER 5 | How Do WLANs Work?

KEY CONCEPTS AND TERMS

Active scanning
Ad hoc mode
Autonomous access points
Basic service set (BSS)
Collocation model
Direct sequence spread spectrum (DSSS)
Distribution medium
Distribution service (DS)
Distribution system service (DSS)
Extended service set (ESS)
Frequency hopping spread spectrum (FHSS)
Full duplex
Half duplex
Highly directional antennas

Independent basic service set (IBSS)
Infrastructure mode
Integration service (IS)
ISM unlicensed spectrum
Jitter
Latency
Mesh basic service set (MBSS)
Multipath
Multiple input/multiple output (MIMO) antennas
Multi-user MIMO (MU-MIMO)
Narrowband
Nomadic roaming
Nonce
Omnidirectional antenna

Orthogonal frequency-division multiple access (OFDMA)
Passive scanning
Quality of service (QoS)
Seamless roaming
Semi-directional antennas
Service set identifier (SSID)
Simplex communication
Single-user MIMO (SU-MIMO)
Spread spectrum
Station (STA)
Thin access points
Very high throughput (VHT)
Wireless access point (WAP)
Wireless distribution system (WDS)

CHAPTER 5 ASSESSMENT

1. Passive scanning allows a client to find a wireless network for the first time.
   A. True
   B. False

2. An 802.11 wireless client can be which of the following?
   A. A wireless-enabled PC
   B. An access point
   C. A Wi-Fi-enabled phone
   D. A wireless thermostat
   E. All of the above

3. A basic service set is comprised of which of the following?
   A. Several access points operating as a single network
   B. An access point and several wireless clients
   C. Several clients connected together
   D. A WLAN and the back-end wired LAN
   E. None of the above

4. What is the effect of increasing the gain on an omnidirectional antenna?
   A. It does nothing
   B. It focuses the beam
   C. It increases horizontal coverage while decreasing vertical coverage
   D. It increases vertical coverage while decreasing horizontal coverage

5. Thin access points have basic switching capabilities.
   A. True
   B. False

6. Grid and parabolic antennas are examples of which of the following?
   A. MIMO antennas
   B. Omnidirectional antennas
   C. Directional antennas
   D. All of the above

7. Site surveys help determine RF coverage only.
   A. True
   B. False

8. Passive surveys automatically collect and assess connection-quality information.
    A. True
    B. False

9. Which of the following are sources of RF interference?
    A. Microwave ovens
    B. Bluetooth radios
    C. Other wireless networks
    D. Malicious jammers
    E. All of the above

10. Self-organizing WLANs do which of the following?
    A. Automatically place themselves in a building
    B. Vote for their own leader
    C. Adjust power levels and channels via a controller to ensure peak performance
    D. Allow for seamless roaming

WLAN and IP Networking Threat and Vulnerability Analysis

CHAPTER 6

THIS CHAPTER CONSIDERS the threats and vulnerabilities that are directly associated with 802.11 wireless networks, their various topologies, and devices. It also looks at wireless local area network (WLAN) vulnerabilities within the context of Internet Protocol (IP) networking and how cyberthieves and hackers leverage WLANs as the "crack in the door" to the corporate networks they attack.

Wireless networks consist of radio cards that work together in a variety of configurations to form peer-to-peer WLAN infrastructures, bridges, repeaters, extenders, backhaul networks, and distribution integration services. These are used to connect to different mediums, such as Ethernet. The radio waves used in wireless networks are an unbounded medium, open to far greater interference, corruption, and eavesdropping than their wired predecessors were. This inherent feature of all wireless networks makes these networks far more vulnerable to data interception, manipulation, and theft.

Poor wireless design and the careless deployment of access points with regard to radio frequency (RF) coverage typically enlarge the attack footprint. Far too often, in-house technicians with little knowledge of RF or site-survey best practices install wireless networks. The results are what one might expect: good to excellent RF coverage indoors that meets capacity and coverage expectations, but massive areas of RF leakage outside the building. This leakage typically goes largely unnoticed by all but the wrong people. RF leakage does not just cause unnecessary interference with neighbor networks. It's a huge security backdoor into the company's network.

An understanding of this—as well as of the human behaviors of both careless or unsuspecting employees and malicious attackers—is a key aspect of the information security professional's craft. This chapter weaves elements of relevant human behavior into its discussion of the technical vulnerabilities.

Chapter 6 Topics

This chapter covers the following concepts and topics:

- What the different types of attackers are
- What the two main types of targets are

© Cherezoff/Shutterstock

- How a hacker might scout for a targeted attack
- How physical security applies to wireless networks
- How hackers use social engineering to access wired and wireless networks
- What wardriving is
- How rogue access points work
- What Bluetooth vulnerabilities and threats exist
- What packet analysis is
- How thieves steal information from wireless networks
- What the dangers of malicious data insertion on wireless networks are
- What a denial of service attack is
- What the security risks of peer-to-peer hacking over ad hoc networks are
- What happens when an attacker gains unauthorized control of a network

Chapter 6 Goals

When you complete this chapter, you will be able to:

- Understand how the skill level and intent of hackers and data thieves determine risk levels
- Describe social engineering and provide examples of how hackers use different methods to obtain their goals
- Describe wardriving and how it can be used to exploit poor access point configurations
- Describe the difference between rogue access points and evil twins
- Understand how evil twins work and describe how they are used
- Understand how Bluetooth connections are made and exploited
- Explain why wireless is vulnerable to packet analysis and describe how hackers exploit this vulnerability

Types of Attackers

Before going into details on specific vulnerabilities, let's take an instructive look at the differences between types of attackers and the range of skill levels that different attackers bring to bear. In doing so, it will become clear that the vast majority of threats can be reduced if not eliminated outright.

Skilled versus Unskilled Attackers

Typically, the more skilled an attacker, the more risk there is associated with an attack. But this is not always the case. By way of analogy, there is a difference between a skilled cat burglar and a simple thug who smashes a glass jewelry counter with a hammer. At the end of the day, however, if the bad guys get away with the jewels, it's of little consequence which technique they used.

This point is made because many aspects of sound IT security are overlooked simply because they seem so obvious. All too often, IT security teams focus their energy and resources on preventing highly sophisticated attacks, but fail to employ simple tried-and-true procedures to prevent more primitive breaches. Remember that the charter of IT security is to reduce risk as much as possible in accordance with the needs of the business and the resources at hand. Employee training and locked doors may not be as exciting as setting up multilayered data-center defenses, but they will prevent many possible (indeed, probable) attacks.

Having said that, it's also important to recognize that the skill level of some hackers makes the task of 100-percent security all but impossible over the long haul. Even worse is the phenomenon of nation state–backed hacker consortiums. Not only are these teams exceptionally skilled and well-funded they also have an enormous advantage in that their targets have to be right 100 percent of the time, while they need to be successful only once.

These groups are also known as **advanced persistent threats (APTs)**. They launch multi-phased attacks to break into networks to harvest valuable information while avoiding detection. These highly complex, long-term infiltration attacks present a significant risk to financial institutions and government agencies, among others.

Insiders versus Outsiders

An authorized insider, especially a knowledgeable one, who turns on their employer is a nightmare—possibly the worst-case scenario for IT security teams. Any insider can cause significant harm if for no other reason than that they are allowed in the building. But insiders are not the only threat. If a wireless network is not properly designed and controlled, an attacker no longer needs to get inside the building to gain access to the network. With wireless, hackers and cybercriminals can overcome one of the biggest hurdles they face: lack of physical access to the network.

FIGURE 6-1 illustrates some general risk levels of skilled versus unskilled outsiders and insiders. It's important to note here that you can prevent scenarios in the bottom half of the chart through basic security best practices and employee training. Again, these are not the most glamorous aspects of IT security, but they may very well be the most important and impactful. Later in this chapter, you will read about a low-level, unskilled employee who managed a significant (and embarrassing) data theft that could easily have been prevented.

This brings up one of the sayings you may hear in your career: "You can't stop stupid." This refrain is often spoken by IT security specialists in response to users who, through their lack of knowledge or awareness, do something that raises the vulnerability level or results in an actual security breach.

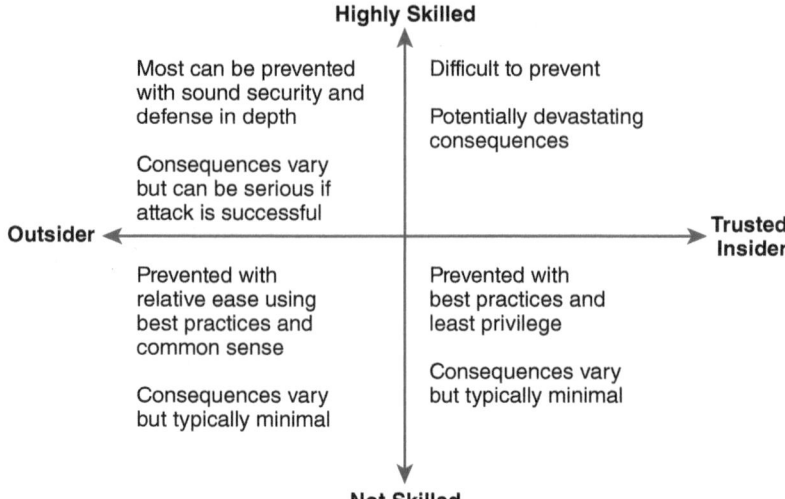

FIGURE 6-1

Comparing the risks of skilled and unskilled internal and external attackers.

While the phrase is funny, it's also a copout. Education on security matters is a critical responsibility of the IT security team. If employees don't understand basic security best practices (at least from a user behavior standpoint), it's the fault of the security team for not training them and of company management for not mandating and supporting said training. What may seem obvious to you as a security professional may not be obvious to someone in the human resources, legal, or finance department, just as you, as a security professional, are likely unaware of generally accepted accounting principles. Indeed, education may well be your best security tool.

Targets of Opportunity versus Specific Targets

In general terms, there are two types of targets:

- **Targets of opportunity**—A target of opportunity is a target that has not previously been identified or considered but that becomes available due to circumstances outside the hacker's control. For example, if someone were to leave a smartphone or tablet at a coffee shop, that might constitute a target of opportunity for a hacker. Other targets of opportunity might include:
    - An unsecured Wi-Fi network in an executive's home
    - An unsecured wiring closet
    - An unsecured rogue access point installed by an employee or an employee attaching to a nonverified free Wi-Fi in a public place to avoid a pay-per-use service

  A target of opportunity can even include cases in which a hacker has a specific company in mind but not a specific goal.
- **Specific targets**—With a specific target, the hacker has a specific goal in mind. This might be the disruption of a business, the theft of customer financial data, or even theft of information for market advantage (corporate espionage).

CHAPTER 6 | WLAN and IP Networking Threat and Vulnerability Analysis

FIGURE 6-2

Comparing the risks of targets of opportunity versus specific targets from skilled and unskilled attackers.

Like Figure 6-1, **FIGURE 6-2** shows a two-by-two matrix—this time with the dangers associated with targets of opportunity and specific targets. As shown in this chart, you can greatly reduce the threat level in three of four sectors with employee training and the application of best practices. In the fourth sector—in which a highly skilled attacker goes after a specific target—the risk level is higher. Although that is cause for concern, most organizations that would likely be the target of such an attack (financial and retail companies, government organizations, political groups) would typically be aware of their higher risk profile and would hopefully have appropriate policies and staff in place.

Scouting for a Targeted Attack

When planning a cyberattack, attackers will spend a great deal of time profiling the network and organization, probing for clues to the systems and devices deployed on the network. Typically, attackers look for any weak spots or backdoors that will enable them to gain access. The first step in the execution of any attack is to gain a foothold.

When assessing an organization's network, many attackers start by using wireless scanning software to detect the presence of a poorly deployed 802.11 wireless network. If the designers of the wireless network have contained the RF to within the company's physical boundaries, then this will be difficult for the casual attacker. However, if RF frequency coverage has been less diligently controlled, then the opposite is true. The attackers will likely be able to detect the network from a safe distance. In fact, using a cheap and easily accessible directional antenna, an attacker could be hundreds of yards away and still penetrate the network.

By gaining access to the wireless signal and the beacons emitted from the wireless network, attackers can capture and analyze packets traversing the air interface. Gaining access

 NOTE

Even in cases when power settings have not been changed, the reach of RF signals can vary greatly, depending on external factors. These include seemingly trivial factors such as rearranged furniture, plant and tree trimming, or changes in season.

to the WLAN at the Physical Layer in this way will enable the attacker to capture packets to map the network and determine the device vendors and operating systems. Using a simple Layer 2 network analysis tool such as Airshark, an open source program that can be downloaded for free, the attackers can capture and eavesdrop on the client station/access point communication crossing the network.

The attackers' goal during this phase is to gather Layer 2 network information, such as the following:

- Media access control (MAC) addresses
- Service set identifiers (SSIDs)
- Basic service set identifiers (BSSIDs)
- Device types deployed
- Authentication use and type
- Encryption use and type
- Channels in use
- Default configurations
- Radio interference from other Wi-Fi neighbors (in multitenant buildings)

With this information in hand, the attackers can plan the actual tactics of the attack that will allow them to associate with the access point.

The attackers may also scout individual employees to see where they live and spend their free time. A common practice is to follow employees home and test the security of their home Wi-Fi networks. Attackers may also follow employees to a public location such as a coffee shop to see if they connect to an unsecured Wi-Fi network. Some attackers may even go so far as to profile these locations and set up an evil twin (discussed later in this chapter).

Why go through all this trouble? Unless the planned attack is a denial of service (DoS) attack, the attacker needs access. In most cases, it's pretty hard to just walk through the door of an organization and get access to an open port. The next best thing is to find an unsecured or weak wireless network or to gain access to an employee's authorized device.

This brings us back to Figures 6-1 and 6-2. Even the most sophisticated attacks have to start somewhere, and that "somewhere" typically involves getting access to the network. Once again, using security best practices and educating employees are key to defense. While taking these steps may deter a skilled hacker for only a limited time, it's still important to do so.

Physical Security and Wireless Networks

The best way to prevent casual eavesdropping, not to mention an attacker performing a Layer 2 packet analysis of the wireless network, is to control the radiation of the RF signal outside the premises. However, this assumes that the internal network hasn't been compromised due to a lack of security. It's essential to treat basic physical security as the foundation of the company's security strategy and ensure that policies are in place and enforced.

This not only includes securing physical access to the building or office but also physically securing internal systems. That means locking doors to rooms containing access switches and installing security doors (with audit logs) for data centers or network labs. In addition, security teams should conduct regular sweeps—that is, survey RF power levels to ensure that RF performance has not changed.

On a related note, you should shut down all switch ports on a switch that are not in use. This prevents users from plugging unauthorized devices into the wired network's wall sockets. This simple measure alone can prevent attackers or malicious insiders from installing rogue access points.

Social Engineering

A significant weakness in any technical system, whether it be a server, a wired network, a firewall, a virtual private network (VPN), or some other system, is the system's end users. These end users—even those who are trained and experienced—may fall prey to **social engineering**. Social engineering is the practice of teasing out information from people that should not be shared to use it to one's advantage.

Phishing is an example of social engineering. With phishing, scammers send an email that appears to come from the receiver's bank or some other trusted organization. This email asks the recipient to click a link, which appears legitimate, but in fact directs the recipient to a site owned by the scammer. The site prompts the recipient to enter their user ID and password. When the recipient does so, the scammer is able to collect it.

One would think that as people become more aware of phishing, the practice would stop. But unfortunately, there will always be people who fall for it. This is what these scammers count on. And as skilled as some hackers may be, you should fear the ones who can manipulate people as much as the ones who can manipulate technology.

Social engineering takes advantage of some of the following human tendencies:

- People are generally nice and want to help other (seemingly) nice people.
- People tend to reciprocate favors.
- People are curious.
- People tend to respond to authority figures.
- People tend to be creatures of habit.

Of course, this is not an exhaustive list. But it serves to illustrate the point. It should also be noted that all these traits are generally positive. Regardless, hackers who are adept at social engineering use these traits to their advantage—and for good reason. As one famous hacker stated, "It is much easier to trick someone into giving a password for a system than to spend the effort to crack into the system." (From Mitnick, Kevin, and William L. Simon. *The Art of Deception: Controlling the Human Element of Security* (ISBN: 978-0764542800). John Wiley and Sons, 2007.)

Some additional examples of social engineering may help illustrate how it works:

- **Chat-up scam**—In this scenario, a hacker calls a company in the hopes of reaching someone who is chatty. For example, the hacker might pose as a remote employee. After three or four pleasant phone calls with someone in the office, the hacker might then say they forgot or erased their Wi-Fi password. Often, thanks to the rapport the hacker has established, the in-house employee will simply supply the information. Hackers may also use these types of conversations to collect other information about the company or about employees.

- **Help desk scam**—In this clever scam, a technically savvy hacker calls various employees (at random or not), claiming to be with the help desk and offering to help with some vague system or computer issue. The hacker will then establish a rapport, which enables them to contact these employees in the future to ask them to "help with an issue"—for example, a "test" involving the Wi-Fi password. This also works in reverse—that is, the hacker, either playing dumb or sounding authoritative, calls the help desk to ask for the password to the wireless network.
- **Curiosity kills scam**—In this scam, the hacker drops a few USB sticks in the parking lot of a specific target (typically a company). The idea is to make it look like an employee dropped them by accident. Often, an unsuspecting (or uneducated) employee will pick them up. In an effort to identify the owner of the sticks, the employee will then plug one or more into an office computer, at which point they will typically see a single file. If the employee opens this file, a program will be secretly installed on the computer that gives the hacker access. (The author knows a gray-hat hacker who swears by this technique.)
- **Authority scam**—With this technique, the hacker, claiming to be an executive or some other authority figure, calls a lower-level employee and demands information, such as the Wi-Fi password. Fearful of getting in trouble, the employee surrenders the information. (This works best in large companies, where not everyone knows each other.)
- **Habits scam**—Hackers know that people are creatures of habit, and that many people create passwords based on these habits (or worse, leave the factory default passwords in place, which are easily found online). This knowledge enables hackers to guess employees' passwords. In one classic case, hackers were able to guess people's passwords based on the knowledge that the people were fans of *Star Trek*. The hackers simply tried different combinations of character names and other aspects of the show, and were very successful. Implementing password best practices has helped with this, but it still happens.
- **Tailgating**—Tailgating occurs when an authorized person opens a secure door (usually with a key badge), and someone follows them in. People who let the "tailgater" through rarely ask to see their badge, because it can be socially awkward. Hackers take advantage of this, using it to gain unauthorized access to buildings. One technique is to open an outer (unsecured) door for someone, who then returns the favor on the next door (the one that requires the key badge). This works particularly well on a cold, rainy day. After all, who would ask to see someone's badge if that person just helped you get in from the cold?

> **FYI**
>
> One of the first major network breaches in the early 1990s—of the Los Alamos Nuclear Lab, no less—was made possible by a chat-up scam. A hacker used this type of social engineering to glean the system password. Social engineering was again used on the same lab in 2006. This technique is especially powerful when hackers use it with multiple employees. Often, each employee divulges seemingly innocent information, but when it is pieced together, it can paint a very accurate picture of what's going on in the organization. During World War II, there was a saying created to warn people of this that said "Loose lips sink ships." The same sentiment applies today with networks.

Wardriving

Wardriving is the 802.11 wireless equivalent of *wardialing*, in which **phreakers** would search banks of telephone numbers looking for a modem to answer. In this way, they found computer systems that were connected to external resources by modems. Now, wardriving attackers search for wireless access points (WAPs) in a form of unauthorized and covert reconnaissance.

Wardriving doesn't require special equipment, although it is typically more successful if a high-gain antenna is used. Usually, the wardriver uses a WLAN utility called a *sniffer* to detect access points and their SSIDs by intercepting and capturing their beacons. Examples of sniffers include Kismet and Airshark, as well as the older but still popular NetStumbler. Wardrivers may also use Global Positioning System (GPS) software and mapping applications such as Google Earth to map and correlate their discoveries.

Wardriving is a tool for finding targets of opportunity. It's for those keen to detect unprotected wireless networks in the area. These people are looking for vulnerable networks, where users have not secured their access points (usually unknowingly). A desirable target for a wardriver will be an access point with its out-of-the-box default configuration—one that still has its default password (often "admin"). Although this configuration is the quickest and easiest way to get a new user up and running, which makes it attractive to many home users, it is inherently insecure.

Wardrivers typically look for the following vulnerabilities:

- Default usernames and passwords
- Weak or nonexistent WLAN encryption
- Default SSIDs
- Default crypto keys for authentication and encryption
- Default Simple Network Management Protocol (SNMP) settings
- Default channels
- Enabled Dynamic Host Configuration Protocol (DHCP)

These vulnerabilities enable wardrivers to easily gain access to and control WLANs.

Wardrivers typically stick to passive attacks or eavesdropping. These types of attacks may not be malicious; they may be done simply out of curiosity. These wardrivers may also map the unprotected networks they find. Other wardrivers, however, launch active attacks on the networks they find. These attacks may take one of several forms:

- **Masquerading**—With **masquerading**, the attacker impersonates authorized users to gain their level of privileges.
- **Replay**—In a **replay attack**, the attacker uses a packet analyzer to capture network traffic between hosts. The hacker can then retransmit that traffic as though from a legitimate user. The message is correctly received, but being "random," can cause disruptions or server errors.
- **Message modification**—This is where an attacker alters, deletes, adds, or reorders the contents of a message. It is an attack on the integrity of the data.
- **Denial of service (DoS)**—By constantly transmitting on the Layer 1 level, a client station can deny others access to the network.

> **NOTE**
> Particularly in a business setting (but also in homes), it is imperative that the administrator not leave access points with their default settings. In addition, administrators should defend against wardriving by implementing encryption settings, reset functions, access control lists, shared keys, and preventing radio frequency leakage.

Wardriving is an attack from outside the boundaries of the home or business premises. You can mitigate it by lowering the power of the access point and by moving the access point to reduce the radiation of RF outside the building. In addition, following best practices with regard to default password settings, authentication, and encryption (for example, changing the default password, setting up MAC filtering, and turning on Wi-Fi Protected Access 2 [WPA2] encryption) will help prevent most security incidents. Wardriving is typically opportunistic in nature, primarily targeting home WLANs and small businesses. If a wardriver detects a secure WLAN, they will likely move on because there are likely many unsecure WLANs around.

Rogue Access Points

A more significant concern for medium and larger networks is rogue access points. A rogue access point, or rogue AP, is an unauthorized AP attached to the wired network. These can be installed by hackers (with malicious intent) but can also be installed by well-meaning employees who simply want easier access to the network. In the latter case, hackers can easily exploit these rogue APs. In addition to being unauthorized, rogue APs are unmanaged, which makes them doubly vulnerable to attack. They may be added to the network to serve as an attack vector—a deliberately and maliciously installed device that provides an attacker with a convenient backdoor into the network. An estimated 20 percent of corporations have had rogue APs on their networks at some time.

> **NOTE**
> Rogue access points have become more common since the commoditization of IT and the introduction of wireless-enabled devices into the workplace. To use their own mobile devices at work, users require an access point and often get impatient waiting for IT to set one up.

Rogue Access Point Vulnerabilities

The dangers of rogue APs are twofold in nature. First, if the rogue AP is installed by an employee who simply seeks easier access to the network, it will likely be poorly done, both in terms of power and security. This can lead to leakage of RF signals from the building, which is akin to leaving a local area network (LAN) Ethernet switch accessible from the street. Second, if a hacker installs the rogue AP, it offers easy access into the corporate network from a safe location, usually from outside the physical security perimeter.

A rogue AP creates a significantly larger attack footprint for an attacker. Potential vulnerabilities include the following:

- Scanning and mapping of the WLAN network
- **Man-in-the-middle (MITM) attacks**
- Attacks on the wired network, such as **Address Resolution Protocol (ARP) poisoning**, DHCP attacks, Spanning Tree Protocol (STP) attacks, and DoS attacks

- Free and unauthorized Internet access and all the associated problems that come with that (including, but not limited to, launching phishing and DoS attacks and visiting illicit websites)
- Data leakage and theft due to unauthorized access

Rogue APs are a real concern. In a large network—especially one with many locations and departments—preventing the installation of rogue APs and finding ones that already exist can be very difficult. There are some automated tools for finding them—for example, security appliance plug-ins that scan for devices listed as wireless APs— but it's very easy to hide one if that's your intent.

One way to reduce the risk of rogue APs is to use Network Access Control (NAC) with mutual authentication, which can block unauthorized devices from connecting. Another approach is to maintain strict control over the Ethernet ports on access switches, ensuring that all unused ports are shut down by default. This would prevent a rogue AP from being inserted into a live Ethernet port, although it isn't always practical because it takes a lot of hands-on management by IT and can cause employee frustration. Yet another way is to deploy a wireless intrusion prevention system (WIPS), which can actively hunt and block rogue APs connecting to the corporate LAN. However, if the rogue AP is an evil twin then this can be difficult.

Evil Twins

An evil twin is a rogue AP installed with sinister intent. In the case of an evil twin, an attacker poses as a genuine network service provider but actually eavesdrops on activities conducted on the network and steals information and passwords. Evil twins are the wireless equivalent of the fraudulent phishing websites used to lure people into divulging their personal information.

The evil twin works because it looks like a legitimate access point. But when users connect to it and use it to access websites and perform other tasks, the access point eavesdrops on their every move, stealing credentials, passwords, and anything else of interest. Most banking websites and email clients use Hypertext Transfer Protocol Secure (HTTPS) and are therefore not vulnerable to this sort of attack, but a lot of information can nonetheless be gleaned through the use of an evil twin. Evil twins are especially hard to detect because they are easy to set up and can be run on a laptop, which means they can be shut down and relocated very quickly.

Typically, evil twins are configured with the same SSID as an authorized access point. In fact, an evil twin may even pass data directly through the original access point. Users fall for the trap because 802.11 management packets are easily forged and access points are not required to prove their identity. To compound matters, many laptops, smartphones, and tablets are configured to automatically connect to the access point with the strongest signal.

Client stations form associations by connecting to an access point advertising its presence through the use of beacons. The client listens for these beacons while in passive mode. Alternatively, it can actively send out probe requests when in active mode. A probe request initiates a probe response from all listening access points with a given extended service set identification (ESSID) if one has been named in the probe request. If no ESSID was specified, however, then all access points will reply with a probe response.

Access point beacons and probe responses carry all the specifications, characteristics, and functionality that an access point supports, including the basic service set identifier (BSSID), which is often the access point's MAC address. The client will respond to the access point that has the strongest signal and the correct advertised capabilities with an authenticate request. If the access point is configured for WPA2, there may be a challenge for a shared key.

In most network configurations, however, an authentication response will simply be sent back by the access point. The client and access point then exchange an association request and a response. This forms the association between them, which will be maintained until a disassociate or deauthenticate packet is received. The problem here is that neither Wired Equivalent Privacy (WEP), which is deprecated but still in use in many home networks, nor WPA/WPA2 can prevent associating with an evil twin because encryption only comes into play after association is complete. They cannot prevent ESSID, MAC, or management packet spoofing, which occurs before the association (if formed).

To set up an evil twin, the attacker first listens for an ESSID being broadcast from the genuine access point. They might use an application such as Hotspotter to listen for probes being sent from other clients in the area. Alternatively, the attacker might use Airshark or NetStumbler to capture and analyze packets to identify the ESSID for the WLAN. Once the attacker knows the ESSID, they deploy a fake access point close to the targeted victim client stations. It's important to emphasize that the attacker does not install a hardware access point—rather, the attacker runs a software-based access point on their laptop. Because most client stations will associate with any access point sharing the same ESSID, it is often not even necessary to forge the MAC address of the genuine access point.

The evil twin is conceptually similar to an Ethernet MITM attack (in which attackers insert themselves between two machines to eavesdrop) using ARP poisoning. In this type of attack, the attacker corrupts the victim's ARP cache with the wrong MAC address, thereby diverting the victim's traffic to their own machine. ARP poisoning is easy to do because you only have to affect one host in a subnet to contaminate all hosts because the infected host broadcasts its own IP as the DNS resolver.

Address Resolution Protocol

ARP is essentially a list-creating service to see what devices are on the network. Think of it as a roll call of sorts. It's used when one device tries to locate another. ARP essentially calls out to ask, "Where is address xxx.xxx.xxx.xxx?" All devices whose addresses *do not* match, do not reply. This enables the original client to determine where the device whose address *does* match exists on the network. This info is kept in an ARP table, which is a local address book. The problem with ARP is that it does not verify the response. That means once a hacker is on a network, they can answer any ARP requests, fooling the requester and "poisoning" the resulting ARP table. Traffic intended for a real target can then be sent to a false one, enabling the hacker to view, copy, or modify the data.

Evil twins are commonly used to divert traffic to phishing websites, where fake pages are constructed to steal login names and passwords. However, there are ways to mitigate the risks presented by evil twins. The most effective is a basic awareness of these risks. It's important to educate employees that not all access points can be trusted.

One of the most common indications that a user may be connected to an evil twin is if the user is unable to establish an HTTPS connection. A bank or Internet service provider's (ISP's) security certificate is unlikely to be out of date or out of order; chances are, if such a connection fails, it's due to a problem with the access point. Another less obvious flag is if a location offers free Internet access. It's true that many legitimate businesses do offer free Wi-Fi, but before you attempt to log on to an access point detected by your device while at a coffee shop, hotel, or other public place, it's a good practice to ask an employee if that access point is in fact supported by the business. Given the choice of connecting to the Internet for free or for a fee, most people will choose the free offering. Naturally, hackers are aware of this.

In an office environment, techniques such as 802.1x **Port-based Network Access Control (PNAC)** can be used for robust mutual authentication. Similarly, strong authentication protocols that require server certificates issued by trusted certification authorities should also be used.

Bluetooth Vulnerabilities and Threats

Bluetooth is a short-range RF communications protocol initially developed as a wireless replacement for the serial interface RS-232, which was a popular short-range interface for computer peripherals. Bluetooth operates in the 2.4 GHz frequency spectrum as a low-cost, low-power radio interface used to connect personal wireless devices such as headphones, tablets, mobile music players, and smartphones. Bluetooth is heavily associated with the concept of the **wireless personal area network (WPAN)**, because it enables wearable or mobile wireless devices to peer with each other to form ad hoc wireless networks. Bluetooth is used to connect peripherals to computers and is widely employed in business and home electronic consumer devices, which can be conveniently interconnected without the need for cables.

Bluetooth connects to up to eight devices in a piconet (a network created using a Bluetooth connection) but uses only 1 megawatt of signal. This low-power usage conserves battery life but restricts range to around 10 meters. Moreover, Bluetooth uses the adaptive frequency hopping (AFH) spread spectrum to mitigate the effects of interference and frequency jamming or blocking. Bluetooth's implementation of frequency hopping spectrum spread (FHSS) uses 79 randomly chosen frequencies and changes 1,600 times every second. By using spectrum spread in this manner, many devices can share the same radio frequencies because they are constantly changing and any clash will last only a fraction of a second. However, because Bluetooth works in the 2.4 GHz frequency range, it does not require line of sight. That means the signal can pass between rooms.

Bluetooth Versions

Bluetooth evolved over nearly two decades from a wireless replacement for physical RS-232 cables to a ubiquitous wireless technology used for everything from file sharing and device pairing, to wireless accessories and even low-power Internet of Things (IoT) devices. As a result, there is estimated to be around 10 billion Bluetooth-enabled devices worldwide and deployed across countless technologies.

Bluetooth's success has come about through its ubiquity, flexibility, and user convenience. It has maintained backward compatibility across versions, therefore retaining its dominance in the personal area network (PAN) domain such as in wearables and smartphones. Nonetheless, it has not stood still and is increasingly evolving to address low-power and long-range requirements. Therefore, Bluetooth is branching out into deployments in power-constrained IoT devices, particularly within low-energy communication mesh networks.

Revisions Compared

Most recently, Bluetooth innovations have focused on IoT technologies such as mesh networking, lower energy profiles, and longer radio range. However, in order to maintain backward compatibility whilst continually improving the specification, Bluetooth's core capabilities have been retained. This is despite many additions and improvements to the specification with each revision.

Notable additions to the specification along the evolutionary path include enhanced data rates of up to 3 Mbps in version 2.0, improved high-speed data exchange over Wi-Fi in version 3.0, and the introduction of a low-energy mode in version 4.0.

VERSIONS	MAX CONNECTION SPEED	TYPICAL MAX RANGE
Bluetooth 1.0 (1999)	0.7 Mbps	~10 m (33 ft)
Bluetooth 2.0 + EDR (2004)	1 Mbps core 3 Mbps with EDR	~30 m (100 ft)
Bluetooth 3.0 + HS (2009)	3 Mbps with EDR (24 Mbps over 802.11 link)	~30 m (100 ft)
Bluetooth 4.0 + LE (2013)	3 Mbps with EDR 1 Mbps Low Energy	~60 m (200 ft)
Bluetooth 5 (2017)	3 Mbps with EDR 2 Mbps Low Energy	~240 m (800 ft)

Since version 4.0, Bluetooth has evolved and branched into two distinct paths: a Low Energy path and a Classic path. The direction of the Low Energy path is tailored toward short infrequent burst-type communication, typically used by IoT devices. The Classic path continues to provide for improvements in range and data throughput for devices that work over a constant connection. The latest Bluetooth 5 revision, for example, focuses on boosting the range and throughput of its low-energy and long-range options, while maintaining backward compatibility with the classic protocol used by smartphones.

CHAPTER 6 | WLAN and IP Networking Threat and Vulnerability Analysis

	BLUETOOTH LOW ENERGY (BLE)	**CLASSIC (BASIC RATE & EDR)**
Channels	40 channels with 2 MHz spacing	79 channels with 1 MHz spacing
Data Rate	BLE 5: 2 Mbps BLE 4.2: 1 Mbps BLE 5 Long Range (S=2): 500 Mbps BLE 5 Long Range (S=8): 125 Mbps	EDR (8DPSK): 3 Mbps EDR (π/4 DQPSK): 2 Mbps Basic Data Rate: 1 Mbps
Power Consumption	~0.01x to 0.5x of Classic	Based on radio class
Network Topologies	Point-to-Point (including piconet) Broadcast Mesh	Point-to-Point (including piconet)

Bluetooth Pairing

An interesting feature of the Bluetooth protocol is how it discovers potential peers. Whenever a Bluetooth-enabled device comes near (within 10 meters or 32 feet) another Bluetooth device, they begin to communicate without any user initiation or intervention. During the communication, they check to see if they have information to share and negotiate a master/slave relationship. By forming this ad hoc piconet, the Bluetooth devices synchronize and frequency hop in unison. This automatic peering is great for plug-and-play connectivity within the home but has serious security implications in the wild.

The Bluetooth specifications allow for three levels of security:

- **Authentication**—This is done to verify the Bluetooth device address.
- **Confidentiality**—This mechanism prevents eavesdropping.
- **Authorization**—This ensures a device is authorized to use a service before being permitted to do so.

Bluetooth also supports four security modes, which define—or rather, initiate—security protection. Not all Bluetooth devices are capable of supporting the security features at any given level, however. The Bluetooth security modes are as follows:

- **Security Mode 1**—Devices that use this mode are designed and produced with no security features, making them vulnerable to attack.
- **Security Mode 2**—This mode determines whether authorization is required before a device can have access to certain resources.
- **Security Mode 3**—This mode requires that Bluetooth devices initiate security before the physical network connection can be established. In Security Mode 3, authentication and encryption are mandatory for all connections.
- **Security Mode 4**—This mode, introduced in Bluetooth version 2.1, is a service-level security mode that uses **Secure Simple Pairing (SSP)**. SSP is a secure method of pairing or connecting Bluetooth devices.

Despite there being four modes, there are only two service security types:

- **Trusted**—A trusted device has full access to all services of another trusting device.
- **Untrusted**—Untrusted devices do not have an established relationship and therefore can reach only restricted services.

These distinctions enable Bluetooth devices to exchange data without asking permission. With Bluetooth Security Modes 1 and 3, no service security trust model is applied. In contrast with Bluetooth Security Mode 2, authentication, encryption, and authorization are required. For Security Mode 4, the Bluetooth specifications call out four separate levels of security:

- **Service Level 3**—This requires MITM protection and encryption, and preferably user interaction.
- **Service Level 2**—This requires encryption only.
- **Service Level 1**—This does not require encryption; user interaction is not necessary.
- **Service Level 0**—This requires neither MITM protection and encryption, nor user interaction.

Not all Bluetooth-enabled devices support these security levels. Some devices have a fixed setting, where the manufacturer of the Bluetooth device decides which level of security to apply. With some devices, such as headphones, this is understandable, since they are not exchanging data and are not deemed a security risk. However, early implementations of Bluetooth were set as Level 0 by default, purely for convenience reasons. That made them very vulnerable to attack. This led to a whole host of Bluetooth attacks, such **bluejacking**, **bluesnarfing**, and **bluebugging**.

Bluejacking

Bluejacking came about through the misuse of a Bluetooth feature whereby a mobile phone could exchange a "business card" or messages with another phone in the vicinity. It soon became clear, however, that this was a fine opportunity for interruption marketing and advertising. Typically, a storekeeper in, say, a shopping mall would set up Bluetooth-enabled devices with high gain antennas to spam any Bluetooth-enabled devices passing by.

For this to work, the Bluetooth devices needed to peer before the messages could be communicated. This of course meant that the marketing was actually consent-based advertising, as the passerby (or, more specifically, the passerby's device) had explicitly agreed to accept the message. The flaw was that the passerby didn't know with whom their device was peering (that is, the "agreement" was the result of a default setting), nor did the passerby necessarily know what the message was until after the fact.

This initially worked because not everyone felt inconvenienced by these unsolicited messages. In Europe, where the first Bluetooth-enabled phones were sold, it became almost a fad, called *toothing*, in which young people sent flirtatious messages to each other over this wireless medium. Many were glad to receive these anonymous peering requests. Eventually, however, bluejacking was considered to be an intrusion. That's because after the spammer's initial message was accepted, their Bluetooth device ID was added to the trusted contacts.

The spammer's device was then able to send messages at any time (if within range). Initially, this was merely an annoyance. But as more and more stores adopted this form of advertising, it quickly became a real nuisance.

Bluesnarfing

Bluejacking was relatively benign. Indeed, it was even quite popular among some users. However, a variant that used the same basic exploitation wasn't so friendly: bluesnarfing. Bluesnarfing is a technique whereby an attacker gains access to unauthorized information on a Bluetooth-enabled device such as a laptop or, more commonly, a mobile phone. In the case of a mobile phone, the attacker can then access the contacts, calendar, emails, and text messages. Where bluejacking was a harmless annoyance, bluesnarfing was actually data theft.

For bluesnarfing to work, the victim's phone must have Bluetooth enabled and be in discoverable mode. In this mode, the phone advertises its Bluetooth ID and can be found by other Bluetooth devices also in the same mode. This makes mobile devices susceptible to both bluejacking and bluesnarfing. However, being in discoverable mode is not enough. The Bluetooth devices must also pair, which (per the standards) requires user intervention. In most cases, this means users must take explicit action to allow their mobile phone to pair with another unknown device. However, a bad combination of lack of awareness and greater convenience (that is, vendors shipping the phone set to security level 0 by default) allowed attackers to exploit users with bluesnarfing attacks.

Once the attacker initiates the bluesnarfing attack from their laptop, theft is quite simple. Bluesnarfing uses the same business-card exchange feature as bluejacking with one fundamental difference: Whereas bluejacking uses a software method called *push* to push out the message to the pair device, bluesnarfing uses a get request to pull in from the device. To use the *get* command, the attacker must know the file structure and directory names on the device. This should make things difficult, but unfortunately for users, the mobile telephone industry named these files and directories using standard nomenclature. For example, on all platforms, the phone book file was named telecom/pb.crf and the calendar file was named telecom/cal.crf. This made theft pretty easy.

In addition to their fixed file locations for the phone book, calendar, and other features, what made mobile phones so susceptible to this type of attack was the fact that they didn't require authentication. In the early implementations of Bluetooth and its message push feature, especially on Nokia and Sony Ericsson mobile phones, convenience trumped security, and no authentication was required. In hindsight, this may seem like a major failure of security. Remember, however, that both mobile telephones and Bluetooth were new technologies, so convenience and ease of use were paramount design criteria.

Even until 2004, there was much debate with regard to Bluetooth being a secure protocol. Bluejacking and bluesnarfing had done little to dent its solid reputation. Many experts blamed misuse and a lack of user awareness for these security problems rather than security flaws inherent in Bluetooth's protocols and the 802.15 standard. In 2004, Bluetooth version 2.0 was released. It addressed low data exchange rates, which made pairing laptops to mobile phones for Internet connectivity and data exchange a viable solution. It also made Bluetooth devices much more tempting targets for malicious attackers. Bluejacking and bluesnarfing were nothing compared to the attacks that were to follow.

Bluebugging

Bluebugging was a quantum leap in attack methodology from bluejacking and bluesnarfing. It didn't just push or get data; it enabled an attacker to commandeer the entire handset. Bluebugging works by first gaining trusted device status, typically through the well-known business-card trick. If successful, the next stage is to establish a connection by tricking the victim's phone into believing the attacker device to be a Bluetooth headset or some other innocent-looking peripheral. Once this is accomplished, the attacker can control just about every function of the phone via **AT command codes**, which are specific commands that enable various functions on the device, used by developers and service technicians. (AT stands for *attention*.)

With full control of the phone, attackers can listen in on conversations (hence the name *bluebugging*). They can also redirect calls and even make calls without the owner's knowledge. Fortunately, the vulnerability that allowed bluebugging was addressed in later firmware upgrades, making it obsolete as an attack tool.

Unfortunately, addressing Bluetooth vulnerabilities across versions and upgrades is not easy because not all devices can be upgraded. Consequently, many devices will remain vulnerable throughout their lifetime. In addition, new versions bring new features, but importantly not all vendor implementations of Bluetooth are standard based; they are often software features and these also can introduce further vulnerabilities such as:

- **BlueBorne**—This vulnerability allows any affected device with Bluetooth turned on to be attacked. The flaws aren't in the Bluetooth standard itself, but in its implementation in vendor software. **BlueBorne** has several known flaws that can allow attackers to take control of a Bluetooth-enabled device. This means they could listen in on any phone calls, make calls, or access and steal any data. The attack also spreads to other vulnerable Bluetooth-enabled targets if any are nearby.
- **BleedingBit**—What makes **BleedingBit** potentially very dangerous is that it doesn't need to pair with the victim device. All that BleedingBit requires is that Bluetooth is turned on and is in range. BleedingBit exploits a vulnerability found in a very common set of Bluetooth chips manufactured by Texas Instruments. But worse, BleedingBit is also "contagious," and this means it will spread through the entire network.
- **KNOB (Key Negotiation of Bluetooth)**—In 2019, a vulnerability in Bluetooth classic surfaced. The vulnerability concerned the key negotiation process, which enabled a type of brute-force attack to exploit a flaw in the firmware of a device's Bluetooth chip. KNOB makes a device vulnerable to MITM attacks via packet injection. Essentially, the KNOB vulnerability gives a false sense of security to users because they believe they have a secure connection with a paired device. The KNOB vulnerability works because Bluetooth permits keys of one-character length, and it also fails to check changes in key entropy during negotiation. Further, the exchange is not encrypted and the pairing device has no alternative but to accept the low-entropy (easily guessable) key. Thankfully, launching a successful KNOB attack over the air and in the wild is extremely difficult and requires expensive Bluetooth protocol analyzers.
- **SweynTooth**—In 2020, vulnerabilities were discovered in Bluetooth devices running BLE including some medical equipment and instruments. **SweynTooth** has three types of attacks: the first can crash devices, the second can reboot devices or force

them into a deadlocked state, and the third and most worrying is an attack that can override the device's security. This gives the attacker full control of the Bluetooth low-power device.

Is Bluetooth Vulnerable?

In 2008, the National Institute of Standards and Technology (NIST) published the *Guide to Bluetooth Security*. It noted that Bluetooth has benefits but is susceptible to DoS attacks, eavesdropping, MITM attacks, message modification, and resource misappropriation.

Like Wi-Fi, Bluetooth presents a trade-off between convenience and security. As such, it will always be vulnerable to the limitations of user education and the risk awareness and risk adversity of the end user. Bluetooth vulnerabilities can be summarized in a few key points:

- **Short PINs used during pairing**—Capturing the key exchange as it happens is not easy or even likely without special equipment to force the devices to disconnect and then re-pair. Even so, choosing longer PINs (the access code that allows a connection to be made) makes it more difficult to crack the PINs quickly. This is therefore an advisable way to mitigate eavesdropping of the PIN exchange.
- **Users pairing devices in public**—An attacker must be able to eavesdrop on devices while they are pairing. Allowing pairing only while in the security and privacy of a home or office considerably mitigates risk. However, this requires manual intervention.
- **User convenience**—Bluetooth will always be vulnerable to users who decide convenience trumps security.

Disabling the Bluetooth feature when in public is the best way to mitigate the risk. A less drastic measure is to switch off discovery mode. The Bluetooth device will still operate, but its Bluetooth ID will remain hidden from other Bluetooth devices. An attacker would have to try to determine the target's MAC address to make a connection, which, at 48 bits, is not something that can be done quickly—even with a packet analyzer.

Packet Analysis

Packet analysis is the practice of capturing and deciphering packets being transmitted across the air interface. There are several freely downloadable open source packet analyzers. One of the most popular is a wireless knockoff of the Ethernet Wireshark called Airshark. (Airshark is not affiliated with the Wireshark tool.)

A packet analyzer works in a wireless environment slightly differently than on a wired network segment. On an Ethernet switch, for example, a probe—typically a cable plugged into the network interface card on a PC or laptop—captures all the packets on that port that are destined for that MAC address and all broadcasts for the entire segment. However, this isn't of much use because the network card must be programmable to work in promiscuous mode. In this mode, the network interface card captures all traffic crossing the wire, regardless of the MAC address. Unfortunately, this isn't much of an improvement, because a switch is intelligent, only sending data out on the specific port associated with that MAC address.

To see all the traffic on an entire segment, a network administrator must enable **port mirroring**. With port mirroring, all traffic on the ports specified is replicated on the port where the mirroring function has been set up. This enables the administrator to see all traffic of interest across many ports on a single port. This feature requires both physical access to the port and administrative access to the switch configuration commands. Obviously, if a hacker has that level of access, wireless security is of little concern.

On a wireless network, all traffic is visible on the same frequencies and channels, making it easily intercepted and captured. All that is needed with Airshark is to run the software. The wireless network card, which enters promiscuous mode, will then capture all the packets it sees cross the airwaves, regardless of protocol or destination. This makes sniffing or eavesdropping on a wireless network far easier than it is on a wired segment, and it's one of the reasons wireless networks are inherently less secure.

Wireless Networks and Information Theft

Wireless networks are susceptible to information theft because, by their nature, they transmit to anyone who cares to listen. All that is necessary is an antenna and a receiver tuned to the correct frequency. A client station need only listen for an access point's beacon before sending an authentication request and forming an association.

But that's not the only way in which information can be stolen. An attacker can install a rogue access point or evil twin to steal user credentials. Attackers can also obtain this information via social engineering. In short, there are many ways an attacker can exploit the natural vulnerabilities of wireless communications to steal information.

For this reason, wireless networks were initially deemed unsecure. Common consensus suggested that they be used only for network access at the perimeter of the network and behind the firewall. Network designers deployed wireless networks in zones built on security interfaces on the Internet perimeter firewall. This design ensured that the company's wired network was behind a firewall and protected from the client stations on the wireless network. Designers considered this to be a secure and practical design, believing that wireless client stations should access only the Internet and the intranet, and not have access to secure areas of the network.

Times and business requirements have changed, however. What was a good design in the late 1990s no longer fits the purpose by the mid-2000s. Wireless user devices started to creep into the workplace in the form of laptops, Wi-Fi–enabled mobile phones, organizers, and BlackBerry devices, all of which were Internet-capable.

At the same time, there was a trend in application development toward web service–based software. These applications resided on web servers and used back-end databases for their dynamic content. Neither the applications nor their clients needed to install anything on a local computer, because they were accessible through a browser using HTTP. When PCs were only located in an office, this was not a big security risk. But with smartphones and tablets, which are easily lost and always connected, a great deal of corporate data could be placed at risk.

By 2010, laptops, smartphones, and tablets were common work tools in the office, and network designers had to cater for them. This meant an expansion of the wireless network into all areas of the workplace. Wi-Fi was now not only for guests; it was all pervasive,

CHAPTER 6 | WLAN and IP Networking Threat and Vulnerability Analysis

penetrating deep into the most secure areas of the network. This was not without serious repercussions with regard to data security and information theft, however. Protected corporate data "walks" outside the building every day, with almost every employee.

Information theft has become a real problem since the acceptance of mobile devices within secure networks. Authorized users now download and store information on mobile devices that leave the company premises. An authorized user downloading information onto a mobile device negates all the controls and security techniques enforced within the network. To compound this problem, the device may or may not be authorized.

The advent of Mobile IP—which is the convergence of WPANs, WLANs, and wireless wide area networks (WWANs) into one coherent administrative entity—has enabled seamless roaming within and between networks. However, it has also led to serious security threats with regard to information theft. There are simply no border checks between these autonomous wireless networks.

As an example of this vulnerability, consider this simple scenario: An authorized user (or more likely, someone using an authorized user's credentials), seeking to steal information, accesses it from company databases and downloads it to a local drive. The application rarely checks to determine what device it is downloading to; it simply transfers the data to the requesting device, regardless of whether it is a PC, laptop, or smartphone.

High-Profile Cases of Information Theft

Information theft is a huge threat to modern networks. Even the most secure networks— or what should be the most secure—are at risk. As an example, consider the National Security Agency (NSA), which was caught short by information theft. In 2013, an NSA contractor named Edward Snowden downloaded thousands of documents from NSA data archives and leaked them to the international media. The loss of such vast amounts of confidential information was distressing and embarrassing not only for NSA officials (after all, they are the professional eavesdroppers and data thieves) but also for foreign governments, major telecom and tier-one Internet service providers, and security software and equipment vendors around the world.

The fact that the NSA was vulnerable to such a low-tech attack, and that no alarms or flags were raised when thousands of classified documents were downloaded by a single entity, is almost beyond belief. However, the NSA was not the first (nor will it be the last) to fall prey to information theft by a low-level contractor or employee. During the Iraq War, Army Specialist Bradley Manning (now known as Chelsea Manning) downloaded hundreds of thousands of secret communications between U.S. embassies around the world. The subsequent release of these documents on the Internet, after they were uploaded via satellite to WikiLeaks, created a huge embarrassment for the U.S. government, especially the State Department. Ironically, WikiLeaks itself would eventually cry foul when it appeared that the hoard of documents was discovered to be freely available for download on BitTorrent. WikiLeaks blamed its media partners for the breach and the subsequent release of the encryption key.

What you can learn from this is that a malicious insider can easily breach networks at even the most security-conscious organization. This brings us back to the issue of the skilled attacker versus the unskilled attacker. These devastating breaches are preventable through the use of policies and rights management tools, neither of which are cutting-edge technology.

After the user has downloaded the information, they can then circumvent all internal security measures by connecting to a mobile (cellular) network provider and uploading the data from the laptop or smartphone with no border checks whatsoever. The person can then delete the data on the local device, secure in the knowledge that the information they wanted is now resting happily in a Dropbox folder or some other cloud-based information storage depository. The bottom line? It's crucial to protect the network from intruders and to keep information private and safe from theft.

Malicious Data Insertion on Wireless Networks

Information theft is an obvious threat to any network. Even with Bluetooth devices, it can be a problem. Stealing digital information is not quite the same as stealing a physical object, however. It does not restrict, curtail, or prevent the rightful owner from accessing the information. That is, the data thieves don't remove the information; they simply copy it. Although this is a crime, the original asset is left intact.

What may be of more concern on some occasions is when attackers modify data in some way for their own gain. For example, a student might change their grade by modifying database fields in a school's system. Or an attacker with a criminal record might attempt to corrupt or delete police files. Here, the intention is not to steal but to modify information to benefit the attacker.

Wireless networks are particularly vulnerable to this type of activity, sometimes called *malicious data injection*. This is because these networks transmit 802.11 frames across an open-air interface. Using a packet analyzer such as Wireshark, an attacker can capture and replay these 802.11 frames. That means an attacker can intercept packets between another client station on the same segment and a network server and inject a compromised payload. This works because by default, 802.11 encrypts only the payload, leaving the MAC address and other headers in cleartext. Capturing a conversation stream using a packet analyzer is easy, and by doing so, the attacker can analyze, modify, and then replay it at their convenience. In effect, an attacker injects a genuine frame with a compromised payload back into the network.

If the network does not use encryption, then inserting malicious data on a wireless network is a trivial task. Using a packet analyzer, the attacker can read the contents of the payload in cleartext and modify them as desired. For example, if a student wanted to modify their grade, that student could eavesdrop on sessions until they discovered a client station conversation with a server requesting a database field update. By capturing that conversation, the attacker could analyze and tailor the request to identify the results database, their name, and the grades field. The attacker could then inject the reconstructed frame as part of a replayed conversation.

Without encryption, data insertion and manipulation are not just feasible but very easy for a reasonably skilled hacker to do. To make the task a bit harder, you can encrypt the network. One option at the advent of Wi-Fi used to be WEP, but it's easy to crack with the right tools and now we need more robust encryption. WPA3 is the modern choice. Using strong encryption will prevent all but the most determined professional attacker. Without the keys, the attacker cannot decrypt the payload, analyze the conversation, or modify it.

Some attackers may be more interested in being a nuisance by modifying and injecting control messages, which have no payload and are therefore still in cleartext. This is more a DoS attack than a data insertion, even though it uses the same methodology.

Denial of Service Attacks

Wireless networks are particularly vulnerable to DoS attacks because they operate on a half-duplex collision-detection medium. Only one radio client station can talk at a time, including the access point. Devices listen for traffic before transmitting, and if there is something else transmitting, they will wait. Therefore, if a faulty transmitter is constantly transmitting, no other client station will be able to communicate.

If a DoS is unintended, a packet analyzer such as Wireshark is a great tool for quickly identifying the culprit and resolving the problem. Identifying the failing transmitter is especially easy if it resides on your own network. If the failing transmitter is on a neighboring network, it might be easy to detect but not so easy to resolve. Interference can also result in an unintentional DoS. Resolving a DoS that is the result of neighboring sources of interference can be very problematic. This is because the spectrum is unlicensed. There is no one to adjudicate in disputes; everybody has equal rights to the frequency spectrum.

Unfortunately, not all DoS scenarios are accidental. Some are the result of an attacker's deliberate handiwork. DoS attacks are the weapon of choice for the less skilled, the less imaginative, and the just plain malicious.

On wireless networks, DoS attacks can be categorized into several groups. These include the following:

- **Application Layer attacks**—These attacks are common on both wired and wireless networks. In Application Layer attacks, the idea is to overwhelm the application server to prevent it from handling requests. This is typically accomplished by sending thousands of requests per second, such as HTTP get requests. For example, the Mydoom worm issued 64 requests per second. When thousands of infected systems issued it at once, it quickly overwhelmed both the server and the network capacity.
- **Transport Layer attacks**—The goal of a Transport Layer attack is to consume the finite number of Transmission Control Protocol (TCP) sockets available on the connecting device, be it a server or a firewall. Called a *TCP SYN flood attack*, this attack sends the synchronization (SYN) packet in the first part of the TCP handshake but leaves the connection open by never sending the acknowledgment (ACK) packet.
- **Network Layer attacks**—In a Network Layer attack, large amounts of data are sent to the wireless network. This floods the bandwidth capacity and overwhelms the target device so it is unable to respond quickly enough to reduce the deluge of traffic. An example of this type of attack is the basic **Internet Control Message Protocol (ICMP)** echo request flood. ICMP is a protocol used by network devices to send error messages. If performed in conjunction with other hosts, this can bring down servers and consume bandwidth, denying other hosts service. Many firewalls block ICMP packets to prevent this type of attack, and many intrusion prevention systems can dynamically change firewall rules to block ICMP packets when they detect an attack.

- **MAC Layer attacks**—A DoS unique to wireless networks is the authenticate/associate request flood. When a client wants to join a wireless network, it sends an authenticate request followed by an associate request to the access point. In most networks, this is a trivial affair. However, if an attacker spoofs the MAC address and sends continual authenticate and associate requests, the target access point has no way of telling whether they are legitimate, so it attempts to process them anyway. This, like the Transport Layer attack, consumes all the access point's memory and exhausts its processing capabilities because it cannot complete the half-open processes. Another technique that attackers may use is the deauthenticate/disassociate flood attack, which forces clients to disassociate and then try to reauthenticate and associate. If done in sufficient numbers, this also consumes the resources of the access point and exhausts the capacity of the network.
- **Physical Layer attacks**—These are attacks against the frequency spectrum used by the wireless network. An attacker can simply cause enough interference using high-gain antennas to create an unacceptable level of background noise. This affects the signal-to-noise ratio, thereby degrading communications.

Peer-to-Peer Hacking over Ad Hoc Networks

Wireless networks configured using the 802.11 standard take two forms:

- **Infrastructure networks**—These require a central access point to serve as the hub of all access and communications. Client stations must communicate with each other via the access point. The same is true for communication beyond the access point via the distribution medium.
- **Ad hoc networks**—With ad hoc networking, peer-to-peer relationships are formed directly between client stations to form informal networks.

Ad hoc networks are typically used to connect peripherals to devices—for example, a printer to a laptop. However, they can also be used to temporarily connect laptops to create an impromptu network—for example, during a presentation or a workshop. Some users also use mobile devices to set up hotspots, enabling other mobile devices to connect directly to the corporate network (which is an ad hoc rogue access point). Ad hoc client stations peer with each other to form the network using the same frequencies and channels as the infrastructure network. Therefore, it is important that they do not clash or cause interference with the infrastructure network, or the main wireless network, by using the same channel.

Ad hoc networks have a few performance and security issues. First, they frequently cause interference and degradation of service to both the ad hoc and corporate networks. Second, they have no way of authenticating clients, so any 802.11 device configured in ad hoc mode can connect to any other ad hoc station to form a network.

This is another case of convenience trumping security. The idea was for ad hoc networks to offer a quick and simple way to interface peripherals and devices. The term *ad hoc*, meaning "for this," signifies that the network is a solution to a particular problem, and is not intended to be adapted for other purposes. Therefore, the 802.11 standards deemed that authentication was not required.

Some IT departments have attempted to ban the use of ad hoc networks. However, it is a common sight to see management use ad hoc networks to peer with partners and clients in meetings to share documents and slides. As a result, the use of ad hoc networks continues, despite IT's best efforts to eradicate them from the workplace. This is a problem with user awareness and lack of understanding regarding the potential for unauthorized access and control. The coming wave of near-field communication (NFC) applications will put further pressure on IT security teams because many devices are expected to have this technology available in the near future (and some already do). It's a good bet that hackers will learn to exploit this technology ahead of the general population's understanding of the risks.

When an Attacker Gains Unauthorized Control

As you have seen with ad hoc networks, it's possible for an attacker to gain unauthorized access to and control of an 802.11 network simply because device authentication is not robust. You can secure access points from unauthorized authentication and association by using WEP and WPA (for casual "free" access) or the more robust WPA2. WEP and WPA can be cracked with off-the-shelf tools and should never be used if WPA2 is available. However, WPA2 is not the answer for hotel or café hotspots. In these cases, device authentication and association must take place at Layer 1 and Layer 2, and user authentication must follow at the Network or Application Layer, driven by web browser username/password challenges.

In a business environment, you generally will not wish to advertise the presence of your WLAN. One way to achieve this is to not advertise the SSID in the beacon. This practice, called **network cloaking**, is not very effective, but it will defeat most casual attacks. Not broadcasting the SSID effectively prevents casual snoopers from identifying the SSID of the network. However, it will not deter a more determined attack because when an authorized device configured to join the network actively probes for the access point, it will send probe requests that contain the SSID. If the attacker uses a tool such as Kismet, which listens for all client station requests and access point responses, they will be able to correlate the request from a station and the response from an access point. When the client station joins the network, it inadvertently reveals the SSID to the attacker. In addition, the access point response contains the SSID and the BSID, which is often its MAC address.

In the end, network cloaking is not a good way to secure or even obfuscate a network. That's because preventing the access point from advertising the SSID in its beacons causes every client device to probe for it. On a very busy network segment with many different networks—for example, in a condominium complex—the network may remain hidden. But even then, a determined attacker will eventually match client probes to access point responses, and that will reveal all that they need. That being said, it should be stressed that network cloaking and using an encryption key such as WPA or WPA2 may not be business-class security, but they are much better than an open network.

CHAPTER SUMMARY

This chapter discussed the inherent vulnerabilities of wireless networking and how data thieves and hackers exploit these vulnerabilities. The very nature of radio-based communication makes this medium vulnerable. As such, it's often the "way in" to networks when the goal is more than petty theft or eavesdropping.

It's not just the vulnerabilities of wireless networks that put them at risk, however. The inherent vulnerabilities of the people who use them are also factors. These vulnerabilities include a lack of training and a lack of awareness, both of which hackers may take advantage of. After all, it's easier to have someone give you a key than it is to pick a lock.

The good news is that many—perhaps even most—of these vulnerabilities can be mitigated if not eliminated through a combination of sound policy, best practices, and employee training and education.

KEY CONCEPTS AND TERMS

- Advanced persistent threats (APTs)
- Address Resolution Protocol (ARP) poisoning
- AT command codes
- BleedingBit
- BlueBorne
- Bluebugging
- Bluejacking
- Bluesnarfing
- Internet Control Message Protocol (ICMP)
- Man-in-the-middle (MITM) attack
- Masquerading
- Network cloaking
- Phreakers
- Port mirroring
- Replay attack
- Secure Simple Pairing (SSP)
- Social engineering
- SweynTooth
- Wireless personal area network (WPAN)

CHAPTER 6 ASSESSMENT

1. Unskilled attackers are not a threat and can be disregarded.
   A. True
   B. False

2. An organization can greatly reduce risk by doing which of the following?
   A. Educating employees
   B. Deploying simple best practices
   C. Adopting least-privilege policies
   D. All of the above

3. Unauthorized wireless access is often a means of access for sophisticated attacks.
   A. True
   B. False

4. Why does social engineering tend to work?
   A. People are dumb
   B. Hackers know mind-control techniques
   C. Hackers know how to take advantage of human behaviors and tendencies
   D. Security is weak

5. Which of the following describes the act of wardriving?
   A. Mounting a battering ram on your car
   B. Searching for unsecured wireless networks while driving around
   C. Jamming other people's wireless networks
   D. Taking over someone else's Bluetooth connection

6. Wardrivers look for which of the following vulnerabilities?
    A. The use of default administrative usernames and passwords
    B. No or weak encryption
    C. The use of default SSID settings
    D. All of the above

7. Which of the following describes an evil twin?
    A. A version of a rogue AP in which the device masquerades as a legitimate access point
    B. A social engineering scam
    C. A Bluetooth hack that takes over another device
    D. A peer-to-peer hack

8. Most Bluetooth vulnerabilities are based on how they connect, or peer, with each other, and can be mitigated by disabling connectivity while out of the office.
    A. True
    B. False

9. Why is packet analysis particularly problematic on wireless networks?
    A. You can "listen" to traffic without a physical connection
    B. Unlike wired networks, you don't need port mirroring to see all the traffic
    C. Packets can be modified and reinserted without authentication
    D. It can be used to initiate a local DoS attack
    E. All of the above

10. Wireless-based DoS attacks only happen at Layer 1.
    A. True
    B. False

CHAPTER 7

Basic WLAN Security Measures

WIRELESS NETWORKS COME IN MANY SHAPES AND FORMS, from small-office, single-access-point networks to vast enterprise networks composed of hundreds, if not thousands, of interconnected access points. However, regardless of the size and scope, the basic principles of wireless security remain the same: to safeguard data privacy, ensure availability of service, and protect against the theft or manipulation of information.

That said, when you consider wireless local area network (WLAN) security measures used to mitigate the various risks and vulnerabilities, you must align your objectives with the actual threats the organization faces. These measures range from basic considerations and affordable solutions to very robust advanced security schemes. All organizations and home users should employ some basic security measures. As the size of the organization or the risk profile increases, the needs of the organization may dictate more advanced security measures. However, these needs do not eliminate the need for getting the basics right.

This chapter focuses on the basic security measures that, generally speaking, satisfy the needs of small office/home office (SOHO) networks. Of course, there are exceptions, such as a boutique trading firm that requires advanced security or a large commodity business that has little in the way of protected data. For the purposes of discussion, however, this chapter considers the typical SOHO model, which typically features a single access point supporting a number of manually manageable clients.

Chapter 7 Topics

This chapter covers the following concepts and topics:

- What the design and implementation considerations for basic security are
- What the basic authentication and access considerations are
- What data protection techniques are available
- What some ongoing management security considerations are

© Cherezoff/Shutterstock

Chapter 7 Goals

When you complete this chapter, you will be able to:

- Understand how proper design and installation contribute to basic security
- Describe methods of radio frequency design, layering, and security management
- Understand the security implications of basic authentication and access
- Describe access methods such as SSID masking, MAC filtering, VPNs, and VLANs
- Understand the importance of data protection on wireless networks
- List the common methods of data protection
- Understand how ongoing management affects security
- Describe best practices for periodic security checks and physical sweeps

Design and Implementation Considerations for Basic Security

Before tackling more technical topics such as authentication and encryption, every basic security discussion should begin with the design considerations that support basic security. Regardless of the scale, scope, or risk profile of a network, designing the network with security in mind will provide the foundation upon which all other security efforts are built.

Radio Frequency Design

Radio frequency (RF) waves can travel through walls and windows and leak into the outside world. Allowing this leakage is akin to leaving unguarded Ethernet switches lying around outside the business premises. A wireless broadcast is by nature available to anyone in the vicinity who wishes to receive its signal. Therefore, it is crucial to restrict the RF coverage to the premise's boundaries. This is not just good security; it is also good manners. Radio pollution broadcast beyond the realm of a property is a major factor in the degradation of performance among neighboring wireless networks.

> **FYI**
>
> The home office setup has not traditionally been a concern of the corporate IT specialist. As the Bring Your Own Device (BYOD) trend becomes the norm, however, it should be. With BYOD, users bring their work devices home, where they are often used to connect to corporate resources. If an employee's home network is not secure, this

can give hackers an easy path into the corporate network. Given this risk, corporate IT departments may find it in their best interest to concern themselves with the security of employees' home networks. This likely does not warrant investing in active support, but it may be worthwhile to develop an easy-to-follow security setup guide that includes simple instructions for setting up Wi-Fi Protected Access 2 (WPA2) security as part of a BYOD or work-from-home policy.

When attempting to contain RF waves within a building, consider using semi-directional antennas and lowering the power. Manufacturers usually ship access points set to the highest power setting. This is not always the best setting, however. Not only does this sometimes cause the access point to broadcast beyond the property's perimeter but it can also be a source of interference with neighboring wireless networks that share the same frequency band. (Remember that a collision of radio waves of the same frequency does not amplify the signal. Rather, it attenuates it.) This is a common mistake, particularly in home offices. People often add access points to their home in an attempt to boost the signal. If extra coverage is needed, however, wireless repeaters are a better solution.

Equipment Configuration and Placement

When deciding on the placement of an access point in a SOHO environment, you must take one factor into consideration: Most manufacturers ship access points with an omnidirectional antenna, which transmits in all directions to form a 360-degree coverage area. In most cases, this is the best solution, because it supports a large area. However, because most buildings have internal walls and floors that can hinder the passage of the RF waves, the theoretical coverage is not normally achievable. **FIGURE 7.1** illustrates the horizontal and vertical signal distribution of a wireless signal.

> **NOTE**
> The principal reason for performing a site survey before implementation is to ensure there is sufficient signal strength and coverage throughout the workplace. Another reason to perform a site survey is to make sure there are no unnecessary broadcasts outside the premises.

It follows, then, that placement of the access point is a key consideration. For example, if the access point is installed near an external window, you can expect that signal to travel through the window, making itself available hundreds of yards outside the premises—hardly a sound security practice. To mitigate this threat, you should place the access point in a central location and adjust the power to ensure adequate coverage without excessive external radiation.

In SOHO wireless designs—particularly in the home—the access point is typically located close to the DSL router or the cable wall socket. Often, this is against an external wall, which is not an optimal location. Because no long Ethernet cables will be required to connect it, however, it may be the most convenient. (Remember that RF will travel through walls and windows.)

In addition to finding a good, central location for the access point, you should also consider the antenna type and coverage pattern. This helps ensure that the necessary coverage can be provided at the lowest power setting and with the least amount of leakage/noise.

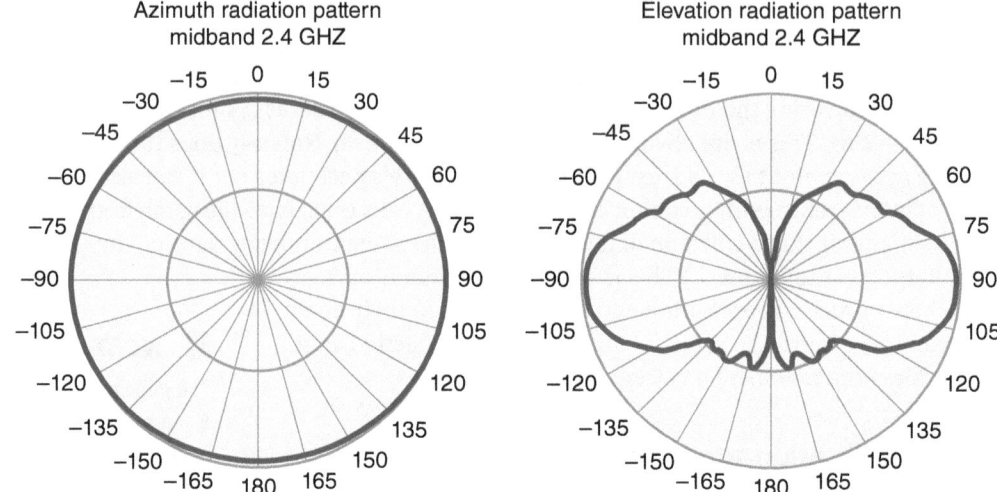

FIGURE 7-1

Example of both azimuth (horizontal) and elevation (vertical) signal distributions for a wireless antenna.

Data from Ruckus®/CommScope.

Interoperability and Layering

Even within the smallest of properties, it can be difficult to get good RF coverage throughout the premises. This section discusses the use of wireless bridging, extensions, and layering. Even in larger and more complex networks, these basic considerations are relevant.

Delivering RF coverage in a building can be a difficult task, prone to unpredictability. Walls, ceilings, floors, and corridors are all obstacles for RF waves. As discussed, the best approach is to place the access point in a central point in the building. Ideally, the access point's omnidirectional antenna will cover the required area both horizontally and vertically (where applicable). The most common way to check for this is to walk the premises using a tablet or laptop to measure the received signal strength. If there are **dead spots**—that is, spots where the signal does not reach—then you might increase the power settings or adjust the position of the access point to try to compensate.

Sometimes, parts of the building may experience a persistent low signal, even when you increase the power setting. Or, it may be that increasing the power setting solves the problem but results in excessive RF leakage. In these cases, a wireless extender or repeater may be the solution. A **wireless extender** or **wireless repeater** is a device that uses the same frequency and channel to overlap the original basic service set (BSS) coverage area by 50 percent, increasing the reach of the access point by half. Unlike additional access points, wireless extenders can use the same frequency and channel as the main access point without creating attenuation.

If the problem is more lack of throughput and capacity than coverage (a common occurrence if there are too many wireless connections on a low-end access point), then an overlay may be required. An overlay uses a twin access point that overlaps the original access point by 100 percent, operating on the same service set identifier (SSID). By setting the two access points to noninterfering channels, you can effectively double the capacity and throughput of wireless connections by spreading users over two separate access points. Note that any one client is connected to only one of the access points at any given time. This was a common design solution during the early days of wireless, when total throughput was less than 10 Mbps. With the advent of high throughput (HT) 802.11n and the latest very high throughput (VHT) 802.11ab, this configuration is required less often. But it still works when needed.

Either of these network topologies is feasible within a SOHO environment. Network designers often deploy them to solve coverage or capacity problems. However, you must never lose sight of the relationship between a coverage area and an attack footprint. That is, the less RF coverage, the less opportunity an attacker has to eavesdrop on, intercept, and manipulate the wireless data on your network, and the more tightly you can secure your network through the application of strategic security management.

Security Management

When managing wireless networks from a security perspective, it is important to have detailed knowledge of all the available security tools and techniques. In 2004, the Institute of Electrical and Electronics Engineers (IEEE) deprecated (openly disapproved of) some techniques that were at one point approved, such as the Wired Equivalent Privacy (WEP) security algorithm.

However, despite newer and better techniques being available, it's not uncommon to see this and other deprecated techniques still in use, particularly in SOHO networks. This is not ideal, but using these defunct security practices is generally better than using nothing at all. They will not offer much resistance against a hacker of reasonable skill, but they will do a decent job of keeping casual trespassers off the network. That being said, it's better to upgrade to more modern approved security measures. Many of these are free to implement and easy to configure and may even already be integrated into the access point and radio network card.

SOHO networks tend to be simple in nature, usually consisting of a single BSS. However, the businesses supported by these networks—in particular, the business requirements—may be very diverse. As a result, the supporting network design may have quite specific requirements and characteristics. For example, consider a public relations/Internet marketing agency that is heavily reliant on the wireless network to support **Voice over WLAN (VoWLAN)**—offering and supporting Skype or Microsoft as well as convenient Wi-Fi access for clients. For this business, the reliability, quality, and availability of these voice calls are the key performance and design criteria. In contrast, another SOHO company, such as a mobile telephone reseller, might not require VoWLAN support but may require very high security, especially with regard to authentication and encryption. For this type of firm, having a high level of security is the overriding goal, because it would likely store hundreds of thousands of dollars of prepaid voucher codes on its servers. Based on these cases, you can visualize two SOHO networks, each with contrasting security and performance design criteria. The bottom line is that while there are some best practices that can be broadly applied, there is no one-size-fits-all design.

Basic Security Best Practices

Basic security best practices include the following:

- Limiting RF leakage by lowering the transmission power of access points
- SSID cloaking
- WPA2 encryption with a strong passphrase
- Authentication
- MAC filtering
- Keeping access points in a locked closet
- Regularly checking for and installing software or firmware patches

Of course, when performing an audit on an existing wireless network or gathering facts and requirements for a new installation, you should address the individual business's requirements. During the initial gathering of requirements and specifications, you should be aware of the design criteria and the relevant threats and vulnerabilities. These should be documented in a process called a *risk assessment*, discussed in a later section. For now, it's sufficient to know that it's good design practice to identify and rank the essential design characteristics as well as the inherent risks, threats, and vulnerabilities, and to lay out their corresponding solutions and mitigating practices. The output of the risk assessment is a document that is an important input into the official security policy.

Despite each network being potentially different with regard to layout, coverage, performance, and security, there are some industry best practices that will help you. Security best practices for a generic SOHO network will satisfy the majority of small business and home requirements. These best practices cover the basic security essentials, such as authentication, encryption, and access control. With regard to basic security practices, you will next examine how security is commonly implemented in SOHO networks and consider what should be considered legacy practice and what is best practice.

Authentication and Access Restriction

TIP

Integrating an SSID with identity-based access techniques such as RADIUS authentication is a good practice.

After implementing a security-centric design, you will want to control access to the network. This can be accomplished in a number of ways, many of which are mutually reinforcing in much the same way as defense in depth. The first step is simply to hide the network. If the network is discovered, authentication plays a key role in keeping unauthorized users off the network. Access restriction is not limited to outsiders, however. There is often a need to segment internal users as well. When used together, these techniques greatly reduce the chances of unauthorized access, whatever the source.

SSID Obfuscation

SSID segmentation was commonly used on older, pre–**Robust Security Network (RSN)** networks to provide security as defined in the 02.11-2007 standard. (RSN is the IEEE 802.11i security standard.) To implement this, the administrator created and assigned different SSIDs for different types of users, protocols, functions, or departments. By assigning an SSID to a **virtual LAN (VLAN),** the administrator provided a method to segment users by SSID/VLAN pairs. Therefore, users who connected to the same access point, but were members of different departments, could be logically grouped and segregated by SSID/VLAN pairs, which provided the necessary security. In addition, each SSID could be configured with different security parameters, making the security model scalable.

SSID segmentation is still used in SOHO networks where there is a digital subscriber line (DSL) router present to switch between the VLANs and provide the trunk backhaul to the access point. Typically, three SSID/VLAN pairs are created, one each for voice, data, and guests. The guest SSID will have no authentication and have access only to the Internet. The voice SSID will provide separation for voice traffic and support quality of service (QoS), a protocol that prioritizes time-sensitive traffic such as voice or live video. Data, on the other hand, will pass over the data SSID for stronger security, authentication, and encryption.

Using SSID/VLAN pairs is a common strategy. However, there is a downside: the amount of work required to configure each individual access point. Fortunately, in a SOHO scenario, this is not an insurmountable issue. Even in a small to medium business (SMB) scenario, with perhaps 20 or 30 access points, it is still manageable. Another possible concern is the number of management frames produced, because each SSID now acts as a virtual access point and reproduces the same number of management frames and beacons per SSID. Obviously, the Media Access Control (MAC) Layer overload generated by using too many SSIDs will affect throughput. For this reason, it's best to avoid overusing the technique. Although some access points can support up to 16 SSIDs, doing so would likely degrade performance in a noticeable way.

Another SSID-related technique is **SSID cloaking**, whereby the Broadcast SSID option is disabled during configuration. This technique works because the client must find, authenticate, and associate with an access point before it can connect and exchange traffic. To authenticate, the client must first find the access point's SSID or MAC address. Without knowledge of either, the client cannot authenticate to the network. Typically, an access point broadcasts its SSID in the beacon that it transmits frequently over the air to all listening client stations.

SSID cloaking can effectively hide the WLAN from unauthorized client stations by inhibiting the advertising of the SSID. By transmitting a null in place of the SSID in the beacon frame, the network remains cloaked from view. When a client passively scans for the network, the SSID will not be revealed. If actively scanning for access points, the client can transmit search probes with the null field set. With this configuration, all access points are required to respond to the client station. However, when the cloaked access point responds to the client, it does so with a probe response with the SSID set to null. Therefore, even an active scan does not reveal the network's cloaked SSID.

SSID cloaking is sufficient to defeat passive or active scanning and even tools such as inSSIDer, which uses a modified active scanning process. This process is not invulnerable, however; it can be defeated using a protocol analyzer or tools such as Kismet. There is an inherent weakness in that authorized client stations must be able to connect to the access point. To do this, they must be preconfigured with the SSID. When a client with authorized access sends out its probe request, it contains the specific SSID, and the access point is required to respond. If an attacker is eavesdropping, they can easily read the MAC address, SSID, and basic service set ID (BSSID) contained in the probe response frame. Additionally, the administrator or user must physically enter the SSID in the client wireless configuration. This makes it very susceptible to social engineering.

Despite not being perfect (you will find there is no such thing), SSID cloaking is a best practice for avoiding casual or opportunistic access to the network. One drawback is that people will need to ask for the SSID to gain access. Basic security layers such as this are an essential part of keeping out less skilled attackers or trespassers who perpetrate the vast majority of casual opportunistic attacks. It should be noted, however, that this method will do nothing to prevent even a moderately skilled hacker from accessing the network.

MAC Filters

When it comes to basic security, MAC filtering comes high on the list. As discussed, each MAC address is a unique six-byte number that is hard-coded into the network interface. MAC addresses are used in OSI Layer 2 communications to identify end station hosts and are the identifiers for the source and destination in Layer 2 frames. Therefore, 802.11 WLANs, which are Layer 2 networks, heavily rely on MAC addresses. Unlike Internet Protocol (IP) addresses, which are logical assignments, MAC addresses represent the physical address of a machine or device.

MAC addressing is the fundamental way in which devices communicate using frames at Layer 2, whereas IP addressing only has Layer 3 relevance. In switched and wireless networks (which are Layer 2 networks), MAC addressing is king, which makes it all the more important to filter based on these unique identifiers.

In theory and principle, MAC filters are used in a "deny by default, permit by exception" scheme, where only those MAC addresses that are listed are permitted access. In very large or public networks, this is not always practical. In contrast, in SMB and SOHO networks that are stable environments with a relatively small number of users/devices (in the hundreds as opposed to thousands) and few guests, this is a manageable practice.

Critics of MAC filtering will point out that MAC addresses can be spoofed, making filtering not especially effective against a skilled attacker. While this is true, MAC filtering (much like SSID cloaking) remains an effective technique against casual opportunistic attacks.

Authentication and Association

For a client station to be able to join a network, it must go through the initial mandatory process of authentication and association. There are many standard mechanisms for this listed under 802.11. Two of these are as follows:

- **Open System Authentication**—One process of connecting to a wireless network is **Open System Authentication (OSA)**. As long as the SSID is known, the client can access the network and receive nonencrypted information.

CHAPTER 7 | Basic WLAN Security Measures

- **Shared Key Authentication**—**Shared Key Authentication (SKA)** is part of WEP encryption. With SKA, a client can access the wireless network and send and receive encrypted data by matching the encryption key on the access point.

While OSA is still in use and approved, SKA (along with WEP) has been deprecated and is not recommended. However, some manufacturers still support them in 802.11 products for backward compatibility. The 802.11-2012 standard defines the RSN methods meant to replace legacy techniques. In practice, however, it is common for SOHO networks to use deprecated methods.

The most common implementation is OSA, which requires a minimal exchange between client stations and authenticating access points. In OSA, the devices exchange hellos, which simply confirm that both parties are 802.11 devices and can use and understand 802.11 frames. An access point using OSA will authenticate any 802.11 client. OSA is typical, not just in SOHO environments, but also in large-scale networks where guest clients are the norm and preconfiguration of the guest wireless client isn't convenient or feasible. Therefore, OSA may be considered legacy and pre-RSN, but it is not deprecated. It is still a perfectly valid and accepted (but not secure) method of Layer 2 authentication.

SKA is viewed as an improvement over OSA in that it involves an additional step that requires the exchange of a matching shared key. This is not a more secure method, however, as encryption does not take place until after authentication and association are complete. That means the preshared-key challenge issued by the access point is in cleartext in the authentication response frame, which is part of the WEP four-way authentication handshake. If an attacker eavesdrops on that handshake, they can capture the access point's cleartext challenge and the client's subsequent encrypted challenge response. The attacker can then use the challenge response in a replay attack or determine the static shared key in order to break the authentication mechanism.

NOTE

There are many types of four-way handshakes associated with different authentication schemes. For example, WPA2 also uses a four-way handshake—one that is not vulnerable.

Worse is the fact that the same static shared key is used not just to authenticate but also to encrypt the payload. Therefore, if the attacker gains the key, encryption is easily broken. Static shared keys are suitable only for SOHO networks, because the effort to preconfigure all clients and keep the static shared key secret is all but impossible in larger networks.

Be aware that WEP is a legacy method for authentication and encryption. Now considered easily cracked, it has been superseded by more robust encryption protocols such as WPA2. However, WEP is still better than nothing at all, despite the fact that it is outdated.

VPN over Wireless

Prior to the 802.11-2007 standard, virtual private network (VPN) over wireless was a commonly used technique for securing user connections. VPNs were particularly useful when configuring inter-building bridges and secure point-to-point links. In security-conscious environments, VPNs were also used for client station access, but that is now discouraged due to high overhead and performance issues.

After 802.11-2007, clearly defined Layer 2 security solutions were provided, which has made VPN usage in the WLAN somewhat redundant. Layer 3 VPNs are still useful in remote point-to-point bridges and links to secure traffic between access points, but network designers rarely use them in client stations to access point security anymore. VPNs are widely used for secure remote connections ranging from public networks or home offices to corporate resources located behind the company firewall.

One downside of using VPNs for secure Wi-Fi access is that they operate at Layer 3, which means an attacker can get access to both the Layer 2 and Layer 3 connections before the VPN tunnel is established. This represents quite a foothold. To prevent an attacker from getting this far, some administrators employed WEP encryption, which encrypts at Layer 2, to protect the Layer 3 information. This double encryption created further overhead and had a significant impact on performance and throughput. Furthermore, WEP is easily broken, so the extra layer of protection had to be balanced against any potential impact on performance.

TIP

It's always a good idea to verify that a Wi-Fi network is in fact offered by an establishment before logging on. Otherwise, you may find yourself on a hacker's "free" Wi-Fi. Free is good, of course, but only if it's legitimate!

VPNs are very good idea whenever using a public (nonsecured) Wi-Fi. Many hotpots in coffee shops, malls, and restaurants now offer free Wi-Fi, but it's not secure. In these cases, establishing a VPN connection is a best practice.

Virtual Local Area Networks

On wired Layer 2 switched networks, network administrators use VLANs to segment the Layer 2 broadcast domain to improve scalability and performance. A VLAN is a logical network segmentation. All members of the same VLAN are treated as if they were connected to the switch, even though they may be on different switches or in different locations. VLANs are useful for providing logical segmentation that can be based on protocol, MAC address, function, or application. VLANs specify **broadcast domains**, which define segments of the network that receive the same broadcast messages over the shared medium. By creating VLANs, an administrator restricts broadcasts to individual VLAN members and provides isolation from other VLANs. This has security and performance benefits for each individual VLAN. For a host in one VLAN to communicate with a host in a different VLAN, the VLANs must be bridged via a router. On a network diagram, a VLAN is represented as a separate cable on a separate switch port, even though in reality they share the same physical cable and transmission medium (see **FIGURE 7-2**).

VLANs also work with wireless networks. By logically segregating the client stations into a common VLAN group membership, an administrator can inhibit the broadcast domain and provide security and performance enhancements. You can identify individual packets as being in a particular VLAN in both wired and wireless scenarios by inserting a tag with a VLAN identifier into the packet header. This method of VLAN ID is called **802.1Q tagging** in wireless domains.

On a wireless network, an administrator assigns a user's traffic to a particular VLAN to separate and segregate traffic. Quality of service (QoS) can also be defined on a per-VLAN basis, thereby giving priority to certain classes of user or traffic type (for example, voice or video). VLANs also play a role in security when you team them with SSIDs. By using SSIDs and VLANs, the administrator can securely segment the wireless network.

CHAPTER 7 | Basic WLAN Security Measures

FIGURE 7-2

VLANs are logical partitions on a network that define groupings of hosts based on logical associations rather than physical connections to a switch. Members of the same VLAN receive the same broadcast messages even if they are connected to different switches.

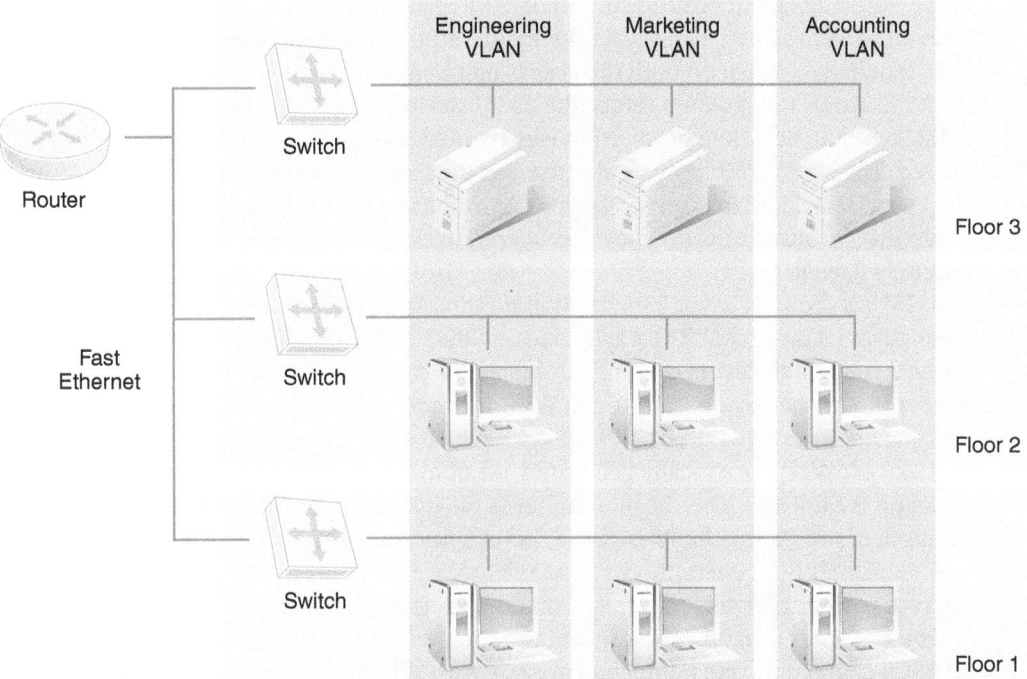

Data Protection

Preventing unauthorized access to the network is a critical concern in any environment. As mentioned, however, wireless traffic is available to anyone within the reach of the radio signals transmitted to and from client devices. Therefore, security professionals must also be concerned with the protection of both the payload information (the data) and the authentication credentials. This is accomplished through use of encryption algorithms.

Over the years, there have been a number of different methods of encryption and encryption implementation. Some of these methods have since been deprecated by the IEEE and other standards bodies because of weaknesses found. However, many devices that use these deprecated methods are still in service and must be acknowledged, if not dealt with. It's an important distinction that while these deprecated encryption schemes have been proven to be vulnerable, it still takes some of amount of know-how and ill intent to hack them. These methods are therefore sufficient for blocking casual access.

This section begins by discussing the older (least desirable) methods and builds up to the method now approved and recommended for secure Wi-Fi access.

Wired Equivalent Privacy

The primary goal of the WEP protocol was to provide confidentiality, integrity, and access control for wireless networks. WEP was defined in the original IEEE 802.11 standard in 1997 as a mechanism to provide for data privacy through encryption, access control via a static-key form of authentication, and data integrity through a checksum to ensure that data had not been modified. In early deployments, static WEP keys were used as authentication keys that had to match on both the access point and the client station. If there was no match, then the access point would refuse the client station permission to associate. If the static WEP keys matched, then permission was granted, and authentication and association occurred. The static WEP key was also used as the Layer 2 encryption mechanism to encrypt the Layer 3 payload, so it served both as an authentication key and a data-protection key. Given that the authentication key was shared in cleartext, however, the solution was fundamentally flawed.

WEP is a pre-RSN Layer 2 encryption method that protects information in the payload from Layers 3 through 7. The payload in an 802.11 frame is simply the IP packet with some Layer 2 (Data Link Layer) control packets and is called the **MAC service data unit (MSDU)**. Encryption of the payload is through WEP. It can be either Layer 2 64-bit WEP, which uses a secret static key of 40 bits, or 128-bit WEP, which supports a 104-bit static key. Both 64-bit and 128-bit WEP use a locally generated, random 24-bit number to add to the static key to make up the full size. This 24-bit number is the initialization vector (IV). The effective key strength of combining the IV with the secret static key is $40 + 24 = 64$-bit encryption and $104 + 24 = 128$-bit encryption, respectively. WEP was the standard way to authenticate and encrypt payloads for several years until faults began to appear in its mechanism.

You can enter a static WEP key into the configuration of each device using hexadecimal or ASCII characters. One nontechnical issue with WEP was that in residential settings, many people did not understand base-16 numbers. As a result, they simply did not turn on security. It's not uncommon for manufacturers to set up equipment to work out of the box with no security by default. If the security setup is perceived as difficult, many users will simply skip that step. Fortunately, today's standard method (WPA2, discussed momentarily) uses a passphrase to generate the encryption key, making it much more user-friendly.

Another problem with static WEP keys was that there was confusion regarding the choice of keys. For example, a typical access point might permit four static keys to be entered. However, only one can be the transmission key, which is the key used to encrypt traffic transmitted from the device. However, the keys must match at both ends of the link. That is, the transmitter and the receiver must both use the same key to encrypt and decrypt. Indicators of which key encrypted the data and which key to use to decrypt it are included as part of the 24-bit IV that is sent in cleartext.

WEP is also prone to what is called an IV collision attack. Essentially, an IV collision occurs when the IV is reused, which results in a full key stream that has also been used. By analyzing packets from the same keys, an attacker could break WEP-encrypted keys in less than five minutes. After the attacker had deciphered and retrieved the static key, that person could then decrypt any data frame they wanted.

> **NOTE**
> Extensible Authentication Protocol (EAP) is an authentication framework that specifies more than 40 methods for WLAN and point-to-point connection authentication. Five of these methods have been approved for both WPA and WPA2.

For these reasons, WEP is no longer considered a viable method for authentication, encryption, or maintaining data integrity. However, as with most legacy security measures, if WEP is all you have—which may be the case on older equipment—it is better than no encryption at all. That being said, WPA2 is much, much better, and should always be used if available.

Wi-Fi Protected Access

The Wi-Fi Alliance introduced WPA certification in 2003, which supported **Temporal Key Integrity Protocol (TKIP)**/Rivest Cipher 4 (RC4) dynamic encryption key generation. WPA (802.11i) was viewed as an intermediate solution to address the serious weaknesses in WEP until the more secure WPA2 was available. WPA uses passphrase-based authentication in SOHO environments and supports strong 802.1X/EAP authentication in the enterprise.

As part of the interim WPA solution, the Wi-Fi Alliance endorsed TKIP as a stopgap security protocol to address the weaknesses in WEP. TKIP (now deprecated) is a stream cipher that uses the same encryption algorithm as WEP, which allowed for firmware upgrades that supported legacy hardware. This was an important consideration given the investment that had been made in Wi-Fi (with WEP). TKIP and the associated WPA offer three basic improvements over WEP:

- TKIP combined the secret encryption key with an IV before initiating the RC4 initialization process. In contrast, WEP simply added the IV to the end of the root key and passed this value to the RC4 routine (also known as ARC4, a widely used software stream cipher). Most WEP attacks exploited that weakness, which gave hackers information needed to crack the cipher.
- WPA used a sequence counter to protect against replay attacks, where a hacker captures an encrypted message and resends it at a later time. Because the packet has all the right information in it (that is, it has not been changed), the receiving end will process it and try to implement whatever commands were contained. This can be disruptive at best and can have disastrous effects at worst.
- TKIP implements a 64-bit message integrity check to ensure that messages have not been modified in transit. (Because the keys that ensured message integrity were not encrypted, however, the message integrity could be exploited, as discussed momentarily.)

Despite these improvements, it was not long before security flaws began to reveal themselves in the TKIP encryption technique on which WPA is founded. These flaws soon became public. Attacks followed, such as the Beck-Tews attack, which targeted the integrity protection of the Layer 3 payload. The Ohagi/Morii attack built on the Beck-Tews attack using a man-in-the-middle (MITM) strategy. These attacks focused on disrupting data integrity rather than seeking to reveal the encryption keys. Changing the TKIP settings on the WLAN controller or the access point could thwart the attacks.

Wi-Fi Protected Access 2

From the start, WPA was considered to be a temporary solution to address the failings of WEP while engineers worked on implementing WPA2, a much more secure solution.

Whenever available, you should use WPA2. At the time of publication, WPA2 is the Wi-Fi Alliance's approved method for data protection and the current standard for 802.11 security.

WPA2 is built on the **Advanced Encryption Standard (AES)** algorithm in **Counter Mode Cipher Block Chaining Message Authentication Code Protocol (CCMP)**, which supports 802.1X/EAP authentication with preshared keys (PSKs). (AES and CCMP will be explained later in this chapter.) The 802.11n amendment that defines high throughput (HT) states that "stations should not use WEP or TKIP when communicating with other stations that can support stronger ciphers." The Wi-Fi Alliance began to insist on this requirement when issuing certification for 802.11n in 2009.

In SOHO deployments, WPA2 uses PSKs, which are 64 hexadecimal digits long. When using PSKs, WPA2 is noted as WPA2-PSK to differentiate it from the more demanding enterprise versions that use 802.1X/EAP for authentication. The PSK is a plaintext English passphrase containing up to 133 characters. This passphrase is then used to generate the unique encryption keys for each wireless client.

WPA2-PSK—or WPA2-Personal, as it is sometimes called—does have some drawbacks, but they are operational in nature rather than flaws in the security mechanisms. WPA2-PSK uses the more advanced AES algorithm, which requires additional processing power to keep the network up to speed. Older hardware, even though it can support WPA2-PSK, may suffer reduced throughput and serious speed impairment. Legacy hardware may require firmware upgrades to support WPA2. If you spend any amount of time supporting SOHO environments, you will likely see options for WEP, WPA, and WPA2-PSK. Whenever possible, it is advisable to use WPA2-PSK in SOHO environments and the more robust WPA2-EAP in enterprise environments. When it comes to security, something is almost always better than nothing.

> **NOTE**
> WEP and TKIP have proven to be faulty and easily cracked with downloadable software. Even so, they are better than no encryption at all.

WPA2 with AES

AES is a block cipher algorithm that may be incorporated into many security products. AES encryption is the standard adopted by the U.S. government. It is also used as the encryption algorithm in **Internet Protocol Security (IPSec)** VPNs. IPSec is a set of protocols for securing IP communications by authenticating and encrypting IP packets.

Block Ciphers versus Stream Ciphers

A **stream cypher** encrypts data of arbitrary lengths in an ongoing fashion. A **block cipher** uses defined blocks of data. Stream ciphers tend to have less overhead and better throughput performance. They are, however, prone to interference when noise (improperly encrypted data) is injected into the stream, which can cause synchronization issues. Block ciphers have more overhead but the encryption and decryption of blocks are not interdependent. Therefore, they are more immune to noise injection.

AES supports three key sizes—128, 192, and 256 bits—although it uses a fixed block size of 128 bits. A block cipher such as AES takes a fixed 128-bit chunk of plaintext called a *block* and works on it to produce a 128-bit block of cipher text. The number of rounds repeatedly performed on the block depends on the key size—that is, a key size of AES-128 requires 10 rounds, AES-192 requires 12 rounds, and AES-256 requires 14 rounds. The greater the number of rounds, the stronger the encryption, but the greater the resources required to decrypt it. AES-128 and AES-256 are theoretically crackable, but the required resources and timeline (in the order of trillions of years) make it impractical in the real world. As an example, it is estimated that it would take 2 billion high-end PCs 13,689 trillion trillion trillion trillion years, and that's a very long time when you consider the universe is estimated to being, in comparison, only 15 billion years old.

WPA2 with CCMP

CCMP is the security encryption protocol defined by 802.11i WPA2. CCMP provides security in the following ways:

- Data confidentiality via encryption
- Authentication
- Access control with layer management

CCMP uses a fixed block size of 128 bits and a fixed key size of 128 bits. AES can support other key sizes, but when deployed within CCMP, it remains fixed. CCMP is a Layer 2 protocol that ensures that information in Layers 3 through 7 is also encrypted in the 802.11 data frame. The Layer 3 payload is encrypted using AES and protected from manipulation and tampering by a **message integrity code (MIC)**. The data headers of the frame are not encrypted. However, a technique called **additional authentication data (AAD)** lends some tamper-proof protection.

The CCMP 802.11i protocol replaces ARC4 stream ciphers, WEP, and TKIP. CCMP is considered to be mandatory for RSN compliance. However, due to the fact that the underlying AES encryption algorithm is processor intensive, many older access points and client network cards are unable to support it. Because wholesale hardware upgrades for clients and access points were required to become RSN compliant (something many businesses could not afford), another solution had to be put in place until the normal hardware refresh cycle of 5 to 7 years ran its course. WPA-AES works well and is still in place, as does PSK, although PSK can be hard to manage (and keep secure) on a large scale.

Order of Preference for Wi-Fi Data Protection

When more than one type of Wi-Fi data protection is available, you should choose in this priority order:

- WPA2 + CCMP
- WPA2 + AES
- WPA + AES

- WPA + TKIP
- WEP
- Open network (no security)

This order of preference will change as **Wi-Fi Protected Access 3 (WPA3)** gains adoption in 2020 and beyond.

WPA3

The WPA2 protocol has served the wireless community well over the years but is getting a bit old and needs a refresh. Thus, in 2018, the Wi-Fi Alliance introduced the WPA3 security standard that adds four new features to bolster the original WPA2 specification.

The WPA3 protocol provides some new or enhanced security features for both personal and enterprise use. Some of these enhancements are the addition of a 256-bit Galois/Counter Mode Protocol (GCMP-256) for encryption, 384-bit Hashed Message Authentication Mode (HMAC), and 256-bit Broadcast/Multicast Integrity Protocol (BIP-GMAC-256). In addition, the WPA3 protocol also supports advanced security controls such as perfect forward secrecy.

WPA3 support will not be automatically added to every device but any new WPA3 devices are expected to be backward compatible with devices that use the WPA2 protocol. Importantly, the Wi-Fi Alliance has placed particular emphasis on four key new features. If manufacturers want to market their products as being WPA compliant then they must fully implement the following four features.

Data Privacy over Public Networks

The first of the four additional features introduced to modernize the WPA protocol is with regards to data privacy. WPA3 adds additional data privacy by introducing "individualized data encryption." What this means is that when you connect to an open Wi-Fi network, the traffic between your device and the Wi-Fi access point will be encrypted. This feature is to address the privacy problems of using Wi-Fi over public, open Wi-Fi networks. By introducing individualized data encryption over a public open network, such as a hotel or mall, this will make it impossible for people to snoop without actually cracking the encryption.

Protection against Brute-Force Attacks

The second feature that the Wi-Fi Alliance has turned their attention to is the inherent vulnerability that exists in WPA2 with regards the passphrase handshake vulnerability, which was highlighted by the KRACK attack in 2017. Although there are software patches available for WPA2 to mitigate the vulnerability, the new WPA3 protocol specifically defines a new handshake that "will deliver robust protections even when users choose passwords that fall short of typical complexity recommendations" (Wi-Fi Alliance, 2009). What this means is that even if you should use weak passwords, the WPA3 standard will protect against brute-force attacks.

An Easier Connection Process for Devices Without Displays

Today, there are a plethora of IoT devices that are wireless enabled but do not have displays or any method to input data. For example, the Amazon Echo and all those IoT smart outlets and light bulbs, which can connect to a Wi-Fi network. Connecting these types of devices frequently involves using a second device such as a smartphone app to be able to communicate with the device. This is often tedious, so many IoT devices remain at their default settings. To mitigate this security concern, the WPA3 protocol includes a feature to "simplify the process of configuring security for devices that have limited or no display interface" (Wi-Fi Alliance, 2009).

Higher Security for Government, Defense, and Industrial Applications

The Wi-Fi Alliance also announced WPA3 will include a stronger and longer session key up to 192 bits, a "192-bit security suite, aligned with the Commercial National Security Algorithm (CNSA) Suite from the Committee on National Security Systems" (Wi-Fi Alliance, 2009). This is intended for government, defense, and industrial applications. This feature addresses a request by the U.S. government to provide stronger encryption on critical Wi-Fi networks.

Ongoing Management Security Considerations

While access and data protection are key, it's important to recognize that good, old-fashioned management and upkeep of the network and equipment are critical areas of basic network security. Any aspect of the network that is not actively monitored and managed becomes a potential vulnerability. Attackers targeting an organization will attempt to achieve their objectives by any and all vectors, including (and preferably) the simple ones. After all, why bother knocking down a wall when you can crawl through an open window?

Firmware Upgrades

Upgrading firmware is an essential aspect of wireless security. Despite the enormous effort that vendors go through to design, test, and certify their products' security, the level of complexity within modern networking equipment all but guarantees that flaws will exist. As these bugs are discovered, they are patched, and new firmware is made available. If not already known in the hacker community, these exploits are rapidly exposed and disseminated. A hacker best practice is to look for unpatched equipment. When such devices are found, there is often a "cookbook" approach to exploiting the unfixed issue.

You can perform wireless 802.11 firmware upgrades on access points and radio network cards to bring them up to date with the latest bug fixes and enhancements. In fact, many vendors provided firmware upgrades to enable their hardware to support WPA, which allowed customers to abandon WEP without replacing hardware. The typical method of upgrading a legacy access point or adapter is to use a **File Transfer Protocol (FTP)** server to download the firmware upgrade from the vendor's website using a web browser or graphic user interface (GUI)–based program. (FTP is a nonsecure application for transferring files). Other times, **Telnet**, a network protocol that supports remote nonsecure access to another device, is used. Both FTP and Telnet originally required the use of a **command line interface (CLI)** window. (CLI is a text-based user interface in which you type commands on a line, receive a response back, then type another command, and so forth.) In some cases, a web browser installed on the device allows for automatic firmware upgrades if there is an Internet connection. This latter method is the easiest and least error-prone way to upgrade a wireless device, although some people still prefer the CLI method.

While firmware revisions have become important aspects of wireless device management, ensuring that every access point and adapter is at the correct firmware revision level can be a trying task in large networks. WLAN controllers have helped with this because only one device needs to be upgraded. This is a good example of central device management, which is a necessary and essential element of large network management. In SOHO environments, management is typically achieved via a local browser device manager or administration portal on the access point. The administrator should regularly check for firmware updates and bug fixes on the vendor's website. Some organizations will have this as a checkbox on a periodic maintenance list.

Physical Security

Physical security is an often-overlooked aspect of wireless security policy. Laptops do get lost or stolen, as do smartphones and tablets. If these devices have been preconfigured for wireless security, the person who comes into possession of the wireless-enabled device will be able to access the network.

In enterprise environments, authentication methods make it easy to blacklist the lost or stolen device to prevent unauthorized access to the WLAN. In SOHO environments, however, it's unlikely that such levels of device management will be in force. Therefore, if a device is lost or stolen, the administrator should change all passphrases and PSKs. This should also be done if someone leaves the company (assuming BYOD is in place). When an employee leaves, it is no longer simply a matter of retrieving their laptop. Now you must try to gain access to their personal laptop, smartphone, and tablet to remove the Wi-Fi security configurations. A more secure method in a SOHO environment is to simply change the passphrases and PSKs on the access point and remaining devices. In environments with high employee turnover, changing the PSKs and passphrases on a monthly basis is a good practice.

Periodic Inventory

It is always good practice to keep an up-to-date inventory of all devices authorized to connect to the WLAN. Fortunately, maintaining such an inventory is a handy outcome of using MAC filtering. That is, you must be aware of all devices and their respective MAC addresses to manage the access list.

Even if MAC filtering isn't in place, it is still wise to list all authorized and configured devices and clients. It is very common for device creep to occur, particularly in SOHO environments where security is perhaps not as robust as in an enterprise. Often, this is because authorized users copy the configuration from their authorized devices onto their personal devices. It's not uncommon to find anonymous devices using the WLAN services (typically the Internet). This is one of the reasons that creating a guest VLAN with Internet-only access is a good idea. Access to the more restrictive employee VLAN can be tightly controlled without upsetting employees and guests.

By running periodic inventory checks, a network administrator in a SOHO setup can audit the MAC addresses traversing the WLAN and identify them as known or unknown. An effective way to identify unfamiliar MAC addresses is to perform a "scream test," in which the administrator filters the MAC address on the access point to deny access and then sits

CHAPTER 7 | Basic WLAN Security Measures

back and waits to hear who screams. In larger organizations, it is still a good basic security measure to perform impromptu checks and keep an especially keen eye out for strange or unknown access points.

Identifying Rogue WLANs/Wireless Access Points

The best preventative measure for rogue access points is to conduct regular and frequent audits of all access points on the WLAN. Obviously, in a SOHO deployment, a rogue access point is going to stand out. However, they are not so easily recognized in an SMB or enterprise environment and it would require advanced software and due diligence from the administrator to detect them.

One strategy is to manage Ethernet switch ports and wall sockets to ensure that unused ports are disabled by default. This will prevent rogue access points from getting an Ethernet backhaul connection. In the past, this was not a practical solution because employees tended to move around and connect from different places. Given, however, that most environments are now wireless (except for large workstations, servers, and printers), locking down unused ports has become a viable option.

Additionally, ensuring that RF coverage is limited to the boundaries of the premises will restrict the installation of access points outside the building to eavesdrop on the internal WLAN. If rogue access points continue to appear on the WLAN, consider configuring one of the newer access points as a Remote Authentication Dial-In User Service (RADIUS) authentication server. This robust enterprise technique is usually well out of scope for SOHO environments and even SMBs, but it can be of value if deployed cheaply on an already existing wireless device.

CHAPTER SUMMARY

With most endeavors, having a sound grasp of the fundamentals is essential even (or perhaps especially) when more advanced pursuits are the ultimate aim. Security is no exception. While the topics in this chapter are considered to be basic and focus mostly on SOHO environments, they are important aspects of a comprehensive security plan even for large enterprises. This is especially the case in today's world, where, due to the ubiquitous availability of Wi-Fi, small offices, employees' homes, and hotspots have effectively become extensions of the corporate network.

In the office, good basic security always starts with good design. After the design has been implemented, controlling access to the corporate network is the next key step. Often, different levels of access may be needed for different classes of users.

In all cases, data protection should be a primary consideration, with a strong focus on the use of WPA2 or WPA3, whenever possible. On public networks, creating a secure connection via a VPN is a sound practice. Beyond direct efforts to control access and protect data, maintaining the network via patches and upgrades and conducting periodic audits and RF surveys will help ensure that backdoors (whether real or virtual) are not left open.

KEY CONCEPTS AND TERMS

802.1Q tagging
Additional authentication data (AAD)
Advanced Encryption Standard (AES)
Block cipher
Broadcast domains
Command line interface (CLI)
Counter Mode Cipher Block Chaining Message Authentication Code Protocol (CCMP)
Dead spots
File Transfer Protocol (FTP)
Internet Protocol Security (IPSec)
MAC service data unit (MSDU)
Message integrity code (MIC)
Open System Authentication (OSA)
Robust Security Network (RSN)
Shared Key Authentication (SKA)
SSID cloaking
Stream cipher
Telnet
Temporal Key Integrity Protocol (TKIP)
Virtual local area network (VLAN)
Voice over WLAN (VoWLAN)
Wi-Fi Protected Access 3 (WPA3)
Wireless extender
Wireless repeater

CHAPTER 7 ASSESSMENT

1. It does not matter where you place an access point within a home or a building because you can increase the power to get the needed coverage.
   A. True
   B. False

2. Which of the following is the best way to increase the range of a wireless signal?
   A. Add another access point on the same frequency and channel
   B. Tell employees to move closer
   C. Use a wireless extender
   D. Crank up the power

3. Which of the following best describes SSID segmentation?
   A. It is practical in SOHO environments
   B. It is a good way to apply different policies to different groups
   C. It can give greater throughput to certain users or groups
   D. All of the above

4. Which of the following describes MAC filtering?
   A. It works at Layer 3
   B. It is flawless because MAC addresses are unique
   C. It is an approved data-protection method
   D. None of the above

5. Clients on the same VLAN act as if they are on a common switch with the same policies regardless of where they are located.
   A. True
   B. False

6. WPA has been deprecated and should not be used.
   A. True
   B. False

7. Which of the following describes AES encryption?
   A. It is a block cypher
   B. It is theoretically crackable, although the time and resources required make it a nonissue
   C. It is used with IPSec, WPA, and WPA2
   D. All of the above

8. All 40 of the Extensible Authentication Protocol (EAP) methods of authentication are approved for WPA and WPA2.
   A. True
   B. False

9. Changing the passphrase is a good way to eliminate or identify unknown clients on a WLAN.
   A. True
   B. False

10. Which of the following is a good way to prevent rogue access points?
    A. Fire anyone who installs one
    B. Jam the common frequencies used by access points
    C. Lock down (shut off) unused Ethernet ports
    D. Post a harshly worded sign in the lobby

Advanced WLAN Security Measures

CHAPTER 8

LARGE ORGANIZATIONS—OR SMALLER organizations with high-risk profiles—differ from small office/home office (SOHO) networks both in the complexity of their networks and the lengths to which they must go to maintain security. In a general sense, the best way to secure a wireless infrastructure is through a layered approach similar in concept to that for smaller organizations, albeit broader in scope and more comprehensive. In this chapter, we will look at more advanced concepts in wireless security. Some of these will be unique to the needs of these networks, while others will be extensions of concepts and techniques basic to any security approach.

Chapter 8 Topics

This chapter covers the following concepts and topics:

- How to establish and enforce a comprehensive security policy
- How to implement authentication and access control
- How to protect data
- How to segment wireless users
- How to manage network and user devices

Chapter 8 Goals

When you complete this chapter, you will be able to:

- Understand the importance of a comprehensive security policy for large-scale networks
- Describe the key components of an enterprise security policy
- Describe the RADIUS authentication process
- Describe the difference between intrusion detection and intrusion prevention

© Cherezoff/Shutterstock

- Understand and describe the benefits and risks associated with discovery protocols
- Describe the method of data protection used in enterprise networks
- Understand why user segmentation is needed and how it is achieved
- Describe the benefits and risks of single sign-on

Establishing and Enforcing a Comprehensive Security Policy

One of the ways that small and medium enterprise (SME) organizations differ from their SOHO counterparts is that there is always a network administrator and one or more individuals responsible for the design and security of the network. Before they implement any security techniques or mechanisms, the administrator's task is to analyze the network's security requirements and decide on a security policy.

A security policy for an enterprise will cover every aspect of the organization's information assets. Wireless access is only one component of a security policy document, but because it is relevant here, it will be examined in an initial high-level overview. The policy breadth and level of detail should be in line with the company's overall goals, its available resources, and its internal security requirements, as well as any external or regulatory requirements. Several of the more common policy points are described in the following sections.

Centralized versus Distributed Design and Management

When you're designing a new wireless network, or when redesigning an existing wireless network due to new technologies or requirements, the first step is to consider whether it's better to have a centralized or distributed security architecture.

With a distributed architecture, each access point must be configured separately both for performance and security reasons. The access point will also work with other network devices to ensure an end-to-end secure service. For example, the access point may provide encryption, but a centralized Remote Authentication Dial-In User Service (RADIUS) server might be responsible for authentication. A single firewall might ensure secure access control from multiple access points.

A centralized architecture, on the other hand, will use authentication, encryption, and access-control servers to administer and manage security. A centralized solution is characterized by the deployment of thin access points (access points consisting of a radio and antenna connected to a wireless switch) and one or more centralized controllers. For larger networks or campuses, centralized control tends to work better because of the time savings it offers for maintenance. A centralized approach can also greatly simplify design and control.

This simplicity is a key consideration. Complexity tends to breed mistakes in implementation, visibility, and control, and the "bad guys" look for mistakes to exploit. In many cases—especially targeted attacks—cybercriminals use **bots**, or programs that perform automated tasks, to look for holes in network defenses. Generally speaking, the simplest design that meets the mission objectives is the best choice—all other considerations being equal.

> **NOTE**
> Organizations sometimes operate a hybrid model, using thin access points with centralized control in some areas (the main employee campus, for example) and separate, individually controlled access points for guest access or other areas.

Remote Access Policies

Today, Internet Protocol (IP) mobility is a major influence in wireless network design. Users require access to network resources everywhere—even when traveling between locations. IP mobility has transformed the way users connect to the network by employing a much wider range of wireless-enabled devices. The challenge to the network designer and administrator is to ensure that the network security is robust and uniform across the entire network.

The first step in fulfilling that requirement is ensuring that there is a uniform, centralized authentication system throughout the network. Unless a device can be authenticated, access should be limited to guest Wi-Fi access outside the corporate firewall. Any applications that connect, such as email or customer relationship management (CRM) apps, will require sign-on credentials, thus allowing access without compromising network security.

The second step is to consider whether to permit resource-intensive apps such as Voice over IP (VoIP) or Voice over WLAN (VoWLAN). These require unique performance and throughput characteristics and key performance indicators (KPIs) such as packet loss, latency (the delay in a network), jitter (a measure of delay variability), and availability. If these applications are not supported, this should be communicated to users. This will not prevent users from raising support issues, but at least it will give support personnel a policy they can point to. If resource-intensive apps become an issue—for example, creating performance problems for apps that *are* approved—they can be blocked by using protocol-shaping filters.

The third step is to ensure that telecommuters and traveling employees can connect through secure virtual private network (VPN) connections. This is imperative when employees access corporate resources via hotels, shopping mall hotspots, and insecure wireless networks.

Guest Policies

Larger organizations often need to accommodate guests, whether they be vendors, clients, or suppliers. There is no "right" policy for guest access as long as whichever policy you do implement is clearly stated and understood. In the best case, there is also a stated reason for the policy, particularly if the policy is extreme (offering either full access or no access at all, although these are both rare in practice).

One way to handle guest access to the network is to establish rules for visitor authentication and policy control. The task here is to allow genuine visitors and guests access to the Internet and perhaps some intranet services, but restrict them from the corporate local area network (LAN).

Some organizations may not offer any Wi-Fi access at all. This policy has ebbed and flowed over the years. At first, it was rare for companies to offer guests Wi-Fi access. As wireless access became the norm, however, this began to change to a point where it was rare *not* to offer guest access via wireless. With the growing popularity of mobile-connected devices, however, it is no longer a great inconvenience not to offer wireless access; many people can check email or connect to the Internet with a smartphone or tablet. Again, it's not necessarily important what the policy is, just that there be a policy that people understand and (hopefully) that was put in place with some purpose or reasoning behind it.

Quarantining

Quarantining is the process of isolating a device from the network until there is some level of assurance that the device is both authorized to connect and free from malware. Whenever a device attempts to connect to a wireless network, it will be allowed access only if it complies with some defined wireless network policy. This can include basic authentication credentials, configuration checks to ensure patches are installed, and even malware scans. If a device does meet the connection criteria, it is connected to a restricted IP subnet with access only to certain services—for example, an antivirus update server.

Another form of quarantine is to use a **walled garden**, which restricts internal access for the suspect device but allows access to instructions and services to remediate the outstanding compliance issues. A walled garden may also allow external access to the Internet. Another option is a **captive portal**, in which an HTTP session is forced to a landing page prior to gaining access to the Internet. This is a common technique used by hotspots either for payment or for the acknowledgment of a user agreement but can also be used as a form of authentication or to check credentials.

Compliance Considerations

Regulatory compliance considerations are a complete topic unto themselves. A full treatment of the topic and its implications for wireless access is beyond the scope of this text. That said, it is worthwhile to provide a high-level summary given the far-reaching implications these regulations have on security teams.

Unlike internal policies, data-privacy regulations are externally driven. However well-intentioned they are, these requirements are by definition forced upon organizations. In many cases, this isn't necessarily a bad thing. Often, regulations specify best practices that many companies would have adhered to anyway (although this is not always the case). The issue that many people have with regulatory compliance is not with compliance itself but rather the proof of compliance that various governing bodies require. This can often place serious demands on resources.

There are also instances where an internal policy or implementation of some security solution or process is often more effective (at least in the organization's eyes) than a required policy or solution. Unfortunately, it's rare that a home-grown solution supersedes a regulatory mandate.

All of this is not to say that government and industry security regulations are bad. It is true, however, that these regulations *do* have an impact with respect to policy and resources. Therefore, they must be accounted for in the overall security strategy and operation.

CHAPTER 8 | Advanced WLAN Security Measures

Employee Training and Education

Any company security policy that does not include employee training should be considered incomplete and inadequate. Security Awareness Training for employee training on security matters has always been a good idea, but even as recently as 10 years ago, the lines between work and not work were much clearer. Back then, typically only a few employees had remote and/or wireless access (executives, knowledge workers, full-time remote workers, and so on). Today's environment, of course, is quite different. It's not an unreasonable expectation that nearly every employee will have a wireless-capable device—either their own or one provided by the company.

It's easy to assume that because wireless connectivity is such a common aspect of people's lives, education on security matters is perhaps not critical—the "everyone already knows" fallacy. This could not be further from the truth. In fact, one could argue the opposite. Because it is so easy to connect, most people are not mindful of the dangers. This is compounded by the ease with which applications can be added (from many sources) and the availability of easy-to-follow tutorials on customization and user hacks. (In this case, *hack* refers to a user modification or the shortcutting of some aspect of a device for the user's benefit.)

All this points to a critical need for organizations to create and support policies regarding user education on security matters. Employees should be briefed on the risks of wireless security and given the training they need to protect both themselves and the organization from cybercriminals. This is a very inexpensive yet highly effective part of a comprehensive wireless security policy.

Implementing Authentication and Access Control

The standard specifying Port-based Network Access Control (NAC) for LANs and wireless LANs (WLANs) is the Institute of Electrical and Electronics Engineers (IEEE) 802.1X standard. 802.1X is a standalone authentication specification (as opposed to an amendment of another specification) and is therefore noted with a capital X, per the IEEE naming standards. This IEEE specification addresses authentication mechanisms for environments that require robust security and access control.

Central to the 802.1X specification is the mechanism for per-user and per-device authentication. The process calls out three entities:

- **Supplicant**—A client device looking to connect to the network
- **Authenticator**—A network device such as a switch or access point
- **Authentication server**—A server supporting an authentication protocol such as Extensible Authentication Protocol (EAP) or RADIUS

In this system, the authenticator acts as a gatekeeper, prohibiting any device access to the network unless and until the device has been properly authenticated. To be authenticated, the client must provide some type of credential. Depending on the authentication protocol, this could be a name and password combination or a digital certificate. These credentials are encapsulated in an EAP over LAN (EAPoL) frame.

The authenticator passes these credentials to the authentication server, which validates (or invalidates) them. The authentication server then notifies the authenticator if access is allowed. If so, the client can connect and communicate with the network. If not, access is blocked (see **FIGURE 8-1**).

Extensible Authentication Protocol

EAP is a method of encapsulation used to securely transport keying material for encryption over wireless and **Point-to-Point Protocol (PPP) networks**. EAP is also used over LANs between the authenticator and authentication server and is referred to as EAP over LAN (EAPoL).

EAP itself is a generic authentication mechanism that transports authentication requests, challenges, notifications, and so on across the network. EAP works by creating a secure tunnel using Transport Layer Security (TLS). Credentials are passed through the secure tunnel to the authentication server. EAP does not need to know the method of authentication. Therefore, it can accommodate several credential options, such as username and password, certificates, tokens, biometrics, and more. However, due to its long association with PPP and VPNs, EAP is closely associated with RADIUS. For this reason, many access points have a RADIUS client built in.

Prior to the ratification of the 802.11i standard (Wi-Fi Protected Access 2, or WPA2), Cisco Systems developed the Lightweight Extensible Authentication Protocol (LEAP). LEAP was developed as a stopgap measure and does not protect credentials. Therefore, it is not recommended. However, it is likely that you will still see access points that support LEAP, given its broad adoption in the industry.

Remote Authentication Dial-In User Service

RADIUS is a network protocol that provides authentication, authorization, and accounting (AAA) services for devices or users connecting to a network. RADIUS was developed around the time of dial-up connections so was a form of authentication for user's dial-in access. RADIUS has now adapted to support all the modern forms of network access. However,

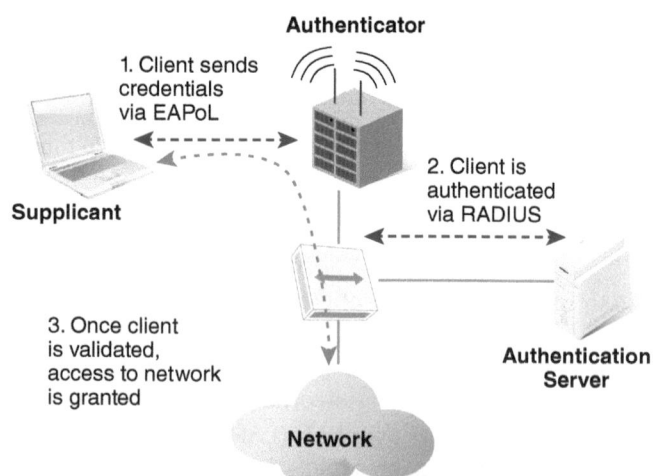

FIGURE 8-1

The 802.1X standard specifies how clients (supplicants) pass credentials to an authentication server via an authenticator, which blocks or allows network access based on authentication verification.

CHAPTER 8 Advanced WLAN Security Measures

the principle has remained the same; when a client attempts to connect to a network, the device (or user) is challenged by a network access server (NAS) device with a request for some type of credential. The NAS passes the user credentials to the RADIUS server. If the credentials are verified (for example, if the username is found and the password is correct), the RADIUS server returns an access accept response. This response also specifies the user's access attributes.

An access point with a built-in RADIUS client can communicate directly with both a client device and a RADIUS server. Consequently, the access point doesn't need to know about users and passwords or certificates. It just needs to form a RADIUS packet from an 802.1X frame received from the Wi-Fi client and pass the authentication request on to the RADIUS server. It is the job of the RADIUS server to handle authentication requests and issue success notifications. The RADIUS server in turn connects to an authentication database (such as Microsoft's Active Directory). This is the central repository for all the authentication data and can issue the success or fail notifications. In this way, the access point is not part of the authentication process but is merely a conduit passing authentication messages between the supplicant and the authentication server. The authentication process using EAP and RADIUS is shown in **FIGURE 8-2**.

FIGURE 8-2

The process shown here outlines the request, challenge, and response process for both EAP and RADIUS.

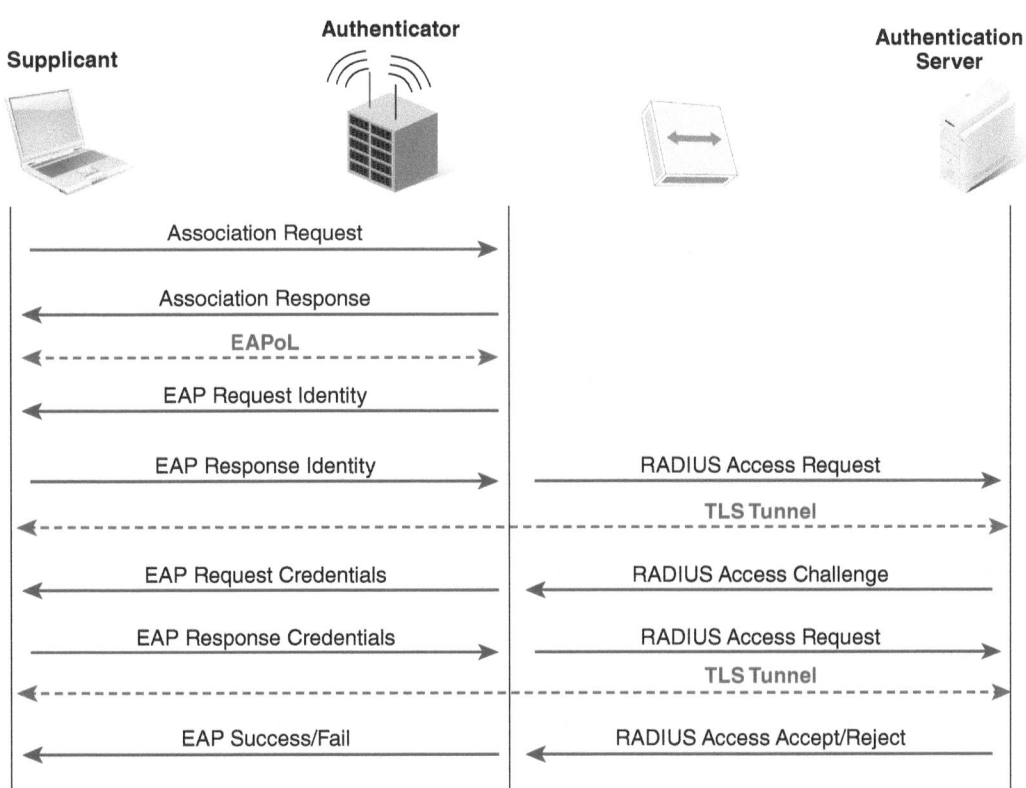

An authentication request is met with an authentication response, which in turn creates an EAPoL tunnel. Once the tunnel is established, the EAP identity request and response are securely shared and the response is forwarded to the RADIUS server as an access request. This request initiates a secure TLS tunnel through which the RADIUS challenge and response are transmitted, after which the tunnel is destroyed and the corresponding accept or reject message is sent to the requesting device.

Intrusion Detection Systems and Intrusion Prevention Systems

Intrusion detection systems (IDSs) and intrusion prevention systems (IPSs) are essential components in today's wired and wireless networks. Both systems work by using deep packet inspection to look inside packets traversing the network. An IDS is purely a detection system. It will raise a flag if it detects suspicious activity on the wire or over the air. An IPS, on the other hand, actively confronts and blocks any suspicious traffic it detects. Both wired and wireless IPSs use known signatures of existing threats to identify attacks. They also monitor data streams to ascertain that the patterns and communication flows are correctly crafted for certain protocols—for example, HTTP requests.

A wireless IDS or IPS is commonly referred to as a WIPS. There are two types of WIPS:

- **Network-based WIPS**—A network-based WIPS consists of sensors that are either in line or configured in promiscuous mode so they can sample and analyze all traffic crossing the network. A centralized server and console analyze and present the results.
- **Host-based WIPS**—A host-based WIPS is an application loaded onto a server or client computer or device that monitors for threats in applications, operating systems, and files, as well as known suspicious behavior.

A WIPS is a crucial element in a wireless environment because it can detect and block suspicious activities coming from an attacker. A WIPS is particularly useful in mitigating man-in-the-middle (MITM) attacks, rogue access points or evil twins, unauthorized associations, MAC spoofing, ad hoc networks, denial of service attacks, and protocol misuse.

Protocol Filtering

Most access points support filtering of media access control (MAC) addresses, various Ethernet frames (EtherTypes), and IP protocols. To implement this, an administrator creates and applies the filters to the access point interfaces in both incoming and outgoing directions. MAC filtering is commonly used for access control and to prevent authentication and association by unknown client stations to the access point. However, it is easily overcome by MAC spoofing. IP protocol filtering, however, can be used to prevent the use of certain protocols on the WLAN, which can help mitigate security threats.

There are two types of protocol filtering:

- **EtherType protocol filtering**—EtherType protocol filtering uses a protocol identifier to identify the protocol that is to be blocked. For example, to block EtherType IPX 802.2, the ISO designator 0x00E0 would be specified. It's not uncommon to find obsolete EtherType

protocols on older networks. Typically transmitted by network-enabled printers or other legacy devices, these protocols should be filtered out at the access point to reduce the potential security footprint.
- **IP protocol filtering**—An administrator can also apply IP protocol filters, which can be configured on the access point by specifying the well-known port number for the specific protocol. By configuring filters to block specific IP protocols, the administrator can lock down the wireless segment to support only desirable IP protocols. By doing so, the administrator restricts the potential range of vulnerabilities for an attacker to exploit.

The administrator can create and assign granular levels of IP protocol filters by specifying a source IP address, a destination address, or both. Therefore, an administrator may allow Telnet but only from specific host addresses or to certain destination addresses. Simple Network Management Protocol (SNMP), for instance, can be filtered and permitted only as a source protocol from a **network management system (NMS) server**, and not from just any wireless client.

Protocol filtering enables low-level granular control of the network protocols allowed or denied on the wireless segment and can be applied directly to the radios or the Ethernet port in both directions. An administrator will typically filter SNMP because, along with the various discovery protocols, it's a prime tool that attackers use to enumerate and map networks.

Authenticated Dynamic Host Configuration Protocol

The goal of authenticated Dynamic Host Configuration Protocol (DHCP) is to supply only an IP address and a network configuration to previously authenticated clients. One way to achieve this is through a captive portal. In this design scenario, a user joining the network for the first time will attempt to connect with his or her device to the wireless network. Once associated with the access point, the user's device will broadcast for a DHCP server over the Layer 2 connection. The DHCP server will hear the request but will have no prior knowledge of the MAC address from which the request is coming. Therefore, it will issue the device an IP address for the captive quarantine portal.

The quarantined device is then automatically directed to a captive portal, which will challenge it for a username and password (often authenticated against either a RADIUS or Active Directory server, but there are other options as well). Once authenticated, the MAC address of the user's device is recorded in the authenticated list of devices/users and is reassigned an authenticated IP address and full network configuration.

Another method commonly used in large organizations is 802.1X port control. With this method, the 802.1X protocol shuts down all traffic coming out of the logical port connecting the user's device with the wireless network except for EAPoL encapsulation authentication messages. By doing so, the protocol blocks all traffic apart from the device's authentication messages to the access point and the RADIUS server. When the RADIUS server authenticates the user or device, the 802.1X protocol opens the logical port to allow the unrestricted flow of traffic, such as DHCP broadcasts.

Data Protection

Data protection is essential to wireless security. It must be implemented correctly to prevent unauthorized access to the system and the replication or theft of valuable data. The IEEE 802.11i standard security amendment (ratified in June 2004) came about as a result of the inherent weaknesses in the data-protection schemes used up to that time. It stated that for any organization with even minor concerns about network and information security, WPA2 must be in place. Further, unlike SOHO offices, where Wi-Fi Protected Access 2–preshared key mode (WPA2-PSK) is sufficient, larger organizations (or smaller organizations with higher security profiles) should use WPA2 Enterprise.

For enterprises and large organizations, the 802.11i standard requires 802.1X for Enterprise mode as the authentication mechanism and Advanced Encryption Standard–Counter Mode Cipher Block Chaining Message Authentication Code Protocol (AES-CCMP) as the confidentiality cipher algorithm. CCMP has built-in integrity mechanisms, so it does not require an additional method of integrity assurance. Therefore, 802.11i requires the use of 802.1X for authentication and AES-CCMP for confidentiality and integrity.

WPA2 Personal and Enterprise Modes

For small offices and home users, using WPA2-PSK is a good solution. The main benefit of using WPA2-PSK is that it is easy to set up and use. In addition, given the relatively small number of users, it's easy to manage. Even if the passphrase must be changed from time to time, the small number of users keeps this task from being too onerous. Remember, however, that with WPA2-PSK, both the clients and the access point must be updated if a change is needed.

In larger organizations, this fact makes the use of PSKs unmanageable given the large number of employees and the frequency with which people come and go. If the PSK had to be changed every time there was a personnel change, it would overwhelm the IT support team. Worse, if the PSK was *not* changed, it would create a security vulnerability because there would be nonemployees (some of whom could be disgruntled after being fired) who still had a valid pass key.

To solve this issue, WPA2 Enterprise takes advantage of RADIUS-based authentication. This can be done with 802.1X devices with a Network Policy Server (NPS). With RADIUS, each user is authenticated on an individual basis from a single server, regardless of which device is used to access the network. One of the benefits of this is that when an employee leaves, it's a simple matter to update that user's credentials in the RADIUS server. Additionally, RADIUS authentication is achieved through a secure tunnel connection. Therefore, after a client has been verified, the session encryption key can be securely passed through the tunnel established in the authentication process.

WPA3

WPA3 was first released in 2018 as an update to the WPA2 protocol. Up until then, WPA2 was the recommended wireless protected access protocol. However, with its introduction, WPA3 brings some essential security improvements. For example, WPA3 finally addresses the long overdue issue with insecure passwords.

WPA3 has four major security improvements:

1. A secure handshake for establishing connections
2. An easy method to securely add new devices to a network
3. Some basic protection when using open hotspots
4. Increased key sizes

However, amongst these new features, only the new handshake is mandatory; the other features are discretionary, but it is likely that most vendors will implement them as optional (nondefault) settings.

WPA3 has two distinct modes: Personal (128-bit encryption) and Enterprise (192-bit encryption). In addition to forward secrecy, WPA3 also replaces PSKs with a new more secure exchange mechanism called Simultaneous Authentication of Equals.

Internet Protocol Security

Internet Protocol Security (IPSec), is an open-standard suite of protocols designed to secure every packet in an IP stream traversing a network between partnering endpoints. A secure session must be initiated and set up before IPSec can secure the traffic. Therefore, it is predominantly used in point-to-point or client-server configurations.
IPSec consists of several security protocols:

- **Authentication Header (AH)**—The **Authentication Header (AH)** provides authentication for data origin and integrity while also providing protection against replay attacks.
- **Encapsulation Security Payload (ESP)**—The **Encapsulation Security Payload (ESP)** provides confidentiality as well as authentication of the data's origin and integrity.
- **Security Associations (SA)**— Composed of the algorithms that provide the security parameters enabling AH and ESP to operate, **Security Associations (SA)** provide the framework for secure key exchanges.

IPSec operates in one of two modes:

- **Transport mode**—In transport mode, only the payload is encrypted and authenticated. If AH is used, the IP addresses must remain fixed or the hash value for the headers will be invalidated.
- **Tunnel mode**—In tunnel mode, the entire packet, including the headers, is encrypted. Tunnel mode is used most often when configuring fixed-link VPNs between backhaul access point links, between routers over point-to-point links, and between campus buildings.

IPSec is a very secure protocol suite that's easy to implement with a shared secret, which is built into Internet Protocol version 6 (IPv6). But it can be tricky to configure in larger Internet Protocol version 4 (IPv4) networks due to its complexity and range of individual components. IPSec is commonly used in fixed-link VPNs and secure remote access client/server VPNs. IPSec's complexity increases at an exponential rate, such that $n \times (n - 2)$ tunnels must be configured, where n equals the number of sites to be connected. Because many enterprise protocols require any-to-any connectivity (VoIP, for example), using IPSec can represent a trade-off between security and performance. IPSec is built into IPv6, making it easier to implement.

Virtual Private Networks

VPNs are an essential component of any wireless network, especially for remote client access over unsecured networks. VPNs provide an extension of the corporate private network over the unsecured Internet. A VPN is created by establishing a secure virtual point-to-point link with virtual tunneling protocols and secure private key exchange.

A VPN can be classified by how it connects. Some VPNs allow employees to communicate remotely with the work network when traveling or working on remote client sites (remote access VPN). Others connect satellite offices to the corporate head office network (fixed-point site-to-site VPN). In both cases, VPNs provide security through confidentiality, data integrity, and source authentication.

There are several technologies that can be used to build VPNs, including IPSec, Layer 2 Tunneling Protocol (L2TP), Secure Sockets Layer/Transport Layer Security (SSL/TLS), Secure Shell (SSH), various vendor-specific technologies, and Data Transport Layer Security (DTLS). DTLS is a very important new addition, as it can tunnel over User Datagram Protocol (UDP), which allows applications to send data without having to set up a connection first.

Malware and Application Security

Another important part of a wireless security policy is to address malware and application security. This is not something specific to wireless networks but rather relates to IP mobility. With IP mobility, users can access the network from anywhere, including unsecure hotspots and networks. As a result, they could potentially return to the enterprise network with all manner of malicious software on their devices. Traditionally, malware and virus control was stringent within the network through strong perimeter controls such as firewalls, proxies, and antivirus servers. Now, however, user mobility defeats the static perimeter protection, so other methods are required to mitigate the malware risk.

To help mitigate the risks borne of mobility, various "health checks" can be built into the process of connecting to the enterprise network. This is especially important when mobile clients connect inside the perimeter after having connected to unknown and untrusted outside networks. Some examples include the following:

- **Client integrity control**—This is a client device application that checks for compliance with network policy—for example, by checking for valid antivirus software, revision, and scan dates.
- **Network-based services**—This is a network element that sends traffic through antivirus servers and intrusion detection/protection systems. This component of NAC is particularly useful when a client device does not support client integrity control. This could include voice handsets, printers, scanners, and even customer or guest wireless laptops and phones.
- **Mobile device management (MDM)**—This is the control system that allows a network administrator to manage individual devices. This caters to both individual and corporate devices and the ability to set corporate policy.
- **Mobile application management (MAM)**—This focuses on applications rather than on the control of the device. MAM is particularly useful with regard to operating system (OS) and software versions that may be vulnerable, as well as antivirus and application wrapping.

User Segmentation

Not all users are created equal—and not all should have the same level of access. In fact, the best policy on user access is, "when in doubt, deny." This may seem harsh, but it's a simple matter to grant access to people who actually need it, and any aggravation they may display will quickly subside once they have access. On the other hand, problems caused intentionally or unintentionally by a person with too much access could have lasting consequences. That's where segmentation comes in.

Generally speaking, there are two types of segmentation:

- **Internal user segmentation**—This is most often accomplished via virtual local area networks (VLANs).
- **External user segmentation**—This is most often achieved with either a wireless connection outside the corporate firewall that allows direct access to the public Internet or a VLAN that provides direct access to the Internet.

Virtual Local Area Networks

A VLAN is one way of isolating visitor traffic and confining it to only external, untrusted areas. A VLAN is a Layer 2 technique whereby the network designer logically segregates traffic, assigning it to a specific VLAN using some identifier. A typical scenario involving the use of VLANs is when there is a mixture of employees in an open-plan office—for example, a single access point serves employees in the sales, finance, and engineering departments along with some visitors. On a wired Ethernet network, this would be a straightforward design. The designer would simply assign each port supporting a sales employee to the sales VLAN, each port supporting an engineering employee to the engineering VLAN, and so on. With wireless, there is no port, but there is another way to identify employees: by assigning them different service set identifiers (SSIDs). By creating an SSID for each department—sales, engineering, and finance—and then pairing that SSID with a VLAN, the designer can segregate the traffic into separate VLANs, which are Layer 2 broadcast domains.

The key to using a VLAN as a security mechanism is utilizing **access control lists (ACLs)**. When a client is assigned to a VLAN, all packets coming from that client are tagged with the VLAN number (called 802.1Q tagging or VLAN tagging). An ACL is a simple lookup list that allows access to certain services. Access to restricted areas can then be controlled via the ACL. From a scalability standpoint, this works very well in large networks because VLAN association spans physical switches.

Using this method, a client authenticates via RADIUS. As part of this authentication process, each client is assigned to a particular VLAN based on the credentials used during authentication. Once assigned to a VLAN, all packets originating from that client are tagged with the VLAN ID. Whenever a client attempts to access an area of the network or use some network service or protocol, the VLAN tag is matched against the ACL for that particular area (usually a port on a switch) or service. If the VLAN is allowed access, the service is granted. If not, it is blocked.

Guest Access and Passwords

In larger organizations, there is typically a need (and an expectation) for providing guest access on the wireless network. There are several options available for granting such access. These include the following:

- **Open access**—With open access, guest access is available to anyone who can receive the wireless signal. Open access is the most common approach. It essentially requires no management or provisioning once set up. With this approach, the guest Wi-Fi offers a direct connection to the Internet. If there is a need for a secure connection, users are responsible for securing their connections via a VPN.
- **Common guest password**—This is low-security method that enables all visitors to share a well-known password for user authentication. A common guest password is a good compromise. It allows a secure wireless connection for guests. The password may or may not change on a regular basis. From a management perspective, the key to using a common guest password is to change it often—perhaps even daily— to ensure that it does not become common knowledge. If it does, it's no better than open access. This approach requires some ongoing management but generally does not create a significant burden on support teams.
- **Provisioned guest access**—This method requires that each guest be given a unique, time-limited password. This method provides the best security but is the most inconvenient to set up and manage. Because it's resource-intensive, this approach is rarely used for general guest access in corporate settings where there are a lot of transient guests. It is a good option, however, for "permanent guests" such as on-site contractors or partners or in especially secure environments where all activity must be monitored. Hotels also make good use of this feature, using the guest name and room number, which are easily tracked and tied to the length of stay.

Each of these options has its pros and cons. There is generally no right or wrong way of providing guest access, as long as it meets the needs of the organization and its guests.

Demilitarized Zone Segmentation

Another popular way to secure visitor access is by placing visitors on their own demilitarized zone (DMZ) segment or VLAN. In this context, DMZ describes an area between the Internet and the corporate network. It enables Internet users to access corporate public services such as web servers and external email and Domain Name System (DNS) servers. By placing these public-facing servers in a designated security area, the designer allows access from the Internet to these exposed web servers and services. The designer does this by allowing incoming Hypertext Transfer Protocol (HTTP), Hypertext Transfer Protocol Secure (HTTPS), Simple Mail Transfer Protocol (SMTP), and DNS external requests through the external Internet firewall destined for the web servers, but denying any other incoming traffic. Placing the inner corporate network behind another more secure firewall interface allows for external communication to public-facing servers but blocks all other incoming traffic, thereby restricting Internet-originated traffic to the DMZ zone.

Traffic originating on the guest wireless network should pass through a firewall before entering the corporate network via the Ethernet distribution medium. Therefore, a logical place to locate a visitor access point is within the DMZ. Visitors can connect to an access point located on the DMZ subnet. From there, they can easily access the company website and gain unrestricted access to the Internet while being blocked from the inner corporate LAN.

Another popular method, depending on the firewall architecture, is to place the visitor Wi-Fi network on its own Wi-Fi DMZ subnet. This is particularly appealing if the DMZ uses public IP addressing or hosts strict outbound traffic rules. By placing the visitor wireless network in the DMZ, the designer can ensure visitors still gain unrestricted access to the company website and the Internet beyond.

Managing Network and User Devices

In addition to controlling network access and protecting data, larger organizations must also deal with the management of network (infrastructure) and client (user) devices. Unlike smaller organizations, where the IT staff (or an IT person) may know everyone in the company by name, this is often not the case in larger organizations. Most users are unknown. In addition, the sheer volume of users—plus the fact that many people now have multiple devices that connect—puts extra strain on the IT and security teams. This section focuses on the management of both network and user devices in the context of large or complex organizations.

Simple Network Management Protocol Version 3

SNMP is an Application Layer protocol used to provide a message format for exchanging information between an NMS and a host agent. An access point will support SNMP alerts. In addition, if an agent **message information base (MIB)** is installed and enabled, it will be able to send information about network status back to an NMS. These alerts, or traps, will contain information such as interface status, authentication errors, lost neighbors, and any other significant network event that the NMS should be alerted to right away. The NMS also interrogates the agent MIB by polling the MIB at regular intervals. The MIB is a hierarchal database for storing all sorts of network information such as interface throughput, packet loss, latency, and jitter, among other operational data.

SNMP version 3 (SNMPv3) is the only version of SNMP that supports robust security. Therefore, it must be the only version supported on the network. The security features in SNMPv3 are as follows:

- **Message integrity**—SNMPv3 uses a hash code to ensure that packets have not been tampered with.
- **Authentication**—SNMP determines that the message is coming from a valid device.
- **Encryption**—SNMP scrambles the message to hide it from a possible eavesdropper.

SNMPv3 uses SNMP server groups rather than communities. The administrator must configure the server group to authenticate for members specified in a named access list.

While SNMPv3 is secure, SNMPv2 is common on older access points and will still work in a mixed environment. Ideally, all devices should be configured through firmware upgrades to support the more secure SNMPv3.

Discovery Protocols

Several discovery protocols are essential for wireless networks to work efficiently, especially in lightweight access point/controller implementations or where VoWLAN is a design consideration. Unfortunately, discovery protocols can be problematic with regard to security. This is because the same information that is useful to an administrator or authorized technician is also very useful to an unauthorized intruder.

Cisco Discovery Protocol (CDP) and Link Layer Discovery Protocol (LLDP) are the two main discovery protocols enabled on Layer 2 networks. Both of these protocols provide a means to discover neighboring devices and map a network. The information exchanged between devices can be invaluable to a technician troubleshooting a network. Similarly, it is of great value to an intruder mapping the network and looking for vulnerabilities and paths to servers. In addition, other network and service discovery protocols are considered security risks. Universal Plug and Play (UPnP) is one, as are the IPv6 Neighbor Discovery Protocol (NDP) and the Web Proxy Autodiscovery Protocol (WPAD).

> **TIP**
>
> A good policy for basic security hardening should include disabling or removing all protocols and services that are not needed or in regular use.

IP Services

Modern autonomous access points come enabled with a suite of IP services such as HTTP for accessing an internal web configuration and administration portal. Other common IP services that are available include the following:

- **DHCP server**—DHCP enables automatic configuration of an IP address, DNS server address, and default gateway address.
- **SSL certificate management service**—This provides support for SSL trust certificates used in Extensible Authentication Protocol-Transport Layer Security (EAP-TLS) for authentication.
- **Network Time Protocol (NTP)**—NTP is used to automatically adjust the time to a reference NTP server clock.
- **Quality of service (QoS)**—QoS settings provide a technique for prioritizing traffic and assigning a QoS value to each protocol. It is commonly implemented where voice and video applications are present.
- **VPN**—These IPSec-type VPNs are used to secure remote access or inter-bridge connections in peer-to-peer configurations.

Most of these services are valuable tools for the network administrator, but some can create security issues if not implemented correctly. NTP, DNS, and DHCP are extremely useful tools for both administrators and potential attackers. Both NTP and DNS request or receive updates from external NTP/DNS servers. An attacker can easily spoof the access point's IP address and place numerous NTP/DNS update requests. The Internet servers will respond to the requests by sending far larger response packets to the spoofed address, thereby amplifying the attack, which will result in a denial of service.

DHCP is an extremely efficient and convenient tool for dynamically assigning and managing a client's IP address and default network configuration. However, if enabled on an access point with the basic default setting, it will hand out a valid IP and network configuration to any client that manages to get a Layer 2 association and broadcasts for a DHCP server. The access point's built-in DHCP server will simply respond with the Layer 3 credentials requested, and the client will have full network connectivity. This is why it is best to consider authenticated DHCP.

Coverage Area and Wi-Fi Roaming

In addition to requiring more robust methods of authentication and encryption, large networks differ from their smaller counterparts in other ways that must be addressed. The most obvious of these is basic architecture. For instance, in large organizations, the network topology is based on the extended service set (ESS) as opposed to a single access point. The ESS may incorporate the aggregation of many basic service areas across some part of, or the entire, organization. This extended network area may share an SSID or have different SSIDs, and it may or may not support seamless roaming of users throughout the organization. Therefore, a logical starting point for considering advanced wireless security is at the physical design and layout of the network architecture.

Large-scale wireless network design consists of an external service set, which is made up of a collection of access points sharing a common distribution medium such as 802.3 Ethernet. The most common design is to have overlapping coverage areas so the network can support seamless roaming. To support roaming, the designer must specify overlapping access point coverage of at least 15 to 25 percent. Therefore, access point coverage area (which is directly related to radio frequency signal strength) is a key consideration.

Not all networks require seamless roaming, however. Indeed, in some security applications, it may be undesirable. In these cases, coverage will be driven by a need to ensure segregation and minimal overlap in coverage. Wireless roaming with this model is termed *nomadic roaming*, as the user's connection is lost and then reestablished when crossing over the access point boundaries. A common deployment of nomadic roaming is when different security policies are applied to different network areas. Segregated, nonoverlapping coverage forces the user to disconnect and then attempt to reconnect using the new security criteria when entering a new area.

A third topology is to use collocation, which is a fully overlapping access point used to increase capacity. Once again, coverage area must be taken into consideration to maximize the advantage.

A newer topology that was introduced into the standards with the 802.11s-2011 amendment defines the mesh basic service set (MBSS). In mesh mode, access points act as bridged trunks that link with other mesh mode access points to backhaul traffic from the network back to a distribution medium portal or gateway, which is typically connected directly to an Ethernet switch. MBSS is a common design in larger networks, which have areas inaccessible to wired connectivity. An example could be a campus with several buildings that are linked together via wireless bridge trunks and backhauled to the main building's Ethernet network.

Whichever wireless topology is deployed, it is essential that access point coverage be diligently undertaken and that overlap exists to cater to either seamless or nomadic roaming. Controlling access point coverage area is not just a performance factor, it is essential as a security measure, because it limits the network footprint and potential area of access to an intruder. As with SOHO designs, the following steps should be taken to ensure there is sufficient coverage, signal strength, and capacity, and that the radio signal does not leak beyond the organization's boundaries:

- Power should be turned down.
- Access points should be placed in optimal locations for RF coverage.
- The correct antenna type should be used.

Access point coverage area, roaming, RF power, interference, and leakage should be part of any initial site survey. In addition, any future expansion projects should incorporate a further limited RF survey to ensure the correct design.

Client Security Outside the Perimeter

Securing the user's wireless device—whether it be a laptop, smartphone, or tablet— is an important step, particularly because today's network infrastructure has altered radically from legacy networks. Previously, a large network was designed with a hard perimeter to defend against external threats but little in the way of internal defense, allowing for easy access for insiders. The rationale was that the network's hard exterior protected against real threats from the outside; inside, users posed a much lower risk. This concept is often referred to as an *M&M design*, in reference to an old ad campaign that claimed M&M candies were "crunchy" (hard) on the outside and "chewy" (soft) in the middle.

Experience has taught security professionals that this approach is problematic, and that both interior and exterior threats should be considered equal. One thing the M&M design *did* do well, though, was protect static devices such as PCs and servers from virus contaminations. It succeeded in protecting the inner network from contagions by having robust antivirus (AV) and intrusion protection systems that actively monitored traffic entering the network. By using deep packet inspection and recognizing attack signatures, the AV and IPS applications acted as gatekeepers that sterilized and quarantined suspect files and email attachments. All traffic entering or leaving the network's perimeter gateways went through AV, IDS/IPS, content URL filters, web proxies, and application and web firewalls. In this way, the traffic was kept free of viruses, worms, Trojan horses, rootkits, and all manner of nasty malware circulating the Internet. Any contagion that *did* make it through—typically brought into the network via external hard drive or USB thumb drive—was easily contained and removed by the client host's AV and IPS.

The M&M perimeter security design worked well for many years because PCs and servers stayed behind the walled defenses and traffic flowed over predetermined links and entry and exit points on the network. Unfortunately, that legacy design is no longer viable. Devices are not hidden behind fortified network walls, but freely traverse the boundaries on a daily basis. In doing so, a device is exposed to threats from the outside world, like a thumb drive picking up all sorts of malware before being brought once more into the corporate network. However, a contaminated thumb drive would be scanned and cleaned as soon as it was plugged into a client network device. In contrast, a contaminated mobile phone, tablet, or other device connects to the network through its own interface. Moreover, if the device is not company owned, it might not comply with corporate security policy and may not have AV software installed, let alone an approved vendor and version.

For this reason, client security in wireless networks is an important aspect of overall network security and defense in depth (although defense in depth was developed and put into practice in many networks before mobility and Wi-Fi were major considerations). There must be a policy for allowing user devices to access the network, but at the same time, these devices must comply with best security practices. Therefore, it is vital that the security policy addresses these issues and that methods to police compliance be put in place. Typical mechanisms to enforce policy are MDM and MAM, which have become standard practice in today's modern IP mobile networks. The security measures in place for a company laptop should be no more or less stringent than those for a user's wireless-enabled portable device, whether that is a laptop, smartphone, or tablet.

NOTE

Radio frequency (RF) leakage is not just a major security risk. It's also a performance and throughput inhibitor, because it is likely to interfere with neighboring networks, causing mutual signal degradation.

As discussed, client device security is important due to the changing network architecture within a large or risk-averse organization. No longer are wireless networks merely extensions of the wired corporate network or overlays for the convenience of guests. They are now an integral part of the network. Indeed, WLANs and supporting client devices have become ubiquitous within the corporate network. What is more, they are being configured for seamless roaming and IP mobility. The coming together of the once-maligned wireless network and the client's wireless and mobile devices has produced a disruptive technology that has transformed the design of SME networks.

Device Management and User Logons

In a complex organization, there are bound to be many users, each with different levels of access to different systems and areas of the network, all requiring some set of credentials to be presented prior to access. Multiply this by however many different devices each employee has (which, of course, they will want to be able to connect from), and you run into a big problem. This topic is described generally as identity and access management (IAM).

The answer to this problem is **single sign-on (SSO)**. Using SSO, users can enter their credentials just once to gain access to all network services and locations to which they are authorized to connect, without having to log in to each one individually. What's more, this SSO function works with all of a user's devices, including laptops, tablets, and smartphones. This also means the same wireless credentials, such as SSID, username, and password, should be available to users in all locations that they may visit, including their homes. Using SSO allows the security team to control access to sensitive data or systems by challenging users rather than the device alone. In the era of Bring Your Own Device (BYOD), this greatly reduces the management burden of tracking an ever-changing array of laptops, tablets, and smartphones in favor of allowing access based on user credentials.

SSO is often presented as a positive from a security standpoint because, theoretically, a person can be removed from all access with one action. This is not always the case, however. If high security standards are maintained (user education and requirements for strong passwords, for example), then SSO can enhance security. However, if poor standards are maintained, then SSO can be a detriment to security because a single compromised device or cracked password yields far greater access. In this context, SSO, like many other tools, will be a reflection of the company's overall security posture, given that any poorly managed process or tool can be taken advantage of.

When implementing or allowing SSO, security teams should also establish a best practice of verifying that all accounts have been closed for removed users rather than just deleting the SSO capability. Failing to do this could result in orphaned accounts that can still be accessed directly.

In sum, SSO is a good solution from a convenience standpoint, and some view it as a security measure. A better way to view SSO is as a magnifier of internal security capabilities. If internal security is strong, SSO can enhance it. If internal security is shoddy, SSO can actually increase the organization's risk profile.

> **NOTE**
> One positive of using SSO is that it may lessen the chances of users writing down passwords to remember them all. This sometimes occurs when users are required to use many passwords, and can be a big security risk.

Hard Drive Encryption

Encrypting data on client devices adds another layer of security. However, this should be done with caution. One of the most common causes of lost data is users forgetting their password on hardware-encrypted devices. Because encryption is designed to be irreversible, this essentially turns the device into an expensive paperweight.

For data in motion, protocols such as HTTPS and SSL will secure web-based applications. VPNs work well for protecting non-web–based data in motion. However, should the mobile device be lost or stolen, data that resides on the device will be vulnerable. There are several ways to protect the data in this scenario. One is to have a master access password that the user must enter to gain access to the device. This prevents access to the data without encryption. Another method—usually used in conjunction with the master password—is hard drive encryption. Yet another option is to store all files on a protected server and access them remotely. There are several good commercial options for this, such as Dropbox and Box. Some companies have also created internal versions of these services.

One twist with respect to hardware encryption is a recent scheme in which hackers take control of a device and encrypt its hard drive. The hacker then blackmails the owner of the device, attempting to extort cash in exchange for the encryption key that will unlock the data. An example of this is a scam called CryptoLocker, which has had some success. Although it turned out not to be permanent, it was still distressing for many people. This is all the more reason to ensure that users are trained on security matters. They should know to always connect via a secure method and not to download apps from suspicious websites.

Quarantining

An important feature of most device management and access control systems is the ability to quarantine noncompliant devices and limit their access to the network. Quarantining is a form of NAC that allows some level of access—typically guest access—if certain parameters are not met, such as antivirus patches. By setting device profiles, an administrator can ensure that each device must match a minimum set of criteria (for example, making sure all patches are up to date) before allowing a device full access to the network. Any device that fails the compliance check will be relegated to a quarantined section of the network.

Wi-Fi as a Service

Cloud-based Wi-Fi management systems have come a long way in recent years from their humble beginnings servicing public hotspots, cafes, and the SMB market. Today, **Wi-Fi as a Service (WaaS)** and Wi-Fi management cloud services cater for the enterprise and large organizations that deploy and manage vast wireless networks. This is simply because wireless networking is now the premier way to build networks but deploying and managing large wireless networks securely is a complex business. Consequently, many organizations are looking to outsource the management of their wireless networks to cloud-based service providers who have the skills, resources, and tools to remotely manage their wireless assets.

Modern Wireless as a Service will typically consist of three components:

- **Hardware as a Service**—The provider will supply, configure, and deploy all wireless assets.
- **Software as a Service**—The provider will supply all software and tools, and update firmware and security patches.
- **Management as a Service**—The provider will remotely manage and support the network from their cloud-based Network Operations Center.

Several key benefits to outsourcing wireless management to a WaaS provider are next discussed.

Simplified Network Management

Cloud-managed wireless and networking solutions offer a single point of management for installation, configuration, network management, and diagnosis. With a central interface and one-click self-servicing, cloud networking solutions simplify the task of handling the dynamic networking needs of the business.

Also, in case of any issue, it is easier to diagnose and troubleshoot the issue quickly with minimal help and support. Most of the issues can be resolved by a central wireless access point management dashboard. Not only does this gives a business greater control over its networks but brings down network management cost.

Cost Efficiency

Cloud-managed services typically work on a subscription-based, on-demand model. This paradigm just like other "as a service" models provides businesses with greater control over expenses and reduces capital expenditure. With a cloud-based wireless management solution, organizations can save on capital investment because hardware such as access points and switches can be provided by the service provider. This also reduces maintenance costs of on-premise wireless assets.

On-Demand Scalability

Cloud-managed wireless networks scale well because they are under unified and centralized management. One of the biggest benefits of deploying cloud architecture is that the wireless network can scale securely as the organization grows.

Enhanced Security

Cloud-based wireless services develop and market their own sophisticated security frameworks built on vast experience and expertise, which mitigates security vulnerabilities. Organizations can leverage this knowledge to create their own security framework that matches their security policies and requirements.

Hands-Off Maintenance

With a cloud-based management service, maintenance is transparent to the organization and wireless access points are automatically updated with the latest system firmware and the latest security patches, which have been verified and tested before deployment. Therefore, making your network up-to-date and more secure.

Deeper Analytics and Reporting

Cloud solutions are equipped with analytical and reporting capabilities that help a business to monitor its network to get real-time updates on security incidents, gather network analytics on usage, as well as provide in-depth prescriptive analysis that assists in informed and proactive decision-making.

Easy Backup and Real-Time Alerts

Ease of management and lower maintenance burden are key factors in deciding to outsource to a cloud-managed service. This is because wireless access points are monitored by tools in the cloud and because the access points actually store their network configuration on the cloud, this eliminates the need for local network management systems. Instead, with a cloud-managed wireless management service, you are automatically notified in case of an alert or an anomaly in the network behavior.

A cloud-managed wireless service may not be for everyone because many organizations do like to retain control over their networks and assets. However, for those who do want to optimize and secure their wireless networks with the least capital expenditure and operational burden then outsourcing to a WaaS provider can enrich your existing business operations through advanced analytics and reporting to secure and make the most of your wireless networks.

CHAPTER SUMMARY

While the basic ideas of controlling access, protecting data, and managing devices remain the same across all organizations, larger and more risk-averse organizations require an approach that meets both the heightened need for security and the increased need for scalability and management. This is not only due to the fact that larger organizations by definition have more people, but also that those people tend to have more devices, access more services, and have more interactions. Larger organizations also have higher rates of churn (employees who join and leave the company), which creates the need for processes and policies to ensure that "anonymous" employees are not left to their own devices. (No pun intended.)

All of this drives the need for more advanced approaches to security in general and to wireless security in particular. Controlling access through the use of a RADIUS authentication service is the approach that most enterprises take, and for good reason. In conjunction with EAP, this method of access ensures confidentiality during the authentication process and provides a secure and reliable means to protect data while controlling access to encryption credentials (both the process and the means to secure keys). SSO can also be a great tool to help create a secure environment without putting an undue burden on employees.

As always, device management is a primary consideration—one that is amplified in a large organization simply because there are more opportunities for lost or compromised devices. As a consequence, many businesses are looking to outsourcing their Wi-Fi operations to cloud providers who can manage the networks for them. The rise of WaaS is becoming prominent as the complexity of managing devices and wireless networks grows.

KEY CONCEPTS AND TERMS

Access control lists (ACLs)
Authentication Header (AH)
Bots
Captive portal
Encapsulation Security Payload (ESP)
Intrusion detection systems (IDSs)
Message information base (MIB)
Network management system (NMS) server
Point-to-Point Protocol (PPP) networks
Quarantining
Security Associations (SA)
Single sign-on (SSO)
Walled garden
Wi-Fi as a Service (WaaS)

CHAPTER 8 ASSESSMENT

1. For larger networks or campuses, centralized control greatly simplifies design and tends to work better due the time savings offered with regard to maintenance.

   A. True
   B. False

2. Which of the following KPIs is unaffected by the use of VoIP or VoWLAN?

   A. Packet loss
   B. Latency
   C. Jitter
   D. Security
   E. Availability

3. Wi-Fi guest access can include which of the following types of access?
   A. Open access
   B. Common password
   C. Provisioned password
   D. No guest access allowed
   E. All of the above

4. Extensible Authentication Protocol does which of the following?
   A. Protects authentication credentials
   B. Can be used over a LAN to securely connect to a RADIUS server
   C. Is independent of the authentication method used
   D. All of the above

5. Discovery protocols are great for IT personnel but are also useful to hackers; for this reason, they should be limited and carefully controlled.
   A. True
   B. False

6. The RADIUS server is a central repository for all the authentication data and can issue the success or fail notifications. However, the access point still needs to know the client's authentication credentials.
   A. True
   B. False

7. Which of the following versions of SNMP offer protection in the form of encryption?
   A. SNMPv2
   B. SNMPv3
   C. Both SNMPv2 and SNMPv3
   D. Neither SNMPv2 nor SNMPv3

8. IPSec is a very secure protocol suite that's easy to implement with a shared secret.
   A. True
   B. False

9. Assignment to a VLAN can be determined through the authentication process when joining a network.
   A. True
   B. False

10. Which of the following are true about SSO? (Choose all that apply.)
    A. Users need to remember only one password
    B. It makes a network more secure
    C. It makes a network less secure
    D. It reduces the management burden of IT
    E. It can enhance security on a well-run network or worsen security on a poorly run network

CHAPTER 9

WLAN Auditing Tools

WI-FI NETWORKS REQUIRE AUDITING on an ongoing basis. Indeed, auditing is essential. It conveys to the network administrator the network's precise status at any time. This is especially true for Wi-Fi because of its inherent vulnerability as a broadcast medium—that is, it transmits to anyone who is interested.

In recent years, Wi-Fi has become pervasive both in the workplace and in the home. This is due in large part to its affordability and its ease of installation. This is great for installers and administrators, but there is a downside. Because they are cheap and easy to install, Wi-Fi networks are constantly under threat of unauthorized access. This unauthorized access can occur as a result of misconfiguration, due to a lack of basic security controls, or because of the addition of rogue access points or ad hoc networks to the wireless local area network (WLAN). To counter these vulnerabilities, it is essential to audit and monitor the WLAN on a continuous basis.

Performing network audits is not an easy task because there are many potential attack vectors. There are, however, many tools available to assist in this sometimes-difficult task. WLAN auditing requires a wide variety of both general-purpose and highly specialized tools. These range from basic WLAN discovery tools to wireless protocol analyzers and network management applications.

Wi-Fi network audit tools come in several categories, although some of the paid-for tools may well cover every aspect. Wi-Fi tools can usually be considered to be one of the following, although they share several common features:

- **Wi-Fi Analyzer**—These tools will usually have the following capabilities: scanner, site survey, stumbler, spectrum analyzers, and heat mappers.
- **Auditing and Security**—In addition to the features in a Wi-Fi analyzer, these tools will also feature: traffic analysis, packet sniffer, and pentester tools.
- **Monitoring and Management**—These tools tend to be comprehensive toolkits that cover every aspect of Wi-Fi monitoring, maintenance, and management, so they will typically cover: scanner, inventorying, site survey, stumbler, spectrum analyzers, heat mappers, traffic analysis, and packet sniffer.

Unfortunately, attackers can use these same tools to find security weaknesses and gain access to the network. Therefore, understanding these tools—how they both help network administrators and can be used against them—is a critical aspect of WLAN security. This chapter focuses on these tools, providing a primer on both how to use them and how to protect the network from them.

© Cherezoff/Shutterstock

Chapter 9 Topics

This chapter covers the following concepts and topics:

- What WLAN discovery tools are available
- What penetration testing tools are available
- What password-capture and decryption tools are available
- What network management and control tools are available
- What WLAN hardware audit tools and antennas are available
- What common attack tools exist and what the techniques for using them are
- What the most commonly used network utilities are

Chapter 9 Goals

When you complete this chapter, you will be able to:

- Discuss WLAN discovery tools such as NetStumbler, InSSIDer, Kismet, and HeatMapper
- Describe how penetration testing tools are used by both administrators and attackers
- Understand the Metasploit framework, including password-capture and decryption techniques
- Describe how network management and control tools are used
- Describe how wireless protocol analyzers are used
- Describe how antennas and WLAN hardware audit tools are used
- Describe attack tools and techniques such as frequency jamming, denial of service, and hijacking
- Understand how network utilities are used by both administrators and attackers

WLAN Discovery Tools

There are many Wi-Fi network discovery tools available both as freely available open source tools and at the other end of the spectrum as built-in features within professional suites for Wi-Fi monitoring and management. However, many of these open source tools were designed with the legacy Wi-Fi standards in mind—802.11a/b/g. These are, of course, still perfectly fit for purpose if the WLAN you are monitoring is built to one of those standards. The problem occurs when Wi-Fi networks start to evolve into 802.11n and particularly when managing 802.11ac (Wi-Fi 5) and 802.11ax (Wi-Fi 6) WLANs because these are not compatible with legacy adapters or tools. For example, when you are capturing 802.11n packets, it requires that your management device has an 802.11n or 802.11ac adapter. Similarly, to capture and decode 802.11ac packets requires an 802.11ac adapter. You cannot capture 802.11n packets using an 802.11a/b/g adapter and you cannot capture and decode 802.11ac packets using an 802.11a/b/g or 802.11n adapter. Moreover, it is not just the adapter that needs to be compatible, so does the Wi-Fi monitoring and management software.

The issue of incompatibility comes about because of the introduction of multiple input/multiple output (MIMO) and Transmit Beamforming technology in 802.11n and 802.11ac networks. These technologies provide serious challenges for wireless analyzers. While this is not a problem when you are performing simple tasks such as device discovery, a site survey, or measuring signal strengths, it is a problem when performing traffic analysis and reporting.

Therefore, when we discuss Wi-Fi auditing tools in this chapter, we will generally only discuss tools that are compatible across the full range of Wi-Fi standards up to at least 802.11ac.

Enterprise Wi-Fi Audit Tools

Professional Wi-Fi network administrators monitoring and managing a corporate wireless network and that have sufficient budget will likely go for an integrated suite of tools that covers their requirements. Therefore, they are likely to go for a comprehensive integrated set of tools that encompass Wi-Fi traffic analysis and site survey. Many vendors support their network products with a comprehensive set of audit and management tools, for example Cisco AWS and Aruba Airwave. However, for administrators in multivendor environments, there are many vendor-agnostic tools available.

One such professional suite of tools is **AirMagnet WiFi Analyzer** by NetAlly, which is a comprehensive wireless network analyzer that also covers all the required features to monitor and manage an enterprise Wi-Fi network. For example, AirMagnet WiFi Analyzer has the capability to analyze 802.11a/b/g/n/ac/ax wireless traffic in real time. The suite of tools also provides you with an overview of wireless network connectivity, capacity, coverage, performance, throughput, and network vulnerability and known security issues. Moreover, AirMagnet WiFi Analyzer can also generate full reports to ensure that your network complies with regulatory standards for PCI, SOX, and ISO.

AirMagnet WiFi Analyzer has many inbuilt features that address Wi-Fi issues and can be considered a comprehensive suite of Wi-Fi monitoring and management tools. Its basic functionality extends to being a professional tool as a:

- Wi-Fi traffic analyzer
- Automatic troubleshooting tool
- Wi-Fi interference detection and Wi-Fi analysis
- WLAN client roaming analysis

However, AirMagnet WiFi Analyzer doesn't come cheap and is aimed at professional Wi-Fi network administrators. There are many other professional tools on the market, however, at budget prices such as **Acrylic Wi-Fi**.

Acrylic Wi-Fi

This suite of professional tools is aimed at advanced Wi-Fi scanner and site survey on Windows systems and has the capability to scan 802.11/a/b/g/n/ac/ax networks, display Wi-Fi channels in 2.4 GHz and 5 GHz channels as well as to visualize signal strength and display Wi-Fi graphs.

The user-friendly graphical user interface (GUI) provides an intuitive visual dashboard, which displays all relevant Wi-Fi information in real time. This information is useful for optimizing Wi-Fi performance and troubleshooting any anomalies in the wireless network.

In addition to basic network discovery, Acrylic Wi-Fi, as you would expect, allows you to optimize access point coverage and surface channel interference issues.

The way it works is that the software scans your local radio surroundings and displays nearby Wi-Fi access points. It has the capability to not just discover existing or rogue access points but also obtain information on signal strength levels, service set identifiers (SSIDs), channels, and WLAN security encryption.

However, if your budget cannot stretch to the AirMagnet WiFi Analyzer or the professional version of Acrylic Wi-Fi then there are several open source or free to use alternatives.

Wi-Fi Inspector

This product is a Wi-Fi scanner and site survey tool and is available as free wireless network monitoring software for use on 802.11ac and compatible networks for analysis of traffic, status, and clients via a real-time and easy-to-use GUI.

Xirrus Wi-Fi Inspector provides easy insights into the local radio frequency (RF) spectrum, and the software is supported by Windows and macOS and can monitor the most popular Wi-Fi standards such as 802.11ac Wave 1 and Wave 2 technologies.

Xirrus Wi-Fi Inspector is free to use; therefore, it is viable for home or small to medium business (SMB) use, but it is also highly capable with support for:

- Real-time monitoring of Wi-Fi networks
- Wi-Fi signal strength and coverage
- Scanning Wi-Fi networks and detecting devices (including rogue access points)
- Limiting a device's Wi-Fi connection

Another free to use Wi-Fi discovery and site survey tool is Homedale from the SZ development team. Homedale can scan your Wi-Fi network and all locally accessible access points, monitor their respective signal strength, and display the results on a graph.

Homedale software supports 802.11a/b/g/n/ac wireless networks in the two popular frequency bands (2.4 GHz and 5 GHz). In addition, you can integrate Homedale with other third-party services such as Google Geolocation, Mozilla Location Service, and Open WLAN Map Service via inbuilt access points and connectors.

HeatMapper

The process of heat mapping during a Wi-Fi site survey enables you to locate and map out the RF footprint for each discovered access point against either a default grid or an imported floor plan. Once you have conducted your site survey (simply by walking around), HeatMapper by Ekahau presents the coverage overlaid on the floor plan showing you where gaps—RF dead zones—and RF overlaps exist. Most Wi-Fi survey tools will have some form of heatmapping, although it may not be graphical. The best tools available on the market, such as Ekahau Site Survey, have easy to understand visual displays of the RF spread.

The enterprise version called the Ekahau Site Survey tool offers additional features as well as heatmapping. For example, Ekahau Pro comes as a component of the Ekahau Connect product suite with the following features and capabilities:

- Simultaneous 2.4 GHz and 5 GHz surveys
- Supports 802.11a/b/g/n/ac/ax
- Simultaneous passive, active, and spectrum surveys

Moreover, as a heat mapper, Ekahau Analyze collects data using one of 15 heatmaps to determine the performance of the Wi-Fi network. As a heatmap, Ekahau can determine and visualize:

- Signal strength coverage
- Signal-to-noise ratio
- Data rate
- Throughput
- Packet loss
- Jitter
- Round-trip times
- Maximum channel bandwidth
- Number of access points (overlap)
- Network capacity problems
- Spectrum channel power
- Spectrum utilization Wi-Fi channel width
- Co-channel/adjacent channel interference
- Capacity issues, such as excessive VoIP calls per radio and access point overload

However, if you cannot stretch to the cost of the professional tools then there are alternatives for an SMB or those on a budget. NetSpot, which is a comparable professional tool, is available as a free to use version Wi-Fi survey and analysis software tool for Windows and macOS. NetSpot allows you to scan 802.11 Wi-Fi coverage areas, perform advanced site surveys, and visualize all results in a single dashboard. The software can also create visual Wi-Fi heatmaps, so you can identify where the signal leakages or dead zones are.

NetSpot is available as a free download but also has a professional licensed version that has many inbuilt features, such as:

- Shows individual channel load
- Identifies any rogue access points
- Creates Wi-Fi heatmaps and provides recommendations for adjusting Wi-Fi radio signals

These network discovery tools can be and are used by administrators for authorized auditing purposes and by attackers for mapping out and identifying potential targets. As noted, Kismet can even discover cloaked SSIDs, so an attacker can easily find hidden networks. To mitigate the risk of an attacker discovering and interrogating the network, an administrator can use a number of techniques to reduce the effectiveness of discovery tools. These involve applying advanced security measures such as 802.1X, Extensible Authentication Protocol (EAP), or virtual private networks (VPNs) to securely encrypt and tunnel the packets. Other methods include using fake access points to generate counterfeit beacons. These mask the presence of genuine access points by confusing tools such as NetStumbler and Kismet. Administrators must also conduct regular surveys to ensure that well-meaning employees have not installed their own rogue access points, which are often exploited by wardrivers.

Penetration Testing Tools

Like any other protected network environment that is subject to attack, the Wi-Fi network infrastructure requires security hardening, such as the use of demilitarized zone (DMZ) web servers. There are many tools and techniques available, but simply installing them and assuming they work has proven disastrous for many organizations. A better practice is for companies to test their own defenses using **penetration testing (pentesting)**. Pentesting is the practice of searching for and then attempting to exploit vulnerabilities on networks that an attacker could potentially exploit. In essence, pentesting is attacking the network with permission. Pentesting works on the theory that you are better off looking for weaknesses yourself than having an attacker find them for you (or worse, having one that goes unnoticed by you for years that is exploited without your knowledge).

> **NOTE**
> Pentesting without express permission is hacking and is considered illegal, even if well intended. It's always best (for professional and legal considerations) to get pen-test agreements in writing.

You can perform manual pentests, but there are automated tools and specialized software frameworks available as well. One of the most popular frameworks for pentesting is Metasploit.

Metasploit

Pentesting tools such as Metasploit are used on mature networks to assess their state of defense. Pentesting is typically goal oriented, and this is where it differs from vulnerability assessments. Metasploit does both. It can be used as a pentesting tool to look for specific weaknesses and as a general vulnerability scanner.

With vulnerability scans or assessments, the administrator looks to find all known common vulnerabilities and to remove them. Once the administrator is satisfied that most, if not all, high-risk vulnerabilities have been mitigated, they will rerun the vulnerability assessment until a high score is achieved. (A low score indicates several problems.) When satisfied that the network is secure, the administrator may test that assumption by carrying out a pentest for specific targets and attack vectors.

> **NOTE**
> An important distinction between vulnerability scanning and pentesting is that vulnerability scans are nonintrusive and do not cause damage, whereas pentesting can cause an outage.

The Metasploit framework is open source software built to assist administrators in "attacking" their own networks. It's a great tool, but it represents a classic double-edged sword in that attackers, too, can and do use it. If nothing else, tools such as Metasploit force administrators to adopt pentesting to prevent attackers from gaining asymmetric information about the network (in other words, to prevent attackers from knowing more about aspects of your network than you do).

Security Auditor's Research Assistant

Another pentest tool that also works as a vulnerability scanner is Security Auditor's Research Assistant (SARA). SARA integrates with the National Vulnerabilities Database (NVD)

CHAPTER 9 | WLAN Auditing Tools

and can perform many pentests, including SQL injection and cross-site scripting (XSS) tests. (XSS is a popular hacking technique that takes advantage of vulnerabilities in web-based application codes. The exploit enables a hacker to get clients to transmit end-user information and data, which can be sold or otherwise used by the hacker.) One particularly good feature of SARA is its ability to integrate with Nmap, a Transmission Control Protocol/Internet Protocol (TCP/IP) network utility that allows for operating system (OS) fingerprinting and remote network port scanning for open TCP/User Datagram Protocol (UDP) ports and applications.

Password-Capture and Decryption Tools

When conducting any Wi-Fi network audit, it is wise to check for weak passwords. Aside from unencrypted sessions, weak passwords are often the most common and serious security threat to a network. Passwords that are easy to crack enable an eavesdropper to exploit them, perhaps gaining deep access into a network (depending on the level of the person whose password was cracked). This is especially true when dealing with default administrator passwords on network devices.

There are several tools for auditing and recovering passwords; Nessus and Aircrack-ng are two of the most popular. Nessus is particularly good at spotting default administrator passwords for web applications. Aircrack-ng is also able to crack Wired Equivalent Privacy (WEP), Wi-Fi Protected Access (WPA), and Wi-Fi Protected Access 2–preshared keys (WPA2-PSK) passwords as well as perform packet capture and forced deauthentication and reauthentication. Being able to force a reauthentication handshake is an essential step in capturing the authentication and association process between clients and access points.

WEP and WPA are not secure. To mitigate the risk of an attacker eavesdropping and recovering network passwords, it is important to establish strong security such as WPA2-PSK or WPA2 Enterprise, which uses Advanced Encryption Standard (AES) encryption and Remote Authentication Dial-In User Service (RADIUS) authentication. Aircrack-ng can still break WPA2-PSK if given a large enough packet sample and enough time, but it is not a trivial task.

However, WPA2 is not without its own security issues, as can be seen with its vulnerability to brute force password cracking. It also provides no protection such as encryption and makes devices extremely vulnerable to attack on open public Wi-Fi networks. WPA3 introduced some long overdue security features to mitigate the risk of brute force password attacks and also enforces encryption of traffic across even open public Wi-Fi networks. However, until WPA3 becomes universally deployed, the vulnerabilities associated with WPA2 will likely persist so enforcing a policy of strong passwords should still be a key security criterion.

More important perhaps is that attackers are not always looking for access point or client authentication keys or passwords. In fact, in most cases, attackers are looking for operating system passwords for applications. Tools such as Win Sniffer and Ettercap can promiscuously capture all packets on a network segment. They can then decode File Transfer Protocol (FTP), Post Office Protocol 3 (POP3), Hypertext Transfer Protocol (HTTP), Simple Mail Transfer Protocol (SMTP), and Telnet passwords. Both Win Sniffer and Ettercap can run for days, collecting packets and recovering passwords. This underscores the advantage that the bad guys have: the ability to automate tasks to collect information and probe for weaknesses.

Another favorite target is open shares on client devices. These enable attackers to gain a foothold in the network. **Open share** is a method of sharing files directly between clients over an air interface. L0phtCrack is one tool that is commonly used on Windows-based clients to crack network password hashes on file shares and network logons. (A hash is a number generated from a string of text via a formula used in encryption.)

Password crackers often use dictionary-style attacks. In this type of attack, attackers use freely available online password dictionaries to try out a vast array of words. Some of these dictionaries contain millions of possible passwords using both real words and combinations accumulated from password databases stolen from websites. These password crackers are called **dictionary password crackers**. In a dictionary attack, each item in the dictionary or word list is encrypted in sequence using the same encryption method as the password. The resulting hash code is then compared to the original password's hash code. If they differ, the software tries the next entry. If they match, the password is revealed.

The advantage attackers have here is that many passwords are actually very simple. (Humans have a hard time remembering complex passwords.) As a result, passwords can usually be cracked very quickly using a large dictionary. When used against a captured password database, password crackers can run thousands—even millions—of permutations against the database per second.

Related to a dictionary list is a **rainbow table**, which is a list of password combinations within a certain range and their matching hashes. A rainbow table may contain hundreds of thousands of known or previously discovered passwords and their hash equivalents, making the cracking of passwords much more efficient. These tables, which are collated by hackers and distributed online, are usually used to crack a plaintext password up to a certain length consisting of a limited set of characters. Having captured the hash, the attacker can then compare the hashed password against the table. When a match is found, the attacker can determine the password.

Dictionary password crackers like Aircrack-ng, Cain & Abel, and John the Ripper also use **brute-force attacks** to try to recover passwords. The difference between the dictionary attack and brute-force methods is that a dictionary attack decrypts passwords, whereas a brute-force attack cracks the password by comparing all possible combinations of characters for a given password length.

Brute-force attacks are generally inefficient against strong, complex passwords because they churn through all the possible combinations of characters. For example, if a password uses both uppercase and lowercase letters A to Z, numbers 0 to 9, and 10 special characters, there are 72 possible entries for every character in a password. That means a five-character password would have 1.934 billion possible combinations ($72 \times 72 \times 72 \times 72 \times 72$). This seems like a really large number of combinations, but it would not take that long for a high-end computer to process them all. On the other hand, an eight-character password has a mind-boggling 722,200 billion combinations, which is why most secure websites require passwords of at least eight digits.

Complicating matters is the fact that the attacker does not know how long a password is. This, along with the sheer math involved in working out all the possible permutations, is usually enough to thwart most brute-force password attacks unless the password is less than five characters long.

To be effective, both dictionary and brute-force attacks require a captured hashed sample password for comparison purposes. This is easily achieved using tools like Airodump-ng, which is part of the Aircrack-ng suite of tools, to capture packets in monitor mode over the wireless network. If this is not possible, then they can be used in a script to try against live sites. Now, however, most sites do not permit more than a few unsuccessful logon attempts before logging the user out or challenging the user with a **Completely Automated Public Turing Test to Tell Computers and Humans Apart (CAPTCHA)**. This makes it very inefficient to use these tools in live mode. They must resort to slow attack mode, whereby they attempt only two or three permutations an hour to avoid triggering the lockout or CAPTCHA.

Technical TIP

A good way to create a strong but easy-to-remember password is to compose a sentence that is meaningful to you, write down the first letter of each word in the sentence, and convert it to "code" by using symbols instead of letters where possible. For example, take the sentence "My oldest daughter Samantha was born in 02." By using a dollar sign ($) in lieu of the letter S and an exclamation point (!) instead of the letter I, you get Mod$wb!02, a nine-digit password that is seemingly random, strong, and rather easy to remember.

The only way to mitigate the threat of password cracking by an attacker is to ensure that users create strong passwords, and that they use different passwords for different applications. This is usually not the case, however. Many users employ the same password for everything. This obviously is not a secure method; however, it is a very human one. Passwords are difficult to remember, especially if you have chosen a strong one. Using the same password over and over again is especially dangerous because of breaches such as those revealed at Target, Home Depot, and others, where entire databases of usernames and passwords were stolen. A routine practice by hackers is to set up an automated scheme to check credit card, bank, and other retail sites for the same combinations.

The problem for users comes down to being able to remember the passwords for each application when required. One answer to this is to use a **password management system (PMS).** A PMS will securely store encrypted passwords for all the websites you visit and will automatically log you on using the appropriate encrypted password. This means you can create very difficult-to-remember—but very secure—passwords and store them in the PMS. You need not remember them, as you will be logged in by the PMS. Access to the PMS password database and repository requires a master password, which should be both strong and easy to remember in the event other passwords need to be modified for any reason.

Remember that passwords are gold to hackers. They allow them to log on and gain access to all of someone's data and services.

Network Enumerators

Network enumerators are software programs that scan a network for active hosts. They often list the IP addresses in a subnet and then go a step further to fingerprint each IP. Popular network enumerators include Nessus and Nmap, which provide information on the network infrastructure such as open ports and supported services. For detecting Windows shares, a program like Legion from Rhino9 will quickly scan an entire subnet and return a list of devices with their open file shares.

Typically, attackers will try to determine the OS and the availability of open ports on any client devices they see on the network, a technique called **OS fingerprinting**, or **port scanning**. A popular tool for this used by administrators and attackers alike is LanGuard. LanGuard can quickly fingerprint an entire network and return in its payload information such as the service packs installed, the security patches installed, services in use, open ports, users and groups, and known vulnerabilities and exploits.

To protect against devices running LanGuard, Nessus, or Legion (or at least to detect their presence), a wireless intrusion prevention system (WIPS) can be used. These programs are easy to detect because they are quite noisy on the network, meaning they request a great deal of information from other devices while scanning and probing ports and services. Because of this, periodic scans with a network protocol analyzer—even brief ones—will reveal the presence of a rogue scanning device on the network.

NOTE
Similar to a network enumerator is a *share enumerator*. Share enumerators scan a Windows subnet for open file shares. They look for usernames and for information on groups, shares, and available services.

Network Management and Control Tools

Network administrators and IT security professionals can bring to bear a broad array of tools and solutions to help monitor, troubleshoot, and manage the network. The most common of these are protocol analyzers and network management tools. These not only help optimize the network, but also send alerts when suspicious or unusual activity occurs.

As noted before, however, many of the same tools can be—and are—used by hackers to gain insight into the network or to find holes in security that can then be breached. Generally speaking, tools are neither good nor bad (although there are some that are purpose-built for hacking). It depends on the motives of the person using them. This section covers some of the more common tools and notes where they can be used against you.

Wireless Protocol Analyzers

Network administrators who need to identify and correct deep-seated network faults often use network protocol analyzers to help them. Protocol analyzers are also used for security purposes to reveal and locate devices that are behaving contrary to the rules and protocols of the network. For example, using a protocol analyzer, client devices that are scanning ports and addresses on a subnet can be quickly identified and, if proven to be a problem, can be remediated.

Excessive probing and scanning behavior is often an indication of a virus infection because that is how a worm replicates throughout the network. However, that is not always the case.

It can also be the result of an attacker running Nmap, Nessus, or some other network discovery tool.

Because they provide a good indication of what's going on, network protocol scans should be a normal part of a network administrator's weekly task list. In most cases, anomalous traffic is found to be legitimate, but there are times when "odd" traffic is indicative of an attack or threat. Listening and investigating is a sound practice.

> **NOTE**
>
> Although the names are similar, Airshark and Wireshark are not affiliated.

Aircrack-ng

Some wireless network tools have built-in protocol analyzers. One such tool is Aircrack-ng, which runs on both Windows and Linux. It's a suite of software tools consisting of a network detector, a packet sniffer/collector/injector, and a dictionary-based password cracker. The suite of programs includes the following:

- **Aircrack-ng**—This is a dictionary-attack tool used against passwords on WEP- and WPA-protected networks.
- **Airmon-ng**—This is a tool for configuring a network card in monitor mode, a prerequisite to network discovery.
- **Airdeauth-ng**—This is a tool to force deauthentication that kicks off an attempt by the client to reauthenticate, which is then hijacked. (This is discussed further later in the chapter.)
- **Aireplay-ng**—This is a tool for packet injection.
- **Airodump-ng**—This is a packet sniffer that captures packets and places them into files for analysis. It can be used as a protocol analyzer. It also shows detailed network information, so it is useful as a discovery tool.
- **Packetforge-ng**—This is a tool for creating encrypted packets for injection into the network.
- **Airbase-ng**—This is a tool for attacking clients rather than access points.

Airshark

Another popular and effective protocol analyzer is Airshark, a wireless variant of the hugely popular Wireshark protocol analyzer for wired Ethernet networks. Airshark, which is free, is very capable, making it a good tool for discovering rogue access points, other rogue devices, viruses, and anomalous activity on a wireless network.

Airshark works by running the network interface in promiscuous mode to capture all traffic on the network (in this case, over the WLAN). In doing so, Airshark can pinpoint TCP conversations and threads, presenting them in a color-coded display. This makes it easy to follow the interactions between devices within any given conversation. Additionally, the software looks inside the packets to reveal much more than just headers and ports but also in-depth drilldowns into the payload of each packet within the TCP stream. This can be a very quick way to check for encryption, as it is readily clear whether the payload data has been encrypted. Furthermore, by checking the conversation streams, it will soon become clear if one host in particular is scanning the network or probing for gateways, DNS servers, or open shares.

Like all other tools, wireless protocol analyzers can be used for both good and evil. It's important to recognize that unlike discovery tools, protocol analyzers are passive in their mode of operation. As such, they do not create noise on the network. An administrator will use them to determine the status of the network and to detect and reveal rogue devices or unencrypted wireless traffic. Protocol analyzers are also useful for determining whether an RF cell is oversized or if devices are incorrectly configured—for example, by logging and reporting failed authentication and associations.

Because protocol analyzers reveal critical insights in unencrypted systems, such as Layer 2 and Layer 3 information and packet payloads, they are widely used by eavesdroppers and attackers. With Airshark, packets can be captured, manipulated, and reinjected in a man-in-the-middle (MITM) attack. This makes Airshark a very useful attack tool—one that can be used by both novices (in a limited capacity) and experts (with great effect).

Tools such as Airshark do not have a network signature that would give away their presence on the network. This makes them hard to detect. However, you can detect Airshark and tools like it by checking for network clients running in promiscuous mode. Security measures to reduce the impact of an attacker using a wireless protocol analyzer include applying Layer 2 and Layer 3 encryption using Internet Protocol Security (IPSec), Generic Routing Encapsulation (GRE) tunnels, or Secure Shell (SSH)/Secure Sockets Layer (SSL).

Network Management System

The foundation of network security is the network management system (NMS). An NMS is a requirement for even the smallest of business networks. It provides the single viewpoint from which an administrator can view, plan, and configure the network. NMS applications such as wireless protocol analyzers come in a vast array of sizes, prices, and functional capabilities. There are open source and freely available NMS applications such as OpenNMS, Nagios, and Zenoss, as well as commercial applications such as WhatsUp Gold and SolarWinds, which are well within the price range of most small businesses. Enterprise-class NMS applications such as IBM Tivoli, on the other hand, can cost hundreds of thousands of dollars and take years to fully implement.

Apart from providing a network view and a single point of configuration, NMS also enables the use of Simple Network Management Protocol (SNMP), which is the network protocol used to raise alerts and alarms by network devices. NMS will typically provide a map of the network layout with each device and its interfaces and connections highlighted. Should any device fail, the device icon on the map will change color and/or the system will emit an audible alarm. When used in conjunction with a wallboard monitor in a network operation center, it can be an effective means of flagging a network issue and raising an alarm.

Alerts in SNMP systems are triggered by **SNMP traps**. These traps are generated when some predefined condition has been met on an SNMP agent in a device or software program. Manufacturers code in possible conditions, and system administrators set the thresholds for the alerts. If the prescribed threshold or condition is met, the trap triggers an alert, which is sent to the SNMP dashboard. This system allows for customized monitoring of the network. Depending on the networks' activity and risk profile, different alerts trigger investigations

CHAPTER 9 | WLAN Auditing Tools

by the security team. Even in simple networks, NMS is important; in large and complex networks, it's a critical security tool.

WLAN Hardware Audit Tools and Antennas

Purpose-built tools are also available for auditing networks and, in the case of wireless networks, over-the-air signals. This section looks at hardware-based auditing tools and antennas, which allow the monitoring and auditing of over-the-air signals.

 NOTE

Remember that all audit tools and techniques allow a would-be attacker to audit a network in the same way.

Hardware Audit Tools

For most wireless audits, a network administrator needs only a laptop with a wireless network adapter. Specialist wireless hardware and antennas are not required. This is true even in larger enterprises. Although these organizations may have specialized equipment, a simple laptop with a wireless adapter may be used for less-complicated tasks.

Some tasks, however, require more specialized equipment—that is, equipment that is designed and configured specifically for wireless network auditing and pentesting. One such device is the popular Pineapple. A pentesting and network auditing toolbox, Pineapple is a microcomputer that runs application scripts on a fast processor with sufficient random-access memory (RAM) for data storage. Pineapple comes loaded with preconfigured attack software such as the Aircrack-ng suite, dsniff, Kismet, Karma, Nmap, and TCPdump, among others.

The advantage of using a special tool rather than a laptop becomes evident when conducting audits in a remote location with limited access to alternating current (AC) power. Because audits are processor intensive, performing one can quickly drain a laptop's battery power. Devices like Pineapple are battery-powered but can run over an extended period of time—much longer than a basic laptop—in an isolated and remote location. Additionally, these tools come configured for pentesting and have many preconfigured attacks ready for launch.

Most of these tools do not come preconfigured with an antenna because the type and power of the antenna used will vary with the method and goals of the audit. These tools will typically come with high quality, low loss interfaces for attaching a suitable antenna.

Antennas

When pentesting, the type of antenna you use can have an impact. Wireless access points and network cards feature one of several types of wireless antennas, including the following:

- **Omnidirectional antennas**—Omnidirectional antennas broadcast RF equally in all directions and have 360-degree coverage. This type of antenna is installed by default on access points and network cards. In most cases, omnidirectional antennas are the correct choice for general office or home use.
- **Directional or semi-directional antennas**—These antennas have a narrower broadcast beam—typically around 180 degrees for both indoors and outdoors. Often, narrower broadcast beams of 45 to 90 degrees are aggregated and mounted on the same mast to

create high-capacity segments covering a 360-degree area. This is a common topology in mobile telecom networks. Directional antennas with tight focus are also used to create short, medium, and long-haul point-to-point wireless links. These, too, are popular in mobile telecom backhaul networks, where microwave radio links are used.

The antenna's power, or gain, is another important consideration. This will determine the coverage area, or RF footprint. The higher the gain, the further the RF footprint will spread. Normally, this is something you would curtail to prevent the RF footprint from spreading beyond the borders of the home or workplace. When pentesting, however, you might wish to test the network security from outside the business. Therefore, a high-gain directional antenna would be required to receive and eavesdrop from a distance. In addition, to measure RF, an RF meter is required. This could simply be a visual guide on a laptop's network interface. Or, it could be a specialized tool integrated into a hardware device, such as the Pineapple pentester tool.

Antennas are a very important part of RF transmission. They determine the RF coverage area and the distance signals travel. Choosing the correct antenna type is crucial to good wireless network design and security. It's also critical for network auditing. When choosing an antenna for use in network auditing, you'll typically want to use two types:

- A 16-decibel (dB) Yagi-style directional antenna for eavesdropping over long distances
- A general-purpose upright 10-Db omnidirectional car-aerial–style antenna for indoor use

With both antennas, the auditing program will be able to detect and receive RF signals in the majority of pentesting and network auditing scenarios.

RF signals that are above the noise floor determine the quality of the radio network. This is determined by the signal-to-noise ratio (SNR). If an antenna cannot receive a clean signal above the SNR, it will not be able to determine the signal from the noise. It will therefore be unable to make sense of the transmitted signal. This is how interference and background noise can prove detrimental to a wireless network. It also provides a simple, albeit crude, way to attack one.

Attack Tools and Techniques

Attackers use a wide variety of means and techniques to attack a network. The nature of these attacks is largely determined by the goal of the hackers (such as disruption of service, theft of data, or control of a client) and the opportunities (security vulnerabilities) of the target. This section explores some of the more common attack methods.

Radio Frequency Jamming

While both interference and jamming cause a decrease in the SNR and disrupt communications between transmitters and receivers, there is a difference between the two. Interference is unintentional. Common sources of interference are neighboring 802.11 networks that share the same frequency and channels. Because both are entitled to use the unlicensed spectrum, this is an unintentional disruption of transmission. In contrast, jamming is a deliberate disruption of the transmission. Jamming can be used to block or censor radio broadcasts, usually at a border region.

CHAPTER 9 | WLAN Auditing Tools

Wireless 802.11 networks are very vulnerable to both interference and jamming, making it very easy to launch a denial of service (DoS) attack on a radio network. In its simplest form, all that's required is a device that transmits continuously on a particular channel. Because Wi-Fi is a half-duplex technology—meaning each device on the network can transmit only when no one else is transmitting—this prevents anyone else from transmitting. In a half-duplex system, each device must listen to make sure it is clear to transmit. If one station floods the transmission channel, then no other station can transmit, causing a DoS.

Jamming of the unlicensed radio spectrum's 2.4 GHz and 5 GHz bands represents a major vulnerability. As enterprises shift toward a wireless infrastructure, it has been a cause for concern. After all, interference is a big enough problem without having to contend with the malicious hijacking of the airwaves. To counter this, network card vendors must ensure that their products cannot be configured to transmit continuously. This prevents radio transmitters from being the source of interference through failure or through deliberate misconfiguration to cause RF jamming.

However, specialized products are available that do allow you to flood a frequency channel. Although these tools are marketed specifically for wireless audits and testing, an attacker can easily use one as a frequency jammer if they wish. Furthermore, a crude but effective RF jammer for the unlicensed 2.4 GHz band can easily be constructed from old cordless phone circuit boards, which can be set to transmit continuously. Making a frequency jammer that can block the four main channels is a trivial task—one that can be achieved simply by following instructions available on the Internet.

 NOTE
The only caveat for the deauth technique is that the chipset and software must support packet injection.

Denial of Service

Flooding the RF spectrum is not the only way to jam a wireless 802.11 network. A more elegant way is to send deauthentication (deauth) packets to force access points to deauthenticate and drop connections. Sending a constant stream of these deauth packets with spoofed MAC addresses to access points will cause them to constantly deauthenticate the client connections.

Aircrack-ng supports packet injection. It also features the Airdeauth-ng tool, which causes the network card to hop between channels, finding all the access points on each frequency. It then constructs a stream of deauth packets aimed at each access point on each channel to force them to drop authentication sessions with their existing clients. Thankfully, deauth attacks on enterprise networks can now be easily recognized and blocked via WIPS. However, on networks that do not have the budget to install WIPS, they are a very real threat. A network protocol analyzer will quickly identify the attack, because it can easily recognize the constant stream of deauth packets. There will be little or no other legitimate network traffic.

Although not technically a DoS because all it actually does is request that the access point deauthenticate a session with a client, the fact that the script keeps injecting a stream of deauthenticate requests means that the access point in practice is constantly authenticating and then dropping the connection to the clients, which in effect is a DoS.

Hijacking Devices

As you have seen, it is a trivial task to introduce interference or deliberately jam the frequencies that wireless networks use. This is due to the half-duplex nature of radio communications. This weakness is difficult to mitigate. You have also seen how you can manipulate the behavior of wireless access points by using deauthentication (deauth) management frames to drop connections. However, these are not the only weaknesses.

Another attack first uses deauth packets and then takes advantage of how client devices attempt to reconnect. In most Windows clients, a device will automatically attempt to attach to any network to which it has previously been attached. What's more, it will favor the access point with the highest transmit signal. Hackers take advantage of this by creating a rogue access point (known as an evil twin) set with a high signal strength. They then force a deauthentication. In response, the client connects to the hacker's evil twin, which appears to be a known access point with a strong signal.

Aircrack-ng features a tool called Airbase-ng that attackers can use to convert a standard wireless network card into an access point. Airbase-ng is very useful for performing a client-side hack that will enable a MITM attack or compromise the client's privacy and confidentiality. The goal is to mimic an access point by faking the SSID and MAC address of one of the client's known (trusted) wireless networks. A client device will poll for these when attempting to associate with a network at connection time. The object is then to send a deauthenticate packet to the real access point, which instructs the access point to deauthenticate every client. When a client is deauthenticated—that is, when the connection is dropped—the client is forced into the authentication and association handshake process. When this happens, the client will by default attempt to attach to the access point with the highest power.

This same feature can easily be used when the client has attached to a network that has a cloaked SSID. When an access point with a disabled broadcast (SSID cloaked) has been used, the client will continuously poll for it regardless of where the client is—even if it's already connected to another access point. A hacker can listen for these polling attempts and then quickly configure their own high-power access point using that same SSID. This is to lure the client to automatically switch over to what it determines is a known access point but is in fact an evil twin.

The crucial step is to ensure that the evil twin has a more attractive signal. The attacker does this by boosting the network card's power output to the largest permissible setting. This is typically set but not physically limited to 27 decibel-milliwatts or 500 megawatts in the United States and Europe. Some countries allow much higher rates, however. For example, if, when you configure the card, you specify the region as Bolivia, then the card will allow transmission at 30 decibel-milliwatts at a full 1,000 megawatts of power. This should be enough to overpower most access points in the vicinity and prove to be the most attractive to any client device wishing to join the network.

After the client has authenticated/associated with the evil twin, it may find itself victim of a MITM attack, with the Ettercap tool used to intercept and analyze all the data. Alternatively, it could simply be left in limbo with no Internet or corporate network connection and no way of passing data through the evil twin.

Hijacking a Session

Another form of MITM attack is referred to as *session hijacking* or *session sidejacking*. It does just what the names suggest. The requirement for this form of user hijacking is that the attacker have visibility at the Transport Layer (Layer 4). Data encryption at Layer 2 or Layer 3 will mitigate this attack, making it unlikely to be effective on corporate networks. Encryption is rarely used in guest areas and hotspots, however, so this can be an effective attack in corporate visitor areas. It is certainly something to be wary of in cafés and hotels.

The vulnerability exploited with this attack is an inherent weakness in HTTP and the way web applications handle HTTP requests. Specifically, HTTP is not session-oriented. It does not remember a user from one command to the next. Therefore, each request or command is treated in isolation, as HTTP has no concept of a user session. While the lack of session continuity may seem like a flaw, it is actually a design consideration to improve throughput. After all, if user credentials were required to be sent with every transaction, it would waste bandwidth and hurt performance.

The way around this is to send a session ID instead of user credentials with every transaction. The session ID is created when the user logs on to the application. To secure the transaction, the web server requires encryption through HTTPS or SSL to ensure confidentiality of the user credentials. This prevents anyone eavesdropping on the wireless network from acquiring the username and password. When the logon process is complete and the user is authenticated by the web server or application, a session ID is created. This is used in all subsequent HTTP transactions as a user/session identifier to enable the application to track the user's activities and ensure that the user is who they claim to be.

The problem is that although HTTPS or SSL is used to perform the logon procedure, the application often reverts back to HTTP for efficient loading of page content and other non-confidential data. An eavesdropper on a wireless network can then see the session ID. It is then a trivial task to spoof the captured session ID and substitute it for the attacker's session ID in their HTTP requests to the same server. An easy-to-use tool for just this purpose is Tamper Data. This free Firefox extension enables the attacker to halt and modify HTTP requests (GET/POST) without the need to inject data using JavaScript or to repost webpages. By spoofing the session ID in their own HTTP requests, the attacker tricks the application into treating them as the user who is logged on. In other words, the attacker has hijacked the session, and will have the same rights and access as the session's rightful owner.

Using Transport Layer Security (TLS), a cryptographic protocol, helps mitigate this attack. It ensures authentication, confidentiality, and integrity of the data. This attack remains possible, however, because TLS or encryption is used only at logon and not across the entire site.

While performing an audit, an administrator should check that company websites and servers consistently enforce HTTPS across the entire site. They should also check that other common website vulnerabilities such as the following have been removed or mitigated:

- **Carriage return line feed (CRLF)**—With **carriage return line feed (CRLF)**, a common HTTP vulnerability, an HTTP packet may be split using a carriage return followed by a line feed (hence the name). By splitting a packet in two, with one packet containing legitimate header and protocol information, the attacker can pack a malicious payload into the second packet. HTTP response splitting, as it is also known, can lead to the hijacking of the client's sessions, web browser, and web server and to proxy cache poisoning.

- **SQL injection**—These types of attacks are very common on dynamic websites that front a database. SQL injection attacks can occur when input is not validated.
- **Cross-site scripting (XSS)**—Dynamic websites are vulnerable to XSS vulnerabilities. Attackers embed malicious code that executes when a user performs a specific action. For example, in June 2014, attackers embedded malicious XSS code into a tweet from TweetDeck. When TweetDeck users logged onto Twitter, it automatically retweeted the malicious code from the user's Twitter account. A similar attack on Twitter in 2010 opened up a pop-up window displaying a Japanese porn site when users hovered over the tweet. While these attacks weren't damaging, other XSS attacks steal user cookies, allowing attackers to impersonate users on some sites with weak security. The best protection against XSS attacks is input validation.
- **JavaScript injection**—Using JavaScript enables an attacker to modify existing information in web forms and input tags, as well as to rewrite cookies that are currently set in the browser. With JavaScript, the attacker can change any parameter within the cookie—for example, changing the authenticated setting from *false* to *true*. This would allow an attacker to bypass any authentication test and gain access without needing to authenticate. JavaScript is also an effective way to launch XSS attacks and inject malicious payloads.

> **NOTE**
>
> *Input validation* means that a blank field will only accept certain types of information or characters. If the site asks for a phone number, for example, it should only accept the correct number of characters for a phone number, and these should be numerals and possibly dashes.

These are just some of the most common website vulnerabilities that a site audit should test for. There are many more. The relevant point here is that attacks are always easier on wireless networks because it is easier to sniff the network with a protocol analyzer. At the same time, however, there are also many network tools available for both wired and wireless networks that can be used to troubleshoot events and issues.

Network Utilities

Network utilities are software programs and scripts designed to analyze or configure various aspects of computer networks. Many of these programs are Unix based and are run through command line interface (CLI) screens, but there are many commercial programs that package one or more for popular operating systems. The most commonly used network utilities are as follows:

- **Ping**—This is used to check for network connectivity and to determine whether a host can be reached over the network. Ping can also be used to measure the round-trip delay between hosts on a wired or wireless network.
- **Traceroute and tracert**—These Unix/Linux and Windows utilities, respectively, are used to trace the path taken between hosts on different subnets or networks. They list every router in the path between the sender and the recipient, along with the round-trip delay between stages and for the entire path.
- **Netstat**—This command-line utility is used to display network connections, routing tables, and network interface information—such as protocol performance—on a host computer.

- **Ifconfig and ipconfig**—These are Linux and Windows command-line utilities, respectively, for configuring and viewing network interface configurations.
- **InSSIDer**—This finds Wi-Fi networks that are in range and provides details about each one. It provides the SSID, the vendor make and model, the channel, the signal strength at the present time and over time, and the public name of the network, as well as what security is in place. InSSIDer is useful in an audit capacity for detecting interference from other neighboring networks as well as discovering dead spots in RF coverage.
- **Hotspot Shield**—This provides a lightweight but secure VPN connection. It uses HTTPS to encrypt all data passing through the secure tunnel. This tool is essential if you have no installed VPN client and you regularly use public hotspots.

IT professionals will likely use all these tools throughout their career. Many great resources are available that describe how they can be used. Of all those listed, the one most relevant to this text is Hotspot Shield. In most cases, enterprises do a capable job of protecting the corporate network. With the advent of ubiquitous Wi-Fi, IP mobility, and Bring Your Own Device (BYOD), the mobile user is often the weakest link in the chain. Users today have an expectation of connectivity, and far too many are far too trusting. The best defense against this is ongoing training and the use of tools such as Hotspot Shield. These are inexpensive but effective measures against ongoing threats. The old saying holds true: An ounce of prevention is worth a pound of cure.

CHAPTER SUMMARY

There are numerous open source and commercial tools to help administrators perform wireless audits. Discovery tools such as AirMagnet, Acrylic, or Xirrus Wi-Fi Inspector help verify what is actually in place (as opposed to what's on a diagram), showing the network's true status. The next stage is to check the network for known vulnerabilities by using vulnerability scanners such as Nessus. This software checks network clients against a national database of thousands of contemporary known OS and application vulnerabilities.

After vulnerabilities have been identified and removed, it's time to bring out the big guns: pentesters. Pentesters are fully developed frameworks of tools that enable auditors not only to verify vulnerabilities but also to ensure that clients are hardened against known exploits. Pentest tools such as Aircrack-ng and Metasploit provide the means to launch fully developed attacks on the network to prove its robustness. Remember to always get written permission from the network owner before conducting pentesting.

Auditing is not just about vulnerabilities and exploits—it also helps ensure that the network is well managed and controlled. The tools available for this function are varied according to network size and budget, with options to match all. Part of the management function is using the available tools to view and analyze protocols and traffic traversing the network. Tools such as Airshark provide visibility, data capture, and data manipulation, along with deep packet inspection. This allows auditors to visualize conversations on the network at the packet level. At the radio level, specialized hardware devices and antennas can monitor RF spectrum and signal levels. You can see how interference and jamming can disrupt a network and cause a DoS.

Network analyzers enable administrators to discover more-advanced attacks such as network-deauthentication attacks against access points. Similarly, network analyzers can be used to discover and track rogue access points, or evil twins. It is not just wireless vulnerabilities that require auditing, however. All general vulnerabilities must be discovered and removed, including common web server and application vulnerabilities such as XSS and JavaScript injection. Network utilities and freeware tools can aid administrators in testing general network connectivity and performance. For example, tools such as ping and traceroute can reveal and help resolve problems at the network-packet level. These are good indicators of connectivity and performance, especially when you're verifying the network key performance indicators (KPIs) of latency, packet loss, and jitter.

The downside of this is that in the hands of a skilled attacker, many of these same tools can be used to break into a network, steal data, or even hijack sessions. As is often the case, the tools are not inherently good or bad—that is left to the user's intention.

KEY CONCEPTS AND TERMS

Acrylic Wi-Fi
AirMagnet WiFi Analyzer
Brute-force attacks
Carriage return line feed (CRLF)
Completely Automated Public Turing Test to Tell Computers and Humans Apart (CAPTCHA)
Dictionary password crackers
Open share
OS fingerprinting
Password management system (PMS)
Penetration testing (pentesting)
Port scanning
Rainbow table
SNMP traps
Xirrus Wi-Fi Inspector

CHAPTER 9 ASSESSMENT

1. Network discovery tools are a set of tools made specifically for hackers.
   A. True
   B. False

2. Programs such as Kismet can perform which of the following functions?
   A. Wardriving WLAN discovery
   B. Rogue access point detection
   C. Low-cost intrusion detection system
   D. SSID decloaking
   E. All of the above

3. Which of the following most accurately describes penetration testing, or pentesting?
   A. To truly gauge security capabilities, pentesting should be done without notifying network administrators.
   B. Pentesting is illegal and immoral.
   C. Pentesting is an important aspect of identifying vulnerabilities and hardening defenses.
   D. Pentesting helps you discover Wi-Fi coverage gaps.

4. Which of the following describes brute-force attacks?
   A. They use torture to make people tell you what their passwords are.
   B. They are efficient ways to crack passwords, given the power of modern computers.
   C. They are fairly inefficient.
   D. They work well on most online portals.
   E. All of the above

5. Which of the following describes dictionary attacks?
   A. They take advantage of people's tendency to use actual words for passwords.
   B. They can be done offline and used against captured packets to test without detection.
   C. They can check millions of passwords per second.
   D. All of the above.

6. Password management systems are a great way to implement many strong passwords without having to remember them all.
   A. True
   B. False

7. Regular audits with a protocol analyzer can do which of the following?
   A. Prevent rogue access points from being established
   B. Eliminate potential viruses
   C. Help find and isolate misconfigured clients and rogue access points
   D. All of the above

8. SNMP traps are a common attack used by hackers to fool users.
   A. True
   B. False

9. Session hijacking works because client devices tend to connect to a known SSID with the highest signal strength.
   A. True
   B. False

10. Which of the following is *not* a vulnerability of HTTP carriage return line feed (CRLF) or HTTP response splitting?
    A. It allows attackers to put a malicious payload into the second packet.
    B. It allows attackers to break encryption in the second packet.
    C. It can lead to the hijacking of the client's sessions.
    D. It can lead to proxy cache poisoning.

WLAN and IP Network Risk Assessment

CHAPTER 10

OVER THE PAST FEW YEARS, there has been a huge influx into the workplace of wireless local area network (WLAN) technology and mobile devices. Even before the advent of Bring Your Own Device (BYOD), security departments considered it a risk to introduce new technologies and devices into the workplace. However, the desire of those in business to accept these new technologies and their advantages has outweighed these security fears. This has put the onus for mitigating the risk onto security departments.

Of course, before members of a team can properly mitigate the risk of new technologies or new technology policies, they must first fully understand the risk in the context of their specific environment. This chapter focuses on the risk-assessment procedure as applied to WLANs and Internet Protocol (IP) mobility.

Chapter 10 Topics

This chapter covers the following concepts and topics:

- What risk assessment is
- What activities are involved in IT security management
- What the security risk assessment stages are
- What a security audit is

Chapter 10 Goals

When you complete this chapter, you will be able to:

- Describe general risk categories
- Describe the methodology and legal implications of risk assessment
- Understand the risk-assessment process
- Describe the types of risk assessment

© Cherezoff/Shutterstock

- Describe the risk assessment stages
- Calculate risk based on input factors
- Describe the threat-analysis process
- Understand the risk-analysis process
- Describe the purpose of security audits

Risk Assessment

The National Institute of Technology Standards (NIST, 2012) defines *risk assessment* as "the process of identifying, estimating, and prioritizing risks to organizational operations (including mission, functions, image, reputation), organizational assets, individuals, other organizations, and the nation, resulting from the operation of an information system. Part of risk management incorporates threat and vulnerability analyses, and considers mitigations provided by security controls planned or in place."

Also according to NIST, the purpose of risk assessments is to inform decision makers and support risk responses by identifying:

i. relevant threats to organizations or threats directed through organizations against other organizations;
ii. vulnerabilities both internal and external to organizations;
iii. impact (i.e., harm) to organizations that may occur given the potential for threats exploiting vulnerabilities; and
iv. likelihood that harm will occur.

Risk assessments are performed less frequently than security audits, which you'll read about later. They are used to determine any change in requirements, technologies, or threats since the last risk assessment (assuming there was one). The actual assessment process includes the identification and analysis of the following:

- All assets related to the system or network
- Known threats that could affect the security of the system or network
- System or network vulnerabilities
- Potential impacts and levels of risk associated with threats
- Required measures to mitigate the vulnerability and threat

To determine the risk, you must know the system or network's vulnerabilities. These vulnerabilities can be categorized into three general groups:

- **Interception**—This applies to data that travels over the network. This data sometimes travels over connections or media, such as the Internet, that are out of the company's

control. Data can be intercepted, stolen, or modified, resulting in a loss of confidentiality and integrity.
- **Availability**—Systems, servers, and applications must be available for the purpose they serve under the **service level agreements (SLAs)** to which they must adhere. (An SLA is a document that identifies an expected level of performance.) Today, as IP mobility and remote mobile access become more pervasive and more workers are based outside the office, there is an even greater focus on providing SLA availability goals.
- **Access**—This refers to the points where remote users can join the network and where internal users can communicate with the outside world through email, instant messaging (IM), and all the other myriad services available on the Internet. These access points are the most vulnerable locations in a network.

Of course, identifying network vulnerabilities is only part of the assessment. You must also identify suitable controls. Fortunately, for most network vulnerabilities, there are recommended, tried-and-tested controls. These ensure that most vulnerabilities that are identified will be effectively mitigated. Some of the controls used to address existing vulnerabilities in the three categories listed earlier—interception, availability, and access—include the following:

- **Interception**—The most effective control for combating interception is encrypting data. The purpose of encryption is to devalue any intercepted data, because it requires an attacker to invest a lot more time and effort to decrypt it.
- **Availability**—The most important control for combating availability issues is the use of fault-tolerant and high-availability designs that eliminate single points of failure. This means no single fault can break a network. There will always be an alternative path, server, or application to provide uninterrupted service.
- **Access**—You control access through security devices such as firewalls, intrusion prevention systems, and Secure Sockets Layer/virtual private network (SSL/VPN) concentrators. An administrator implements these controls through network designs and configuration in demilitarized zones (DMZs) or virtual private network (VPN) access networks. Because these points in the network are effectively the gateways into and out of the network, a robust security policy should be in place to authenticate and authorize remote users. There should also be strict rules to filter incoming network protocols and activity.

> **NOTE**
> Reliability is closely related to availability. It is measured in the "five nines" (99.999 percent) for enterprise-class servers and infrastructure devices.

> **NOTE**
> Nessus is noninvasive. It simply checks for existing vulnerabilities. It doesn't test them, but it can still produce a lot of traffic. This can burden the application or web server, so it is always best to coordinate with server and network administrators when using it.

To identify and address all network vulnerabilities, network inventories, network diagrams, and network designs should be available as input into the process. Because there will be far more network vulnerabilities in a real-world scenario where web applications and web servers are open to the Internet, it is important to get an understanding of their requirements and protocols.

This is where gathering information and collaborating with application developers and server administrators becomes relevant. Holding interviews with key parties is the standard way to gather information, along with doing research on vendor websites. Running vulnerability scanners is also critical. A vulnerability scanner such as Nessus can check for thousands of known operating system (OS) vulnerabilities, as well as those of hosts, infrastructure devices, and servers. After you identify all the assets and their vulnerabilities, it is time to suggest or introduce appropriate security measures.

You should conduct security risk assessments at least every 3 years (although more often is better), as they are only snapshots of the time they were performed. This is just a best-practice suggestion but a security risk assessment should be done whenever new technologies or major changes are introduced into the system or network; it is best to undertake a new security risk assessment. An example of this would be recent trends in BYOD and **Bring Your Own Application (BYOA)**. (BYOA includes cloud-based applications such as Dropbox and Google Docs, for example.) These have introduced new vulnerabilities and risks that previous risk assessments did not identify. This scenario would entail performing a fresh, comprehensive assessment. This assessment should have the scope to redefine security policy and asset management as well as identify new vulnerabilities that come with mobile devices, such as differing OSs, applications, malware, and connectivity methods. Once a comprehensive network risk assessment is complete, you typically follow it with a security audit.

Risk Assessment on WLANs

Looking at risk assessment from a wireless perspective, it's critical to identify how a risk assessment on a WLAN differs from one done on a traditional wired network. This means identifying the risks associated with mobile devices in particular. The most common risks on WLANs in a business network are as follows:

- Exposure of critical information to wireless sniffers and eavesdroppers
- Leakage of critical information beyond the network boundaries on unsecure devices
- Theft of data
- Loss, theft, or hijacking of a mobile device
- Fraud caused by disruption, eavesdropping, and copying of data
- Viruses, worms, and Trojans

Although most of these risks also exist on wired and static networks, the nature of mobile and wireless networks greatly increases the probability of the first four items, which are all interrelated. Even so, the key elements necessary to mitigate risks associated with WLANs and mobile devices have not changed from the fundamentals of risk mitigation for wired networks. These elements include the following:

- Access control
- User authentication
- Data encryption
- Intrusion prevention

CHAPTER 10 | WLAN and IP Network Risk Assessment

- Antivirus and antimalware software
- Standards, guidelines, and policies
- Network-perimeter and Internet security
- Transmission security
- Application and web services

The steps taken to assess risk on a wireless network or on a single segment that uses portable mobile devices (laptops, mobiles, and tablets) are no different from the steps taken to perform a traditional risk assessment. Your efforts should be seamlessly incorporated into a single, all-encompassing risk-assessment procedure. To that end, this chapter examines a framework for IT security management that includes a risk-assessment and security-audit model for the entire network infrastructure, whether fixed or mobile.

Other Types of Risk Assessment

Not all risk assessments are comprehensive in scope or scale. In some cases, risk assessments can be limited in scope or depth. Examples include the following:

- **Preproduction assessment**—This type of assessment is comprehensive, but its scope is restricted to the production of the information system before it is rolled out. It concerns itself with the major functions and changes that the system under production will produce.
- **High-level assessment**—This type of assessment involves high-level risk analysis across the entire network, without a deep or fully detailed technical review. High-level assessments are typically applied during the planning phases of network or system development before the design process begins.

IT Security Management

IT security management involves several related activities. These may be performed as part of a cycle of repeating security processes, depending on the organization's size and complexity. Security practices and policies should be appropriate for the size and value of the network and its assets. They should be cost-effective and in proportion to the value of what you wish to protect. The first step, then, logically requires identifying the assets needing protection, as well as their relative value and importance to the company. Subsequently, you must identify the threats to those assets and the possible consequences of those threats.

Methodology

There are two methods for performing a risk assessment:

- **Quantitative assessment**—In this scenario, the assessment assigns a monetary value to the assets. The result is a quantitative statement on risk and the cost of impact.
- **Qualitative assessment**—This method uses a more subjective calculation of risk by assigning it a level—low, medium, or high—and a probability multiplier to determine the risk and impact level.

> **NOTE**
> Because risk is about uncertainty, it can be both negative and positive. Consequently, a risk assessment might well uncover instances of a positive risk, which could bring about a beneficial impact.

The qualitative approach is the most commonly used because it can be used for all types of risks and projects. Quantitative assessments, on the other hand, require specific data to support specific risks and projects so have a more specific use.

Legal Requirements

Laws and regulations require periodic risk assessments in many different enterprise environments. In the United States, both the Sarbanes–Oxley Act (SOX) and the Health Insurance Portability and Accountability Act (HIPAA) require security and control infrastructures to be put in place. Although compliance with these acts does not require specific methods or techniques, it does require that the company be able to prove to an independent auditor that it has effective and reasonable security in place.

Other Justifications for Risk Assessments

There are other rationales behind performing security risk assessments, such as justifying the cost of security. An effective risk assessment identifies threats to valuable company assets and raises awareness within management of the required costs to protect systems and services. Additionally, it can promote communications and productivity and break down interdepartmental barriers because collaboration is required across all the disciplines of the IT department and its business partners. Furthermore, this interdepartmental collaboration raises awareness of security and its requirements, and enables nontechnology departments to provide input into the process. Consequently, a department may revise its work practices or procedures to mitigate the risks it has become aware of thanks to the assessment. This is a form of self-auditing.

Assessing security risk is typically the initial step in security management. When you are assessing network assets (such as servers, hosts, infrastructure, and applications) for risk, the goal is to identify vulnerabilities, prioritize, and apply quantitative or qualitative values to possible impacts or consequences. The network auditors can then use this assessment as the basis for establishing a cost-effective security program, which they can present to the company's management.

Risk assessments vary in scope depending on the complexity of the network and the company assets. For instance, a small company with just one WLAN and a single Internet connection may follow the same framework but will not cover the scope or require the full diligence and rigorous investigation of an enterprise risk assessment. For the purpose of clarity and completeness, the following sections discuss a framework suitable for a company of any size.

Security Risk Assessment Stages

The activities undertaken when performing a risk assessment are as follows:

- Planning
- Information gathering

CHAPTER 10 | WLAN and IP Network Risk Assessment

- Risk analysis
- Identifying and implementing controls
- Monitoring

Before discussing the stages of a risk assessment, it's a good idea to review the following points:

- An *asset* is anything of value, such as people, property, intellectual property, or information. In essence, an asset is what you are trying to protect.
- A *threat* is anything that can damage or compromise an asset. In other words, a threat is what you are trying to protect against.
- A *vulnerability* is a weakness that makes a threat possible or even probable. A vulnerability can also be a gap in the protection measures against a threat.
- *Risk* is the combination of all three, related in the following way:

$$\text{Asset} \times \text{Threat} \times \text{Vulnerability} = \text{Risk}$$

In other words, risk is a function of a threat exploiting a vulnerability to damage an asset. Consequently, if there is no vulnerability, there is no risk. Similarly, if there is no threat, there is no risk. (This is a dangerous justification, however. You must take great care before dismissing a threat when there is a known and defined vulnerability.)

These are important definitions. Understanding the relationship between assets, threats, vulnerabilities, and risk is a crucial step in accurately assessing risk and ultimately securing assets.

FYI

By documenting the project scope, constraints, roles, responsibilities, and potential risks, you ensure that no one can later challenge the authorization or legality of the risk assessment. This is important because some of the methods used in vulnerability assessment and network discovery are similar to or the same as techniques used by hackers. Getting permission up front is critical. If you were to disable a mission-critical server by running a vulnerability scanner, and you did not have documented prior approval from stakeholders, the consequences could be dire. It is vital to document the proposed tests and activities, as well as any potential risks associated with those tests and activities, and to get signoff before starting. In addition, it is often important from a political standpoint to seek the assistance of relevant stakeholders—such as server and network administrators—before running any process or activity that could be viewed as hacking in the event of even a coincidental service failure.

Planning

Planning is an essential stage. It occurs before any security risk assessment can take place. Even when doing small security audits with limited scope, you should complete the planning stage in its entirety. During this stage, stakeholders, roles, and responsibilities are determined. Even more vital, it is when permission is sought, granted, and documented.

Information Gathering

The objective of the information gathering stage is not just to collect information but also to gain an understanding of the existing systems, network, and environment. This knowledge, gleaned through interviews, group discussions, and other research, will enable you to identify risks, as well as controls for those risks.

The information typically gathered during this stage includes the following:

- Security policy and objectives
- System and network architecture
- As-built designs and diagrams
- Physical asset inventories
- Network protocols and services
- Access control procedures
- Firewall deployment and policy
- Identification and authorization mechanisms or systems
- Documented policies and guidelines

There are two general types of information gathering: a general controls review (GCR) and a system review.

General Controls Review

A **general controls review (GCR)** identifies threats existing in the general security processes. GCRs are concerned with the high-level security of an existing system. The review involves collecting data to check whether physical security measures match policy and documentation. Physical access control can also be visibly noted, and policy and procedures checked for compliance and efficiency through group meetings or multilevel interviews.

An auditor carries out a GCR via documentation reviews, site visits at data centers and computer rooms, and stakeholder interviews with staff at different operational levels. These multilevel interviews can prove particularly effective in revealing a failure to implement security policy at the hands-on, operational level, as well as a lack of security awareness.

A general control review is concerned with high-level functions such as the following:

- Physical security
- Change-management control
- Access control/authentication and authorization
- Awareness programs and training
- Roles and responsibilities
- Security policy, guidelines, standards, work practices, and procedures

The GCR is used to check that these features are implemented properly, not just stated on paper in policies and guidelines. If access to computer resources and applications is restricted to roles and authorized persons with privileged access, then that should be demonstrable.

System Review

Unlike a GCR, a system review seeks to identify vulnerabilities at the network and application levels. This type of review focuses its attention on the operating system, administration, and monitoring tools used by administrators of the different platforms. Typically, a system review will look to server logs and system files for baseline vulnerabilities, but it will also consider other information sources such as the following:

- Access control files
- Running services and processes
- Configuration settings
- Security patch levels
- Encryption and authentication tools
- Network management tools
- Logging or intrusion detection tools
- Vulnerability scanners

The review team will attempt to spot abnormal behavior, such as repeated unsuccessful login attempts on secure accounts like administrator or root. In addition, system configuration files and network configurations can be checked against documentation for compliance. Noncompliance can be an indicator of poor change-control procedures, unauthorized configuration changes, or improvised emergency configuration changes.

During a system review, it is often worthwhile to use specialized tools or scripts to collect all the data from the individual systems in the network. These can automate and greatly reduce the time taken to collect configuration files manually. After all the system data has been reviewed, the next stage is the risk analysis.

Risk Analysis

This is the stage where the security team uses risk-analysis techniques to determine the value of assets and identify any associated risks. Risk analysis is a component of risk assessment, a single-process stage. It does involve several subprocesses, however, including:

- Asset identification and valuation
- Threat analysis
- Vulnerability analysis
- Asset, threat, and vulnerability mapping
- Impact and probability assessment
- Risk results analysis

Asset Identification and Valuation

A risk assessor must identify all the assets that fall within the scope of the risk assessment. These assets can be physical, real objects, such as servers, hardware, or network devices, or intangible assets, such as information, services, or reputation.

Classification is an important part of identification. An asset should be classified by category, such as process, application, server, router, or information. (These categories will differ depending on the type of system or network under assessment.)

The purpose here is to show the importance and relevance of the asset to the system under assessment. The assessor can then place a value on the asset in terms of its importance. This should take the form of an inventory checklist of assets and their corresponding values in terms of the following:

- Tangible values, such as the cost of replacing physical devices
- Intangible values, such as reputation or lost data
- Information values
- The name and type of information
- Information flows, whether incoming or outgoing
- The physical location of physical assets
- Software installed on servers
- Indicators showing the importance or value of the asset
- Indicators of services supplied by the asset
- Values assigned to each individual asset

Threat Analysis

When the asset inventory checklist is complete, it's time to start the threat analysis, in which each asset is analyzed to identify any current threats. A *threat* is anything that can exploit a vulnerability, whether accidentally or maliciously, and disrupt or destroy an asset. Sources of threats include human error, hacking, industrial espionage, disgruntled employees, theft, malicious actions, and environmental phenomena.

Threats can be classified into one of three general categories:

- **Social threats**—These threats are related to human actions. They can be intentional or unintentional.
- **Technical threats**—These threats result from technical issues or faults.
- **Environmental threats**—These threats are the result of environmental elements such as storms, floods, fires, and power outages.

The purpose of the threat analysis is to identify and document the threats to each asset. It is not just a case of listing each possible threat, however. You must also indicate the likelihood that the threat may occur and the threat's potential to harm or destroy the asset.

You can identify and evaluate environmental threats using your knowledge about the history of the climate in the geographic area in question. Similarly, historical records will reveal the likelihood of main power supply outages. You can evaluate social threats against human actions such as lack of training or negligence. Again, knowledge of history is a way to qualify these types of threats. In contrast, technical threats can be more specifically identified using a vulnerability scanner. After all, a risk cannot exist without a vulnerability.

Vulnerability Assessment

A *vulnerability* is a weakness or gap in the defenses of an asset that makes it a potential target for exploitation by a threat. The vulnerability could be any number of things—for example, a missing security patch, which would allow an asset to be compromised. It could also be poor security awareness among employees and management or environmental causes such as a poor regional or national electrical supply. Vulnerability analysis is the task of identifying, evaluating, and documenting the existence of these possible asset vulnerabilities.

System or network vulnerability is measured in terms of accessibility and the corresponding number of authorized users. Therefore, a system or closed network with strictly limited authorized users and no external connection will have less inherent vulnerability than a web server DMZ that is open to the public Internet and accessed by tens of thousands of users on an hourly basis. Accessibility and the number of authorized users make systems or networks vulnerable to threat by increasing the threat landscape and the possible number and variety of threat sources and vectors.

Each vulnerability must be considered and evaluated by applying a level of degree of importance, such as low, medium, or high. Mission-critical assets and their vulnerabilities should be identified first. With technical vulnerabilities, tools such as vulnerability scanners are used to identify missing patches or gaps in a system or network's security. A vulnerability scanner uses databases containing thousands of known vulnerabilities to check a server or router for existing problems. Scanners such as OpenVAS 7.0, an open source program from the Nessus Project, are updated daily and can test for more than 35,000 vulnerabilities.

TIP

Having the server administrator present when an asset is checked is a good precautionary strategy. Although vulnerability testing is nonintrusive, it is good practice to involve those responsible for the asset just to be sure nothing goes wrong.

Scanning such a large number of system requests on a server or host can cause it considerable stress, leading to a performance drop as resources are consumed. Therefore, it is best to have the server administrator present when the asset is being checked so they can monitor the server's health during testing.

Running a vulnerability scanner, such as Nessus, Saint, or OpenVAS, will detect many vulnerabilities in operating systems, applications, web servers, and services, including open ports and missing OS and security patches. Not all detected vulnerabilities are genuine, however. Indeed, some may be false positives—for example, when the scanner highlights a known vulnerability that doesn't apply to the device being scanned. There are several types of false positives. Typically, these are caused by one of the following:

- A response header that is hidden or suppressed
- A header rewritten by a firewall
- A version update that is loaded but not active (version updates often activate on a server reboot)
- Failed or prematurely terminated version updates

Researching and evaluating these false positives can waste time. Worse, they can result in testers skipping other vulnerabilities, which may or may not also be false positives.

> **Technical TIP**
>
> It is good practice to tune the vulnerability scanner to the environment being tested and to perform a detailed analysis of the results. To confirm any vulnerabilities discovered, it's best to use secondary tools such as Nmap or something similar.

There is also the real potential for false negatives—that is, when the scanner doesn't highlight a vulnerability, leaving the security flaw still in place. Such false negatives can be very dangerous. Generally speaking, it's worth dealing with false positives to reduce as much as possible the occurrence of false negatives, which are essentially time bombs in the network. Therefore, the accuracy of the vulnerability scanner is of paramount concern.

Vulnerability scanners typically use **version analysis**, in which the scanner sends out requests to the target system and, upon receiving a response, analyzes the headers for the version details. After the scanner has ascertained the hardware or OS version, for example, it assumes that the vulnerabilities known to be associated with that hardware or OS version will be present, and lists them. This can produce a quick and impressive list of vulnerabilities that do not actually exist and, at the same time, miss real vulnerabilities.

The alternative method is to employ a vulnerability scanner that uses **behavior analysis**. Behavior analysis relies not on version data, but on how the system responds to requests, the aim being to find unexpected responses to a query. These unexpected responses enable the scanner to accurately identify that a vulnerability is present. Behavior analysis takes longer, but is more thorough and accurate.

You can quickly rectify most identified vulnerabilities, but you should take care to gain permission for any changes to the system through the proper change-control channels. Again, you must never make changes without proper authority, even if the system is open to severe threat. Permission must always be granted, and the relevant change control process followed.

Asset, Threat, and Vulnerability Mapping

Asset, threat, and vulnerability mapping is the process of documenting or pairing asset vulnerabilities with any potential threats that could expose those vulnerabilities. If you recall the relationship between assets, vulnerability, threat, and risk (Asset × Vulnerability × Threat = Risk), then you can see the purpose of asset, threat, and vulnerability mapping. It reveals that risk is present only if there is both a threat and a vulnerability. By performing this mapping, you can greatly reduce the list of potential risks that require risk results analysis, saving a lot of time and unnecessary effort.

NIST (2012) offers the following guidance on mapping: "Threat-vulnerability pairing (i.e., establishing a one-to-one relationship between threats and vulnerabilities) may be undesirable when assessing likelihood at the mission/business function level, and in many cases, can be problematic even at the information system level due to the potentially large number of threats and vulnerabilities. This approach typically drives the level of detail in identifying threat events and vulnerabilities, rather than allowing organizations to make effective use of threat information and/or to identify threats at a level of detail that is meaningful."

In other words, there may be times when a detailed mapping is too resource-intensive, especially mapping that gets in the way of mitigating a vulnerability or threat. Remember, the object is to reduce the risk, not to provide documentation for its own sake.

Impact and Probability Assessment

As you might guess, an impact and probability assessment is a combination of the following:

> **NOTE**
> When a risk assessor performs an impact and probability assessment, they must make assumptions regarding the overall estimation of risk. The reasoning behind these assumptions should be clearly defined and documented.

- **Impact assessment**—The purpose of an impact assessment is to determine the consequences (including the degree of harm) of a threat exploiting a vulnerability on an asset. Unfortunately, however, due to budgetary, technical, or time constraints, it might not be possible to address the impact of all threats. Therefore, the emphasis should be on addressing only those assets, threats, and vulnerabilities with the most severe potential impact. The impact might be on any number of things, such as availability of service, loss of reputation, profit loss, increased costs, or drops in performance. The more severe the potential impact, the greater the risk; therefore, these items should be addressed first.
- **Probability assessment**—Also called a likelihood assessment, a probability assessment is used to identify greater or lesser threats based on a score rather than a mathematic probability. The probability assessment expresses its results in terms of high, medium, or low based on the ability or motivation of the threat source, the nature of the vulnerability, and the effectiveness of security controls.

Risk Results Analysis

Risk results analysis involves analyzing results and presenting them using any of the following methods:

- Quantitative and qualitative methods
- A risk map
- A matrix approach

Qualitative methods rely on an assessor's subjective assumptions, which are based on experience, research, history, and judgment. They are descriptive and are expressed in words and rankings of severity for comparison. Qualitative methods rely on the categorization of risks and a subjective ranking system, whether that system uses terms like "low," "medium," and "high" or a number-ranking system (for example, 1 to 10) to convey the degree of importance. Assessors typically use qualitative methods when doing high-level assessments prior to a planning phase. Qualitative methods are also used extensively in repeating enterprise risk assessments where there is neither the time nor the budget to use fully quantitative methods.

In contrast, quantitative methods are expressions of financial worth. They are usually presented in the following three terms:

- **Single loss expectancy (SLE)**—The **single loss expectancy (SLE)** is the expected monetary cost of the occurrence of a risk on an asset.

- **Annual rate of occurrence (ARO)**—The **annual rate of occurrence (ARO)** is the probability that a risk will occur in a particular year. It factors in the number of times it will occur in a year if no action is taken.
- **Annualized loss expectancy (ALE)**—The **annualized loss expectancy (ALE)** is the product of the ARO and the SLE.

Quantitative methods require more time and greater effort on the part of the assessor because much of the detailed information is not readily available—either because it's speculative or because it's tightly controlled. In addition, collecting this information results in considerable operational overhead. Because of this, a qualitative approach is used for the majority of the report, with some quantitative methods employed for critical systems or where a large financial investment will be required to mitigate the risk.

A risk map is a way of graphically displaying assets and their risk levels, with the impact on the y-axis and likelihood on the x-axis. Both axes are measured from low to high, which means an asset plotted in the bottom-left corner of the map is a low-impact/low-probability risk. In contrast, one in the top-right corner is a high-impact/high-probability risk. Risk maps are good tools for visualizing the relationship and severity of risk within the scope of the assessment.

A matrix approach is a way to display the results of a qualitative assessment. A table or grid is drawn for each risk, with columns for impact, likelihood, and risk level (Impact × Likelihood); rows for the individual risk categories of confidentiality, integrity, and availability; and a final row for the overall assessment. **FIGURE 10-1** shows an example of a matrix approach.

Identifying and Implementing Controls

After the completion of the risk-analysis stage, it's time to start selecting and implementing appropriate controls, or safeguards, to mitigate the risk for each asset, threat, and vulnerability mapping. Of course, not all risk can be removed. That means there will be times when you must opt to avoid risk, reduce risk to a tolerable level, transfer risk (if possible), or accept risk. Which of these you choose depends on the asset value and the risk culture within the organization. Some companies are risk averse, whereas others live happily with high levels of risk. The purpose here is to find controls that, at a minimum, match the levels of risk that management is willing to tolerate.

In general terms, controls can be classified into three types:

- **Technical controls**—These include access controls, antivirus software, firewalls, and so on.
- **Administrative or management controls**—These include security planning and implementing rules.

FIGURE 10-1

Analysis of risks on unencrypted data on a WLAN.

Risk Category	Impact	Likelihood	Risk Level
Loss of confidentiality	4	3	12
Loss of data integrity	3	2	6
Loss of availability	2	2	4
Overall	3	2	6

- **Operational controls**—These include security personnel, data backups, contingency plans, and system maintenance.

Choosing suitable risk controls requires technical knowledge and skill. The cost of managing risk must be appropriate for the risk exposure.

Monitoring

Risk assessment is a repetitive process, and each assessment represents a snapshot in time. Therefore, the results should be properly documented because they will be the baseline for future reassessments of the environment.

As you have seen with wireless and IP mobility, technology in the enterprise can change quickly. New devices and technologies have quickly become pervasive in the network environment. Reassessments are necessary to address these new technologies. Therefore, having a solid risk assessment baseline is crucial. It can serve as the foundation for introducing new protocols and hardware. This allows the seamless introduction of change without requiring a full, ground-up assessment. Additionally, documenting and monitoring the existing organization's systems and infrastructure by means of the risk assessment also facilitates conducting a security audit.

Security Audits

A security audit follows many of the same steps as a risk assessment. The difference is that a security audit looks for proof that best practices and procedures were followed. In many cases, these audits are internal. However, the same person or team who performed the risk assessment should not carry out a security audit.

An auditor conducting a security audit for WLAN security vulnerabilities should look at the following points (in addition to those items examined on a wired network):

- The potential for unauthorized access to the WLAN
- Radio frequency (RF) leakage beyond the physical perimeter
- The use of weak encryption protocols such as WEP or WPA
- The existence of rogue access points
- If preshared keys are used, how widely the preshared key is known (or the method through which it is shared)
- The knowledge, training, and awareness of security personnel and staff
- The wireless network's conformity to the company's security policy

Security risk assessments and security audits occur at different times in the security management cycle. A security risk assessment is conducted at the beginning of the cycle. It is the means of identifying the assets to be secured and the possible vulnerabilities of those assets. These will determine the security tools or measures required. A security audit, on the other hand, is a repetitive checking process that occurs throughout the cycle. During a security audit, the security measures recommended in the risk assessment can be checked for effectiveness and compliance with the security policy.

Generally speaking, a security audit's purpose is to ensure that the organization adheres to its own security policy. But a company with a poorly thought out and limited security policy can demonstrate a sterling record of compliance if auditors focus simply on checking the box. This may go on for years, with weaknesses exposed only when a breach occurs. Inevitably, the forensic (post-breach) assessment will not only reveal the root cause of the breach but also point out severe lapses in the risk-analysis and risk-mitigation processes.

CHAPTER SUMMARY

Given the complexity of both networks and organizations, the ongoing (and increasing) number and veracity of external threats, and the high cost of a breach, performing comprehensive risk assessments is a must for all organizations.

Regardless of the organization's size, it's imperative to act as one's own worst enemy by diligently and honestly looking at vulnerabilities and potential threats. It's not always a pleasant task to point out areas where your own team may have left a vulnerability open, but it's much better than having a bad guy find it for you.

Once completed, you should follow up the risk assessment with security audits that occur on a regular basis to ensure that mitigating actions and controls have been put in place and are being followed.

Both network infrastructures and the modern threat landscape are very dynamic. For this reason, risk assessments and audits have limited shelf lives. Both should be an ongoing part of standard operations. This is one area where complacency kills.

KEY CONCEPTS AND TERMS

Annualized loss expectancy (ALE)
Annual rate of occurrence (ARO)
Behavior analysis
Bring Your Own Application (BYOA)
General controls review (GCR)
Service level agreements (SLAs)
Single loss expectancy (SLE)
Version analysis

CHAPTER 10 ASSESSMENT

1. Risk assessment on wireless networks is significantly different from risk assessment on wired networks.
   A. True
   B. False

2. Vulnerabilities can be categorized into which of the following three groups?
   A. Data, client, and people
   B. Interception, availability, and access
   C. Compliance, assessment, and audit
   D. Encryption, authentication, and RF power

3. Common risks on WLANs in a business network include which of the following?
   A. Exposure of critical information to wireless sniffers and eavesdroppers
   B. Data leakage of critical information
   C. Theft of data
   D. Loss, theft, or hijacking of a mobile device
   E. Fraud caused by disruption, eavesdropping, or the copying of data
   F. All of the above

4. Which of the following describes preproduction vulnerability assessments?
   A. The manufacturer performs them so you don't have to.
   B. They are the only assessments needed once the network is deployed.
   C. They are performed on new solutions/devices before these are added to the production network.
   D. They are a waste of time.

5. Risk can be calculated as which of the following?
   A. Asset + Vulnerability + Threat
   B. Asset × Vulnerability × Threat
   C. Asset × Threat – Vulnerability
   D. (Asset + Vulnerability)/Threat

6. Vulnerability analysis is the task of identifying, evaluating, and documenting the existence of possible asset vulnerabilities.
   A. True
   B. False

7. Risk analysis results can be calculated using both quantitative and qualitative methods.
   A. True
   B. False

8. The implementation of barriers, hardening, and monitoring are examples of which of the following?
   A. Security controls
   B. Protocols
   C. Security regulations
   D. Assessment techniques
   E. All of the above

9. Which of the following is *not* a point of emphasis when auditing security on a wireless network?
   A. RF leakage beyond the physical perimeter
   B. The use of weak encryption protocols such as WEP or WPA
   C. The completeness of the existing security policy
   D. The wireless network's conformity to the company's security policy

10. The annualized loss expectancy (ALE) is which of the following?
    A. The sum of the ARO and the SLE, which indicates the actual value of loss by combining the likelihood of a security event with the average cost of a security event
    B. The sum of the ARO and the SLE, which indicates the likely loss for a specific year
    C. The product of the ARO and the SLE, which indicates the likely loss for a specific year
    D. The product of the ARO and the SLE, which indicates the actual value of loss by combining the likelihood of a security event with the average cost of a security event

PART III

Mobile Security

CHAPTER 11	Mobile Communication Security Challenges **241**
CHAPTER 12	Mobile Device Security Models **259**
CHAPTER 13	Mobile Wireless Attacks and Remediation **279**
CHAPTER 14	Fingerprinting Mobile Devices 297
CHAPTER 15	Mobile Malware and Application-Based Threats 317

Mobile Communication Security Challenges

CHAPTER 11

IT'S BEEN MORE THAN 15 YEARS since the first recognized occurrence of malware that specifically targeted a mobile phone in 2004. The malware, called Cabir, was released, not into the wild, but to antivirus software developers. This was primarily as a proof of concept that mobile phone operating systems such as Symbian were not invulnerable to the malware that plagued PCs.

Despite this head start, little was done to address this issue. By 2012, malware had become a significant problem. Not surprisingly, the rapid growth in the last several years of malware targeting one or another smartphone operating system (OS) has matched the huge growth in the use of smartphones and tablets. Although malware developers have focused especially on the growth in market share of Android OS devices, that doesn't mean the operating systems and phones made by other smartphone manufacturers are secure. Apple iOS and Windows Phone are also targeted, making them susceptible to malware and other mobile phone specific threats that now pervade the mobile environment. This chapter looks at the vulnerabilities of each of the major mobile OSs as well as other issues that increase risk.

Chapter 11 Topics

This chapter covers the following concepts and topics:

- What general threats and vulnerabilities exist for mobile phones
- What exploits, tools, and techniques exist for mobile phones
- What security challenges exist with Google Android smartphones
- What security challenges exist with Apple iOS smartphones
- What security challenges exist with Windows Phone smartphones

Chapter 11 Goals

When you complete this chapter, you will be able to:

- Describe the security challenges specific to smartphones and tablets
- Describe the vulnerabilities and exploits of mobile operating systems

© Cherezoff/Shutterstock

- Explain why the Android model is prone to exploits
- Describe the Android security architecture
- Describe the Apple iOS security architecture
- Describe the Windows Phone security architecture
- Understand how open or restricted access to applications affects user security

Mobile Phone Threats and Vulnerabilities

Back in 2012, when the smartphone market was booming, it was believed that the nascent technology was under severe threat from malware. This at the time seemed to be a real threat as not only did OS vendors have to contend with more than merely identifying and mitigating the emerging threat of malware, they also had to try and educate their user base. First, patches and fixes had to be made available for a whole array of devices and OS versions. Second, they may have had to address a lack of security awareness among end users. Unaware of the risks involved, or simply not knowledgeable about the need for accepting software updates, a significant number of end users did not regularly download and install available security patches. Consequently, cybercriminals found attacks on mobile phones an attractive and potentially lucrative proposition.

Attacks on mobile phones were growing each month as cybercriminals shifted their focus from PCs to mobile devices. For instance, in 2013, Sophos, a security firm, discovered 1,000 malware samples in the wild that targeted Android devices each day. By 2014, that figure had risen to 2,000 samples per day. This acceleration is consistent with other OSs. Such was the burgeoning growth and diversity in malware targeting the Android OS that it was generally believed to be unacceptably vulnerable and therefore unfit for corporate use. However, as we now know due to hindsight, the threat of malware was at the time greatly exaggerated because there is a huge difference between detected and viable malware. Consequently, despite the numerous alarms coming from the software security firms, the actual real-world successful compromises of Android OS devices were exceedingly rare. This was down to several things, but the most important was that the Android OS was designed to be highly secure and isolated application runtimes in their own virtual **sandbox**. What this means is that even if the user downloaded malware, the application would run in complete isolation from other apps or system resources. Indeed, the user had to take affirmative action to allow the malware app to access resources so the issue was not with the technology; it was with the user education.

This is a recurring insight with IT security because we can look back over the last few years and see that as IT technologies mature, like the smartphone and wireless, the shift has been away from identifying and mitigating technological threats and vulnerabilities to surfacing and rectifying flaws in user processes and procedures.

CHAPTER 11 | Mobile Communication Security Challenges

For example, back in 2014, malware was not the only security challenge. Cybercriminals also focused on other attack vectors for mobile phones, such as OS attacks, side-loaded mobile applications (those downloaded from unauthorized third-party sites), and communication attacks. Today, however, we can see that the most serious security issues are more user based. Indeed, many security practitioners believe the most important vulnerabilities are in data leakage, outdated devices, as well as social engineering. These are general IT security issues not specific mobile or wireless technology vulnerabilities.

Mobile OSs are inherently secure by design; the security issues that arise come from user activity such as rooting Android or jailbreaking iOS. In addition, due to a lack of awareness on the part of end users, system updates often do not get installed (exposing the device to exploits), mobile browsing vulnerabilities exist, and data is not stored in a secure manner. Furthermore, interfacing with other trusted devices such as a PC through a USB or Bluetooth connection brings additional threats, such as USB exploits and USB hijacks. Even new OS features can introduce further attack vectors. For example, a single swipe with a near-field communication (NFC) tag can reveal financial information stored on the device.

In addition to the fact that large numbers of people now use smartphones, cybercriminals are attracted by the inherent weaknesses of user-compromised mobile devices. Vulnerabilities can be introduced as a result of poor technical controls, lack of user awareness, and poor practices. As with any Wi-Fi–enabled device, data communications are an obvious vulnerability, but there are also major security concerns with side-loaded applications that are not approved, certified, or verified by the OS vendor. Moreover, it is not just unsecure application software that can prove to be a vulnerability.

In addition to OS and application vulnerabilities, the logon and authentication of the end user can be problematic as well. Although mobile phones often have security measures to enable the use of passwords, personal identification numbers (PINs), or even biometric tools such as fingerprint readers, end users often disable them or enable only the most basic form such as the four-digit PIN. Using a simple PIN rather than something more complicated makes the phone far more convenient to use, but it also renders the device less secure if it is lost or stolen. In such cases, the person who finds the phone may gain access to the data stored on the device. The unsecure storage of confidential data is therefore another potential weakness.

Something else to consider is the size of smartphones. Although this has been mitigated somewhat by the growth in screen size, smartphones are still small relative to PCs and laptops. Hackers often take advantage of this fact in their attempts at phishing. Because the full URL of a website cannot be seen clearly on a smartphone screen, it makes them particularly vulnerable to phishing attacks.

>  **NOTE**
>
> One could argue that only the most security-conscience users bother to match the URL of a link to the name presented in the text of an email message—by using a mouse-over function or by right-clicking the link—before clicking on it. In this view, the risk from phishing on a smartphone is no worse than on PCs and laptops.

These vulnerabilities are a concern for all mobile smartphones and tablets regardless of OS. However, as mobile phones have become pervasive items for personal and business use—with penetration percentage rates reaching the high 90s and above in some countries—users' security awareness has become a major issue. As a result, there has been much debate in recent years regarding the wisdom of people using their smartphones for both business and pleasure. Unfortunately for security teams, the perceived productivity and employee-satisfaction gains have trumped the legitimate concerns about the increase in risk that these devices and their inherent vulnerabilities represent. Consequently, cybercriminals around the world—particularly in Russia, Eastern Europe, North Korea, Iran, and China—have launched massive attacks targeting smartphones (particularly Android phones) based on the dual promise of exploiting users and finding new pathways into very lucrative corporate networks.

Exploits, Tools, and Techniques

Every mobile phone—even those featuring the early, yet secure, Symbian OS—is vulnerable to basic forms of cyberattack. The early threats and vulnerabilities that come about from communication channels such as Short Message Service (SMS), Multimedia Message Service (MMS), Wi-Fi, and Bluetooth have always been a problem. With the advent of the smartphone, the scope of the risk has grown to encompass all forms of Internet threats including browser attacks, OS attacks such as remote jailbreaking (a method of circumventing provider-based security), **Remote Access Trojans (RATs)**, rootkits, and the myriad types of malicious software already in the wild for PCs.

Mobile phone vulnerabilities and exploits can be categorized as shown in **TABLE 11-1**.

Although Wi-Fi and Bluetooth have real-world vulnerabilities, these vulnerabilities are not specific to smartphones or tablets. That is, they apply to any mobile device or Wi-Fi–enabled PC or router. What *is* specific is the way cybercriminals have developed smartphone-specific OS malware.

The vast majority of malware attacks developed between 2012 and 2014 targeted the Android operating system. This is not only due to Android's popularity, but also because of its open security model. This open approach was in stark contrast to iOS and Windows Mobile that restricted users to only download and use approved applications from the respective app stores. Google's open approach left it more vulnerable to malware. However, malware was not the only thing showing explosive growth in the Android ecosystem. Another threat, **potentially unwanted applications (PUAs)**, was also thriving.

Developers create PUAs in an attempt to monetize their applications through connections to aggressive third-party advertising networks. These are sometimes referred to as mobile adware, or **madware**. On behalf of these third-party advertisers, the madware collects location information and tracks the device, perhaps even harvesting browser histories and contacts. Although PUAs are not strictly malware, their categorization is becoming a gray area because they introduce security risks along with the sometimes-unwanted third-party advertising. Some examples of Android PUAs are AirPush, Adwo, Dowgin, Kuguo, and Wapsx. The use of PUAs was growing exponentially back in 2014 within the Android realm. However, robust regulations combined with increasing user awareness of privacy and the value of their personal data have curtailed this once lucrative market. PUAs are still prevalent but nowhere near as intrusive as they once were and only a fraction, however, compared with the vast ocean of Android-specific malware.

Google Android Security Challenges

The same year that Apple introduced the iPhone (2007), another technology giant, Google, announced its interest in the mobile phone market through the acquisition of a startup

TABLE 11-1 Categories of mobile phone vulnerabilities and exploits.

VULNERABILITY CATEGORY	EXAMPLE OF EXPLOIT	DESCRIPTION OF EXPLOIT
Surveillance vulnerabilities	Audio attack	This involves switching on the microphone to listen in on conversations.
	Camera attack	This involves hijacking the camera to monitor the user or the user's surroundings.
	Location snooping	This involves the activation of Internet Protocol (IP)/browser tracking to monitor location. This is a common malware trick to gain advertising revenue.
	Call logs	This involves recording recent calls and messages, which can be read and/or stolen.
	Global Positioning System (GPS) tracking	This involves the activation of another location and tracking port to monitor location. This can be very accurate.
Financial vulnerabilities	Stealing transaction codes	This technique is commonly used for man-in-the-middle attacks against online banking sites.
	Stealing account numbers	This is possible when the phone is used as a data repository or a mobile wallet with an unsecure data store.
	Making expensive calls	This involves bypassing security measures to make calls, which are then charged to the user's account.
	Sending premium-rate SMS messages	This involves using a mobile handset to pay for services and products. This is a common way for cybercriminals to monetize their attacks.
	Extortion via ransomware	This is a method of extortion where malware is placed on a phone that prevents the phone from being used until a ransom is received. This is another popular method for turning nefarious cyberskills into cash.
Botnet activity	Participating in distributed denial of service (DDoS) attacks	This involves hijacking the phone to participate in mass attacks on a third-party network—for example, by sending out Domain Name System (DNS) or Network Time Protocol (NTP) requests.
	Sending premium-rate SMS messages	Again, this is a way to make money at the owner's expense.
Data theft	Communications	Emails and SMS messages are all open to theft.
	International Mobile Station Equipment Identity (IMEI) number theft	The IMEI number uniquely identifies the mobile phone and can be used for a number of purposes, such as blocking the phone on an operator's network.

(Continues)

	TABLE 11-1 Categories of mobile phone vulnerabilities and exploits.		(Continued)
	Banking data	Unencrypted or poorly protected banking data can be captured and used to fraudulently access a user's on-line account.	
	Credit card data	Credit card details can be extracted from the phone, especially if it has no encryption or is using unsecure NFC.	
	Contacts and phone book	This is another popular target for cybercriminals, which furthers their reach for potential victims.	
	Photographs and video	The attacker can invade the user's privacy by stealing their pictures and videos.	
	Call logs	Tracking call activity is another way a cyberattacker can invade a user's privacy.	
Impersonation	Sending SMS messages	This involves sending false messages to collect information from contacts or to engage in illegal or illicit activities (including harassing the user).	
	Posting to social media sites	This is typically done to harass or embarrass the user.	
	SMS redirection	This is used for eavesdropping and potential extortion.	

company named Android. Google's interest in acquiring the company and in delivering an open source, Linux-based mobile phone OS raised many eyebrows. Technologists viewed it as an interest in acquiring market share and as a direct challenge to Apple and Microsoft.

Although Android was originally developed as a digital camera interface to work with touchscreen mobile devices such as smartphones and mobile tablets, Google developed it further. This development quickly led to Android becoming the most popular OS for a variety of devices—not just mobile phones, but also tablets and game consoles. The first Android phone to become commercially available was the HTC Dream, which was unveiled in 2008.

Android benefited from the open source model. This model attracts an enormous number of community programmers and developers, who continually add features and functions desired by the public. This offers a huge advantage over closed systems, such as those used to develop Apple iOS and Windows Phone, where upgrades and features are added only if they are deemed financially viable, and then by a single team that may be competing for internal resources and priorities within the company. The open source model allows for rapid, multithreaded development driven by customer demand. Many apps are developed, with the popular ones attracting more developers and enjoying rapid innovations and updates.

For this reason, Android-powered mobile phones have dominated the market, despite the popularity of the Apple iPhone. (This is despite the lukewarm reception upon the Android's launch, with companies such as Nokia and Microsoft scoffing at its relevance.) Furthermore, the adoption of the Android OS by Samsung, HTC, and many other vendors has made it the dominant OS in the mobile phone market. It is likely to stay that way, even though most vendors now deliver a mixture of open source and proprietary software.

Despite the fact that Android has focused on building a secure architecture, it is not free of vulnerabilities and threats. Far from it. If you believe the research conducted by antivirus companies, security threats on Android devices were growing at an exponential rate. (Google denies this, claiming that viable malware threats are rare.) At the time of writing, history appears to have come down firmly on Google's side of the argument. However, breaching or compromising Android's security architecture was never going to be easy. A more worrying security issue surfaced in September of 2013, when both the U.S. National Security Agency (NSA) and the British intelligence Government Communications Head Quarters (GCHQ) stated that they had gained access to user data on iPhone, BlackBerry, and Android devices. This was because many popular applications collect personal data and transmit it unsecurely across the Internet. This of course is a vulnerability common to all IT devices not just smartphones. Nonetheless, smartphones were more susceptible to this type of eavesdropping due to a lack of enforced Secure Sockets Layer (SSL)/Transport Security Layer (TLS) controls combined with a high-level of personal usage data. Usually, this information is used for advertising and marketing purposes, but that does not mean that it cannot be (or is not being) used for more sinister aims.

> **FYI**
>
> Different phone providers offer different versions of the Android OS (an inherent risk in any open source model)—although they are all based on a common core OS, which is managed by Google. From a security perspective, managing multiple versions of the OS is more difficult than managing a single version. This is not an insurmountable problem, however.

Criticism of Android

Google Android received a lot of criticism from security experts when it was released due to the dangerous combination of its increasing popularity and its perceived vulnerability to attacks. That being said, although Android is fundamentally open source, an Android OS implemented by a reputable manufacturer (which modifies, tests, and packages the code for its own means) is no more or less vulnerable to malware attacks than other smartphone operating systems—as long as end users show some awareness of the potential security vulnerabilities brought about by rooting their devices. Rooting an Android device gives users privileged control (root access) and the ability to download and upload any software they wish, including security updates, even from third parties. This is great for tech-savvy users who are diligent about security, but it provides cybercriminals with a way to prey on those who are not security literate.

The most lucrative targets for cybercriminals are those who are naïve, are gullible, or just don't want to spend the money for genuine applications rather than download so-called "free apps." In particular, cybercriminals have targeted Android by developing malware that is delivered to the end user via Trojans. This has proved to be an effective delivery mechanism. Typically, the malware payload is bound to an existing genuine application, such as the popular game *Angry Birds*, which is then made available free of charge on pirate sites such

> **TIP**
>
> If you use your Android device as designed and download only applications from Google Play (discussed later in the chapter), then there is little chance of your device being infected by serious malware from legitimate application sites.

as Pirate Bay or on peer-to-peer (P2P) software sharing sites. The unsuspecting end user downloads the "free" software and installs it along with the Trojan. In the end, the $1.99 saved by downloading the pirated version can cost the end user thousands of dollars—not to mention a great deal of time and stress—if the hacker successfully uses the malware to access the user's bank accounts or credit card numbers.

Android Exploitation Tools

Android is an open source form of a widely used version of Linux. Because of this, there are many tools available that developers and programmers can use to decompile, analyze, and study Android OS code and applications. The people who use these tools may be good guys who seek to identify malicious or otherwise problematic code. Or, these same tools may be used by cybercriminals to exploit known vulnerabilities and create their own malware.

> **FYI**
>
> One of the great myths of open source software is that because many people study the code, the chances that a bug will go unnoticed for a considerable amount of time are decreased. This myth has been disproved by the recent **Heartbleed** vulnerability, which was a huge hole in the security code in very popular open source applications. This vulnerability existed for more than a decade and was not widely known until it was publicly disclosed in 2014.

The tools available for evaluating and exploiting security issues with Android include the following:

- **AndroRAT**—This tool, which can be bound to other applications, can read messages and contacts, steal data, view video, record calls, and more. This full framework of open source tools is freely available and is constantly updated.
- **Android SDK**—The Android software development kit (SDK) is the official Android development tool. It enables developers to compile and decompile applications for Android. This is an essential development or research toolkit.
- **DroidBox**—Another application for analyzing Android applications, DroidBox can check for password hashes, check files for read/write data, and record incoming and outgoing communications (SMS messages and phone calls), among other things.
- **Android Framework for Exploitation**—This tool can scan the network, looking for security issues and vulnerabilities on Android devices.
- **RiskInDroid**—This is a handy tool for calculating the inherent risk of Android apps based on their required/requested permissions.

There are many more open source tools available to both secure and exploit Android devices. The fact that Android is built on open source code is both its strength and its weakness. However, the applications that run on the Android architecture are easily viewable and can be analyzed, verified, or modified without difficulty. To gain this capability, all that is required is to download the Android SDK and decompile the code. It is then very simple to modify the code or use tools such as AndroRAT to create "test" code that binds to legitimate applications and will circumvent antimalware software detection.

Android Security Architecture

Android is an open mobile platform built on a robust security architecture. This architecture was designed to ensure the protection of users, data, applications, and devices by providing a secure development environment. The Android approach is to build multilayer security for an open architecture while providing flexibility and protection for users of the platform. However, Android does have developers' interests in mind and has tried to reduce the burden on application developers by introducing many security controls that can be implanted into software.

The Android security platform controls and features include the following:

- **Security at the OS Linux kernel**—This ensures that native code is constrained by the application sandbox.
- **Mandatory sandboxing of applications**—This prevents applications from interacting with each other and limits access to the operating system.
- **Secure interprocess communication**—This provides standard and secure mechanisms for accessing file systems and other resources.
- **Digital signing of applications**—This identifies application authors and deters or prevents malware.
- **User-granted application permissions**—These require applications to obtain express permission from users before accessing resources such as camera functions, contact lists, or GPS.

The Android software stack contains the security measures required to secure applications, with each layer assuming that the components lower in the stack are secure. The top layer is the application layer, which hosts device-based applications such as the dialer, SMS/MMS, browser, camera, and so on. Below that are the application frameworks, which are the services provided. These include the activity manager and the package manager, among others. Below the frameworks are the libraries and the Android runtime virtual machines. This layer is built on the Linux kernel, which provides interprocess communications control and ensures that even native code is constrained by the application sandbox.

Android Application Architecture

Because Android is an open source platform, every application created for Android devices consists of essential building blocks. Therefore, every application can be decompiled and reviewed as blocks of source code. This is made easier because Android consists of basic software components that make up each application. These components are as follows:

- **Activity**—This is a user interface whereby a user can enter data or interact with the application in some other way.
- **Service**—A service performs operations in the background—for example, playing music.
- **Content providers**—These provide information to third-party applications. A content provider can be seen as an interface that processes data in one process and feeds it to another independent process.
- **Broadcast receivers**—These respond to systemwide notifications such as "battery low" or "microphone unplugged." The OS normally initiates these notifications or broadcasts, but trusted applications can also issue broadcasts.

Google Play

Google Play is the digital distribution platform for Android applications. It was launched in 2012 through the merger of Android Market and Google Music. As a result, Google Play is not just an application store but also offers a broad catalog of other products, such as music, books, magazines, games, movies, and TV shows.

Google Play provides a market for users to browse Google and third-party applications for Android devices. Google Play does not, however, install these applications. It merely downloads the application as a package. The PacketManagerService on the device then opens to perform the actual installation in the device's internal storage.

Google does have an approval process for apps. As a result, some restrictions do apply to third-party applications. Google also uses advanced application security techniques backed by its own machine learning algorithms such as running an automated antivirus and malware detection program called Google Play Protect against all uploaded applications as part of the vetting process to remove malicious applications. You can also use Google Play Protect to verify apps downloaded outside of the Google Play store. Another security feature in Google Play displays all the permissions required by an application before it is installed. In theory, this should warn the user of an application's intentions. The user can review the permissions requested and check whether they are suitable or compatible for the type of application. The user can then decide whether to install the application or not.

Google Play is not restrictive. That is, Android device owners are free to access applications from other sources. Therefore, Android devices can download applications from any third party, or even side-load applications from a developer's or corporate website. This lack of restriction is in contrast to the walled garden of Apple and Microsoft. (A *walled garden* is any environment that controls the user's access to web content and services. In the case of Apple and Windows, it refers to their authorized enterprise portals, which are the only sources for downloading applications to Apple iOS or Windows Phone devices, respectively.) Rooting, the process of circumventing these restrictions, is also called *jailbreaking*.

> **NOTE**
> Rooting is also done to allow phones to be used on switch carrier networks other than the one with which they were originally associated. (Carriers often subsidize the cost of a phone and lock the phone to their own network.)

Apple iOS Security Challenges

In 2007, the introduction of the iPhone changed the mobile phone landscape. The arrival of the original iPhone, which was more of a handheld computer with a large touchscreen than merely a phone, sparked changes in mobility, computing, photography, and independent software development, to name just a few areas. Its operating system, called iOS, ran a Safari web browser and offered built-in Wi-Fi and Bluetooth in addition to traditional mobile communications.

The iPhone was one of the most disruptive devices of the new century. It certainly transformed the way we benchmark mobile phones and even the way we work and play. From a security aspect, however, it opened up a whole new way of thinking. The iPhone, along with the other smartphones that were to follow, was not simply a mobile telephone, but a complex computer in a miniaturized format that carried with it a treasure trove of user information beyond what any other device had ever held.

When the iPhone was launched in 2007, and followed a year later by the iPhone 3G, it was clear that it would change the way people interacted with technology. The public embraced this change. Suddenly, mobile data and Internet, along with web access from a mobile phone, became hugely popular. Indeed, it was so successful that within a few years, data usage levels skyrocketed and Internet access on mobile devices became the norm.

It's debatable whether the iPhone sparked the widespread adoption of smart devices or if the iPhone (and, later, the iPad) simply happened to appear at just the right time. Previous attempts at smartphones and tablets had failed due to a lack of applications and connectivity. Perhaps the difference-maker was the creation of the App Store, where users could download thousands of Apple-approved third-party applications. This was a major divergence from previous strategies pursued by the likes of Nokia, BlackBerry, Windows, and even Apple, and it sparked massive user interest and demand.

Before this, manufacturers of so-called smartphones made the development of third-party applications as difficult as possible for independent and small software houses. In contrast, Apple actively encouraged independents to develop applications for its product, resulting in a huge repository of applications available from the App Store. This freed Apple from having to guess which applications would be profitable from a development standpoint. It also kicked off a modern-day gold rush as developers of popular apps became millionaires overnight.

Unlike Android, Apple iOS is closed source and follows a walled garden philosophy. That is, only verified applications from the App Store are available for download. From a development standpoint, this has made the iOS less attractive to cybercriminals. Indeed, to download or side-load unauthorized applications, the user must jailbreak the device, at which point Apple can claim innocence of any damage caused by attacks. From Apple's perspective, the walled garden approach is the more secure choice—and many security pros agree.

Apple has one more security advantage over Android: It typically releases only one new phone per year. In contrast, Android devices (which use open source software) are released by the hundreds. Apple has pushed BlackBerry aside in the corporate boardroom battle and has managed to keep Samsung's Galaxy, an Android device, at bay in corporate settings (at least for now).

> **FYI**
>
> The App Store is something of a double-edged sword. On one side, it makes it easier to secure applications and prevent unauthorized downloads from third parties. This limits the iPhone's vulnerability to malware. Some view this as too restrictive, however—a problem that has been exacerbated by Apple enabling service providers to lock devices to work only on their networks. (This was because service providers were heavily discounting phones purchased with a service agreement.) The result was that users found ways to jailbreak the iOS security features, which of course opened phones to attack. Unfortunately, jailbreaking devices has become very popular. Some users feel limited by the fact that they can only download Apple-verified applications through the App Store. Many want the freedom to download or even build their own applications.

Like the Android OS, the Apple iOS has a component-layered model. The layers consist of the following:

- **System architecture**—This involves the OS platform and hardware used to protect the iOS device. It also relates to sandbox testing and application isolation. It includes a secure boot-chain, system software authorization, a secure enclave, and touch ID.
- **Encryption and data protection**—These are the techniques used to safeguard against theft. They include file data protection, passcodes, keychain data protection, and more.
- **Network security**—These are the techniques used to protect data when it is transmitted across the open Internet. They include SSL and TLS security.
- **Application security**—This includes digital authentication and verification, runtime process security, data protection within applications, sandboxes, and service isolation.
- **Internet services**—These include iMessage, FaceTime, Siri, and iCloud.
- **Device access**—These are the basic security tools such as passwords, PINs, remote wipe, mobile device management (MDM), and even remote access tools.

A key consideration with the iPhone was how it could be secured against theft or loss. After all, the mobile device held private user data, such as account information and passwords. For this, access control is always a good starting point. To that end, the iPhone had a password lock. In addition, it used many other access control techniques, such as application permission requests, which are similar to the permission per process control in Android.

Apple iOS Exploits

Although this chapter has stressed vulnerabilities in Android, the Apple iOS and Windows Phone also suffer from them. In 2014, for example, cybercriminals took advantage of a vulnerability in the Find My iPhone application to perform brute-force attacks on passwords for celebrity accounts on Apple iCloud. This enabled the attackers to steal the celebrities' stored personal data and photos. Fortunately for celebrities (and the rest of us), Apple has since distributed a patch to lock out failed password attempts to prevent cybercriminals from stealing people's personal information.

The following are a few examples of vulnerabilities found in Apple iOS. None of these were fatal or created excessive risk, but they do illustrate that iOS does have vulnerabilities.

- Due to vulnerabilities in 802.1X, a cybercriminal could impersonate a Wi-Fi access point.
- A flaw in iOS enabled cybercriminals to gain access even to sandbox applications to retrieve data from an iCloud account.
- A logic issue existed that prevented the screen from locking.
- The address book used in iOS was vulnerable due to poor encryption.
- Cybercriminals could write to the /tmp directory and install unverified applications.
- Cybercriminals could spoof the validation of updates and developer certificates. In the case of the latter, forging these certificates or obtaining them on the black market enables a cybercriminal to bypass Apple's App Store validation process. By forging these digital certificates, a cybercriminal can produce an application that bypasses all permissions and gains access to trusted features without having to prompt the user for permission.
- Bluetooth was unexpectedly enabled by default, creating a host of security issues.
- Because of a vulnerability in the graphics engine, a maliciously crafted PDF file could lead to application crashes.
- In iMessage, attachments may persist even after being deleted, which could expose sensitive information that was assumed to no longer be present. In newer versions, additional checks are done to ensure attachment deletion.

What is more, Apple created applications, such as IOKit, for those developing device drivers and applications. In the right hands, these tools are a boon for development. But unfortunately, they are also a godsend to cybercriminals.

NOTE

Although Apple issues regular patches to fix many vulnerabilities, many new vulnerabilities are exposed with each new OS release. And of course, these fixes are relevant only if users update their software—and many do not.

Another security issue was the iOS's support for web browsing. Although web browsing became a common feature, it had the same dangers and vulnerabilities with mobile devices as with PCs. The iOS 4 release brought additional security and management features, which were greatly enhanced to permit granular policy control of the device. This was a prerequisite to gaining widespread corporate adoption. Nonetheless, cybercriminal attacks involving the web browser have evolved from simple email phishing to falsely rendering web browsers to execute scripts of their own. An example of this was the JailbreakMe tool, which took advantage of flaws in the Safari web browser to jailbreak iOS devices.

A significant threat to iOS devices is the risk of iOS surveillance—more specifically, **mobile Remote Access Trojans (mRATs)**. This attack route focuses on jailbroken devices that have had all their security mechanisms removed. Of course, after security has been circumvented, it is easy for a cybercriminal to gain control of a device. After the cybercriminal has achieved this, they have access to all the features and data on the phone, including the ability to delete data and take pictures and videos.

Similarly, Apple employs a permission model to secure features and mechanisms. It does this through an iOS profile. If a cybercriminal can forge or fake this profile, that person can gain control of all the features of the phone. All that is required is to persuade the user to download a fake profile. This loads a rogue configuration, rendering the phone open to remote control.

Apple iOS Architecture

The iOS architecture is a layered model. At the highest level, iOS can be considered an intermediary between the underlying hardware and the applications running on the device. Applications do not talk directly to the hardware but rather go through the iOS and device drivers. Therefore, the iOS is built on several layers that stack on each other, providing more sophistication at each subsequent layer. From the top down, the layers are as follows:

- **Cocoa Touch Layer**—This higher level layer provides a level of abstraction from lower levels. It is where application development occurs. This makes it much easier to write code, as it reduces the amount and complexity of the code.
- **Media Layer**—This layer contains the graphics, audio, and video technologies used to implement multimedia features in applications.
- **Core Service Layer**—This layer underpins the system services that applications require. It also supports technologies such as iCloud, social media, and networking.
- **Core OS Layer**—This layer contains the low-level features that are the foundation of all the higher layers and their features.

To assist developers, Apple has supplied a developer library. It contains application programming interface (API) references, programming guides, and many sample code blocks. The lesson for end users is that to remain secure, they should use the App Store to download applications. Apple has created this marketplace for developers to upload and sell verified applications on which end users can rely.

The App Store

The App Store was launched in July 2008 with a library of 500 applications. These were a mixture of business applications and games, 25 percent of which claimed to be free. At the time of this writing, the App Store boasts more than 2 million apps and has had in excess of 75 billion downloads since its launch.

The App Store is a digital distribution platform run by Apple for mobile apps developed for its mobile operating system. The platform provides a repository of applications developed with Apple's SDK. Users can browse these apps and download them directly to their iOS device. The App Store is the only authorized source for third-party applications for iPhones, iPods, and iPads. Apple maintains the App Store and monitors the quality of the applications uploaded by developers. These applications can be offered free or sold for a price, in which case Apple takes 30 percent of the revenue and the developer is paid the remaining 70 percent. Many so-called "free" apps have in-app revenue sources or pull in advertising revenue from other vendors.

By hosting the App Store, Apple was able to create a vast marketplace for trusted third-party apps. This gave their devices a distinct advantage over competitors such as

Microsoft and BlackBerry. Apple approves applications before they are uploaded to the App Store by using basic reliability testing and code analysis. Additionally, Apple rates applications based on content and determines appropriate age groups and categorization.

Because the App Store was designed for consumers, it left business users unable to download or upload their in-house applications to their employee devices. Apple resolved this issue with an extension called the Enterprise App Store, which enabled businesses to publish these applications using the Apple iOS Developer Enterprise Program. These applications are still subject to Apple's control and Apple can terminate the application on the user's device simply by revoking the application's certificate, known as Apple's "kill switch."

Despite Apple's original vision of a global marketplace for iOS applications, national laws and regulations have resulted in the launch of many different App Stores. Furthermore, there are restrictions whereby users can only use the App Store that caters to the country in which they are registered.

Windows Phone Security Challenges

The Windows Phone OS was the replacement for the Windows Mobile 6.5 OS. Although the Mobile 6.5 OS did not achieve huge market share, it was very business oriented. In fact, it was developed for that purpose. The Mobile 6.5 OS had very strong and granular permissions and features that could be controlled by a user or administrator. Unfortunately, its successor, Windows Phone 7, had none of these security and management features required by business network administrators. This was rectified in Windows Phone 8, however. Windows has now added security and management features comparable to the iPhone iOS.

Windows Phone OS Exploits

Microsoft security is notoriously very strong, and the Windows Phone OS is no exception. It uses the same update/patch approach as all other Windows products. Additionally, Windows Phone OS is less likely to be jailbroken than iOS due to the diversity of devices on which it runs. Consequently, exploits on the Windows Phone OS are typically due to breaches of trust rather than breaches of internal security. This is because applications must ask for permission to access phone functions as they use them rather than just at startup. Therefore, the user always has the option to deny the request.

Ironically, the fact that Windows Phone has low market share (and high programming complexity) means it is not an attractive target for cybercriminals compared with Android. As a result, Microsoft can proudly claim to have a negligible amount of successful malware attacks. This is the reverse of the security issues on Windows and Apple computers.

Windows Phone Security Architecture

Windows Phone 8 has a large number of security controls to protect third-party applications. The system is heavily compartmentalized, using a sandboxing approach to applications. This prevents them from interacting with one another. File and protocol handlers exist to assist in app-to-app communication in cases where it is needed, but the interaction remains limited.

In addition, there are other mechanisms for protecting data storage on the device itself. For example, Windows Phone 8 uses BitLocker disk encryption to protect not only the storage areas but also the isolated data storage compartments that applications use.

Windows Phone Architecture

Like iOS, Windows Phone 8.1 is a closed system. The underlying OS code is not available to developers. Only APIs are used along with the Windows development kits. Windows Phone 8.1 is based on the Windows NT kernel and is a stripped-down Windows system that boots, manages hardware and resources, authenticates, and communicates just like any other Windows device. It also contains low-level security features and network components. Where Windows Phone 8.1 differs is that it contains additional mobile phone–specific binaries that form the Mobile Core.

> **NOTE**
> Unlike Apple and Android, Microsoft uses one OS for phones and another for tablets.

The architecture itself is a layered model. Applications run on top of an operating layer, which provides the services and programming frameworks that applications can use to create the user experience. Below the operating layer is the system kernel, which controls the file/system and storage, input-output (I/O) manager, memory manager, and networking and security functions. Below the kernel are the device drivers, which talk directly to the original equipment manufacturer (OEM) hardware. Developers use the Windows Phone SDK 8.0, which contains tools and emulators necessary to create applications that run on the OS.

Windows Store

Windows Store is the successor to Windows Marketplace, which was Microsoft's online software storefront. Windows Store was launched in 2012 with Windows 8. It was designed to provide a platform for users to browse and purchase Windows applications and what used to be called "metro-style apps." Microsoft design guidelines require apps to be tightly sandboxed and constantly monitored for quality, compliance, and security.

Windows Store is the digital distribution platform for Windows applications. It is the primary method for distribution of metro-style apps. Microsoft scans apps for security issues and flaws and to detect and filter malware. Similar to Apple, Microsoft has taken the walled garden approach. Windows Store is the only source of authorized applications.

Windows Store provides a marketplace for users to browse for Windows Phone applications and for third-party developers to showcase their products. For developers, Microsoft provides a portal and tools for tracking sales, financials, adoption, and ratings via a developer dashboard.

Microsoft sees the development of Windows Store as strategic, viewing the lack of external applications as one of the factors restricting the adoption of Windows Phone. The development of Windows Store is believed to be key to making up ground with Google Play and the Apple App Store. Currently, Windows Store offers more than 170,000 applications available for download, with games, entertainment, books, and reference being the largest categories.

CHAPTER 11 | Mobile Communication Security Challenges

CHAPTER SUMMARY

The rapid growth of smartphones and tablet sales since 2007 has proven to be an excellent opportunity for cybercriminals. Not only has it provided a vast array of new attack routes, it has also allowed them to focus on potentially affluent victims. Additionally, these new devices and technologies have become the primary form of Internet browsing and communication. Increasing numbers of smartphone users are taking advantage of their devices to conduct e-commerce and mobile banking, as well as using their phones as mobile wallets. More a lifestyle tool than merely a phone, these devices contain valuable personal and financial information such as bank or credit card details.

While the Android operating system is the most targeted due to its open source nature and multithreaded versions, all the major mobile operating systems, including Apple iOS and Windows Mobile, have proven to be either directly vulnerable or susceptible to vulnerabilities based on user actions and third-party applications. In addition to these platform-specific vulnerabilities, all smartphones are vulnerable to the same browser-based vulnerabilities that plague laptops and PCs.

Compounding the security issue is the fact that thanks to new technologies, users require time to become familiar with the device and the security best practices that can help mitigate risks. It also takes time and experience for the manufacturers of these devices to become aware of security vulnerabilities and to take steps to fix them.

KEY CONCEPTS AND TERMS

Heartbleed
International Mobile Station Equipment Identity (IMEI) number
Madware
Mobile Remote Access Trojans (mRATs)
Potentially unwanted applications (PUAs)
Remote Access Trojans (RATs)
Sandbox

CHAPTER 11 ASSESSMENT

1. The Android OS is less susceptible to attacks than Apple iOS and Windows Phone because it is based on an open source model.

   A. True
   B. False

2. Compared to PCs, mobile devices have which of the following?

   A. Less risk, because they are moving targets
   B. More risk, because they are subject to all the same exploits and many of their own issues
   C. The same amount of risk as nonmobile devices
   D. Less risk, because they are newer

3. Potentially unwanted applications (PUAs) are used for which of the following purposes?

   A. To create competition in Google Play
   B. To monetize applications through connections to aggressive third-party advertising networks
   C. As a category for low-rated applications
   D. To clean up memory on phones

4. The open source nature of Android results in which of the following?
   A. Better security
   B. Fewer code releases
   C. Consistent software versions among vendors
   D. Rapid, multithreaded development driven by customer demand

5. Google Play displays all the permissions required by an application before it is installed. This warns users of all the services the app will use before installation.
   A. True
   B. False

6. Apple's security-based decision to restrict the downloading of apps to those in its App Store is one of the reasons people jailbreak phones, which is one of the biggest security problems in iOS.
   A. True
   B. False

7. Which of the following is the key difference between the Android and Apple iOS security approaches?
   A. Open source models are more secure.
   B. Apple uses a walled garden approach, requiring all apps to go through its system.
   C. Google Play lacks security checks.
   D. Google uses a walled garden approach, requiring all apps to go through its system.

8. Why are Windows Phones less likely to be jailbroken?
   A. Lack of popularity makes them a low priority target.
   B. The OS is offered on a wide variety of devices, giving people more options.
   C. The code design makes it harder to achieve.
   D. There are no third-party applications available.

9. Windows Phone's sandboxing approach does not allow apps to directly interact with each other.
   A. True
   B. False

10. The Heartbleed vulnerability, which existed for years before it was discovered and publicly disclosed, illustrated which of the following?
    A. Developers are lazy.
    B. Open source code is not secure.
    C. Bugs will always exist in complex code, no matter how diligent developers are.
    D. Hackers are slow to exploit vulnerabilities.

CHAPTER 12

Mobile Device Security Models

DUE TO THE IMMENSE POPULARITY OF SMARTPHONES AND SMART DEVICES, and to the very lucrative potential of gains in market share, each of the major smartphone vendors—Google, Apple, and Microsoft—has adopted an approach to improve security and lower risk for users. Interestingly, these vendors' approaches don't significantly differ. All three are based in part on the two main concepts of controlling access to applications (downloads) and compartmentalizing applications and their resources once downloaded.

Recent years have also seen a push to ensure that mobile platforms are well suited for use in enterprise settings, given the growth in acceptance of Bring Your Own Device (BYOD) policies. In response to this, mobile platforms are increasing their support of enterprise management security, monitoring, and control services that allow information technology (IT) teams to efficiently manage large numbers of mobile devices.

This chapter looks at the security models of each of the major mobile platforms—Google Android, Apple iOS, and Windows Phone. After reviewing the similarities and differences between these, the chapter then explores how IT organizations manage the security and control of smart devices on a large scale.

Chapter 12 Topics

This chapter covers the following concepts and topics:

- What the security model and features of Google Android are
- What the security model and features of Apple iOS are
- What the security model and features of Windows Phone are
- What the security challenges of Handoff-type features are
- How BYOD affects security
- How enterprise mobility management can boost security

Chapter 12 Goals

When you complete this chapter, you will be able to:

- Describe the Android approach to sandboxing
- Understand how the open source model affects Android security
- Discuss the concept of application provenance
- Describe the Apple iOS approach to sandboxing
- Describe the Windows Phone approach to security
- Identify emerging trends in mobile security
- Discuss how enterprise mobility management works

Google Android Security

The Android operating system (OS) is built on the Linux open source OS. However, the applications used on Android devices are developed in Java. In particular, Android uses the Dalvik Java platform. Developers typically write their apps in Java and then use the Google Android software development kit (SDK) tools to convert their applications to run on the proprietary Dalvik platform on all Android devices.

The Android security model is based on an open system. As an open system, Android allows owners to download applications and software from any website. Therefore, to fully benefit from Android's security model, it is up to the owner to vet the trustworthiness of any download's source.

The Android Security Model

Android is built on the Linux kernel, which has been used for many years in security-sensitive environments. Android applications use *process sandboxing* for security. This means that each Android application runs in its own Dalvik virtual machine (VM), and each VM is isolated within its own Linux process. Although the Java and Dalvik VMs are secure, Android does not rely on Java VM to enforce security. Instead, it looks to the Linux kernel. The Linux kernel is, therefore, the foundation for Android security and provides key security features. These include a user-based permission model, process sandboxing, extensive mechanisms for inter-process control, and the ability to remove unsecure elements of the kernel. This is made possible through Linux's extensive multiuser features, which isolate and prevent User A from reading User B's files, applications, resources, or memory. Android builds on these concepts to create what is known as the **Android sandbox**.

The Android Sandbox

Android takes advantage of Linux's multiuser environment by adapting the multiuser-based protection as a means for identifying and isolating Android applications. The Android security system assigns a unique user ID to each Android application and then runs the application as a separate user process with its own permissions. This configuration isolates applications and their files, resources, devices, and memory. Furthermore, because the application kernel is within the OS kernel, this isolation extends to OS applications, libraries, application frameworks, and the application runtime.

One of the security benefits of application isolation within a sandbox is that not only are inter-process communications controlled but the resources and memory are as well. One benefit, for example, is that a memory crash in one application will not create a security issue that compromises the overall security of the device. In a sandbox environment, a memory crash will only allow arbitrary code to be executed within the confines of the affected application, and under the same permissions. That is, there is no leakage between applications when one crashes or is compromised.

File-System Permissions

Permissions are how the Linux file system ensures that one user cannot access the files of other users or, in the case of Android phones, how it prevents one application from accessing other applications. In Android, each application runs as a user with its own set of permissions. Unless a developer explicitly grants permission to expose files to other applications, files created by one application cannot be read or used by another application.

Android SDK Security Features

Notable features in the Android SDK make memory-corruption issues much harder to exploit. One of these features is ProPolice, which prevents stack overflows. There are also tools to protect against leaking kernel addresses and integer overflows during memory allocation. In addition, there are techniques for format string protection and a hardware-based "No-eXecute" parameter (control) to prevent code execution on the stack or heap.

Android can also encrypt data through cryptographic application programming interfaces (APIs), which support **crypto primitives**. These are low-level cryptographic algorithms used to build cryptographic protocols such as Advanced Encryption Standard (AES), RSA (an acronym of the last names of its developers, Ron Rivest, Adi Shamir, and Leonard Adelman), Digital Signature Algorithm (DSA), and Secure Hash Algorithm (SHA). Android 3.0 and later also support full file-system encryption where the encryption key is protected with AES-128 by a key derived from the user password. This protects the user's data from unauthorized access. Later versions of Android such as Marshmallow (Android 6) provided full file-system encryption at AES-256 but was not set by default. Consequently, only around 25 percent of smartphones running on Android 8.0 or below were encrypted and secure. Google's insistence that later versions of Android such as Nougat (Android 7) were to be encrypted by default led to a vastly improved 80 percent of smartphones being secured. The problem, however, persists that cheaper phones do not have the capacity and processing power to effectively run encryption algorithms so have been exempt.

It is not just cheaper low-end smartphones that are exempt from Google's insistence on encryption by default. Android is open source and has been widely adopted as the OS of choice for many Internet of Things (IoT) devices. However, IoT devices are often resource-constrained in so much because they have little memory or processing power. Google has been working to address this anomaly in the Android ecosystem by introducing low-power versions of their Android encryption called Adiantum. The aim is to get full-file system encryption running seamlessly on even low-end IoT devices and cheap smartphones by using highly efficient encryption algorithms.

Rooting and Unlocking Devices

In the Android security model, only the lowest layer, the OS kernel, and a small subset of the core applications run with root permissions. Root (super-user) permissions give a user or application unrestricted access to the operating system, kernel, and any application. *Rooting* a device gives someone privileged root permissions, enabling them to override the Android OS security and do anything on the device. Changing the permissions of an application to root, however, significantly increases the device's vulnerability to malware. Unfortunately, some versions of Android enable the user to change the boot sequence through the boot loader, which enables them to upload an alternative version of the OS or an application with root permissions.

Android Permission Model

Android applications run in sandboxes. By default, these have only limited access to system resources. This approach restricts access to features and resources that may, if used maliciously or incorrectly, cause the device to be compromised. These restrictions are implemented by a variety of methods. Some are protected by a deliberate lack of an API. Others have protected API status. Only trusted applications can access these protected APIs, which are protected by a mechanism called *permissions*.

> **NOTE**
> Examples of protected APIs include the camera function, location data, Bluetooth functions, and Short Message Service (SMS) and Multimedia Messaging Service (MMS) functions, but there are many more.

To use these APIs, an application must receive the permission of the device's owner. This is done through a request for permission to access features and resources at the time of the application's installation. These requests for permission are listed in the application manifest and are supplied to the user for inspection prior to installing the application. It is the owner's responsibility to grant or deny permission to access certain protected features and to allow the application's installation. After the user has given permission to the list of requests in the manifest, the application will be installed with those permissions.

> **NOTE**
> The onus is on the owner to verify that the application indeed requires the permissions requested.

The application will retain those permissions for as long as it remains installed. Although the owner can view and check the permissions for any application, they cannot revoke an installed application's individual permissions. Permissions can be revoked or removed only when the application is uninstalled, although individual features can be globally turned off—for example, Bluetooth or Wi-Fi.

> **FYI**
>
> The major difference between Android and other mobile operating systems—apart from its open system model—is that once the owner gives permission to an application, that permission is binding for the lifetime of the application. The device will not prompt for permission again. The application will work seamlessly without any interruption to the service. Other operating systems take an alternative approach, asking the owner for permission when the application is executed or as the feature is accessed. There is little doubt that from a user-experience perspective, the Android method is preferable. The application works uninterruptedly, without prompting for permissions. Users can switch freely from one application to another. That is Android's vision. From a security perspective, however, it may be inferior. The owner is asked for permission only once. The list of requests may be substantial, or it may be unclear which capabilities the application is requesting access to. For the security conscious and the technically adept, this isn't a problem. Those who are less security or technologically savvy, however, may not fully realize which permissions they are actually granting in their haste to install the application and get it working.

Another security concern with rooted Android devices is that Android natively supports active content such as Flash, Java, JavaScript, and HTML5. This can open attack routes for malware. Additionally, there are new features, such as **Always on Listening** and **Auto Content Update**, which can be vulnerable to misuse because they can dynamically enable or disable features, provide location information, or record conversations.

Apple iOS Security

The Apple iOS is a slimmed-down version of Apple's OS X, which is used in the company's line of Mac computers. OS X is actually a derivative of the **FreeBSD** variant of UNIX code maintained by a large open source community. Its security model is based on access control, application security, encryption, and isolation.

The Apple Security Model

Apple uses a walled garden approach to security, restricting access to and downloads from websites other than its own App Store. This focus on **application provenance** through its iOS Developer Program and iOS Developer Enterprise Program is to ensure that users can trust the authenticity and integrity of applications in the App Store. Those who produce apps must register and pay to become official iOS developers. When they do, they are given a digital certificate with which to sign their applications. Only digitally signed applications from authorized developers can be uploaded to the App Store. The digital certificate also ensures the integrity of each application, as the code cannot be tampered with after release. In addition, Apple's application provenance requires developers who wish to publish consumer applications to distribute and sell through the Apple App Store. By doing so, Apple can police the quality and safety of each application submitted before supplying the certificate and posting it to the App Store.

Application Provenance

The application provenance approach employed by Apple does not guarantee that applications in the App Store are safe. It does, however, greatly increase the odds that application developers will be held accountable for their work. This can deter developers from publishing malicious code. That said, the certification process is not entirely foolproof. For example, a developer could use a stolen identity to register for an account and attempt to upload malicious applications (spyware or mobile adware).

> **NOTE**
> The digital certificate is issued to the developer. The developer uses this certificate to sign their application. The certificate is then verified by the certificate authority. In the case of the App Store, Apple itself acts as the certificate authority.

Despite the possibility of spoofing, application provenance has proven to be a successful approach for Apple, judging by the lack of malware in the wild that targets non-jailbroken iOS devices (that is, those in which the user has not circumvented security features). There are several reasons for this:

- Developers must register and pay to gain access to the App Store and obtain a digital certificate.
- Apple thoroughly tests all applications for security and malware, and violations are likely to be identified. That means rogue developers will be identified, banned, and possibly prosecuted—a daunting prospect given Apple's financial might and strong economic and technical leadership.
- Apple's digital certificate for approved products makes it impossible for cybercriminals to modify or tamper with released applications. That means hackers must create new applications rather than embed malware into existing code.

> **FYI**
>
> iPhone owners are responsible for doing their own due diligence when vetting the trustworthiness of the sites from which they download applications. This of course works in Apple's favor, because the company can disavow any problems that occur on phones whose users have circumvented their restrictions.

The main limitation of application provenance is that it works only on iOS devices that have not been jailbroken. If an owner jailbreaks their device to download apps from any other source, that person circumvents Apple's application provenance model. Once the provenance model has been disabled, an iOS device is vulnerable to malware attacks.

iOS Sandbox

Apple also isolates applications by running each one in its own sandbox. As noted, a sandbox isolates an application's data, resources, and services from other applications. Additionally, apps cannot determine whether other applications are even present on the device. Further isolation is achieved by preventing an application from gaining access to the OS kernel or from obtaining root privileges through the installation of drivers. This segregation approach provides a high level of separation between applications, and between applications and the OS kernel. The result is that the Apple iOS provides very strong protection from malware.

Apple's sandbox vision differs from Android's from a conceptual point of view. Apple allows applications free access to system resources and features. In contrast, Android requires the owner's permission before access is granted to an application or system feature. With iOS, an application may access contacts and calendars in addition to music, phone, and video files. Similarly, iOS allows direct access to Safari search history, YouTube playlists, and the device's microphone and video camera without the need for the owner's permission. This, of course, has created privacy issues. Although Apple's provenance approach has been praised, with regard to the handling of access to services, it has drawn criticism.

Security Concerns

Although applications are isolated from each other, they are still theoretically vulnerable to web- or network-based attacks. For example, an attack on the device's Safari web browser via a malicious webpage could gain control of the browser's logic—although because it could not spread to any other application, the attack would be contained. Nonetheless, the malicious payload—although segregated—is free to obtain system-wide access to the calendar, address book, and even the cameras. Moreover, it can steal any data that passes through the browser application, including web passwords and login credentials.

That being said, ensuring that apps on the device cannot be modified by other apps limits the impact of malware and prevents acts such as loading drivers or rootkits into the kernel. Sandboxed applications also have restricted access to SMS and to the telephone function, which severely limits the potential effect of Trojans. Such forms of malware would have unrestricted access to the Internet, which could result in malicious data loss given free access to contacts, addresses, and so on.

Permission-Based Access

The Apple iOS version of permission-based access control is built on a much more limited restriction list than Android's. In iOS, an application requires an owner's permission only when doing the following:

- Accessing location data from the GPS
- Receiving remote notification alerts from the Internet
- Initiating an outgoing call
- Sending an SMS or email message

Besides being much more limited than Android's extensive permission-based control of system resources, this permission model is longer lasting as well. For example, if the owner of a device grants a Global Positioning System (GPS) application permission to access location-based features, then that permission is granted permanently. This need for permission is similar to the Android model, but different. With Android permissions, a request is required every time an application attempts to initiate a call or send an SMS or email. In contrast, with Apple, the request is made once and not required again. This is great from a convenience standpoint but it does lead to a higher risk of abuse because few people read all the fine print when accepting a user agreement. This can lead to users inadvertently granting permission to services that they might not allow, not realizing those permissions are excessive. Indeed, in iOS 13,

Apple changed some of the permission structures to try and address the lack of granularity in user permissions. For example, with regards to the location feature, iOS 13 allows the user to select one of three revised options whilst dispensing with the previous choices of only when using the app, always allow, and never allow. The 3 new choices are:

1. **Allow once**—Allows the app to collect location data for the current session
2. **Don't Allow**—Forbids the app to collect data at any time
3. **Only while using the app**—This allows the app to collect data when running and active in the foreground but importantly also when running in the background! Effectively, this is a disguised "Always Allow."

This does not really mitigate the issue because the updated permissions only add to users' confusion. As an example, "Only while using the App" is ambiguous and that the user will be continually prompted to accept "Always Allow." Instead of helping users choose a secure setting, the revised app permissions simply add another layer of complexity and confusion. Really, Apple is distancing itself from the inconvenient truth that ultimately the security of the app and the device is not with them but with the user.

Encryption

Apple relies on more than application provenance, isolation, and strong access control. It also provides a high level of protection through full-device encryption. However, Apple's implementation of full encryption requires that a key resides on the device. That's because it continuously runs apps and services in the background, even when the user is not logged into the device. Therefore, data encrypted with this technique would be readable to someone with physical access to the phone, even if they did not know the user password. To mitigate this, iOS uses a layered approach, with certain data—typically email files and messages and other personal data—double encrypted. Layering the encryption in this manner secures the files from unauthorized access. Only someone with the master pass key can read the data.

Jailbreaking iOS

Apple iOS security measures are, by design, restrictive, and are robustly enforced by Apple's security strategy. Many owners have found this too restrictive, however. They want to download their own applications without interference from Apple. Jailbreaking, or unlocking the device, has become a popular way to free the device so it could be used to download applications from any source. Jailbreaking is also used by those who wish to unlock carrier restrictions, which bind carrier-subsidized phones to the carrier networks.

Jailbreaking the device gives owners root (super-user) privileges. This enables them to override the application provenance model, which is the foundation of the iOS security framework. By jailbreaking a device, owners can download applications, music, and content from wherever they like. Unfortunately, this freedom comes at the cost of security, undermining the layers of protection behind Apple's walled garden security strategy. One would assume that users capable of successfully jailbreaking a phone would know enough to protect against malware, but given the ease with which cookbook-type instructions can be found and followed, many users get in over their heads.

Windows Phone 8 Security

Windows Phone 8.1 was the last specific version of the Microsoft smartphone OS. With the 8.1 release in 2014, Microsoft had pursued corporate acceptance with features that provide security and management control, mitigate data leakage, and offer protection from malware. Microsoft has also integrated security features such as BitLocker, Windows Defender, SmartScreen Filter, and user access control into the security architecture for its mobile devices. In 2016, Microsoft introduced Windows 10 Mobile as the successor to Windows Phone 8.1 but it was a version of the PC OS and was itself made obsolete when Microsoft withdrew support for the mobile version in 2020. However, because there were sufficient numbers of devices running Windows Phone 8.1, it is interesting to view Microsoft's approach to mobile phone security architecture.

Platform Application Security

Microsoft's strategy was similar to Apple's iOS strategy in the way they both worked to protect applications through isolation. Like Apple iOS, Windows Phone uses strict application provenance. Consequently, Microsoft has adopted the secure App Store model, naming its version the Windows Store. It follows the same basic model, whereby only strictly vetted applications are released and downloaded to devices running Windows Phone 8 (WP8).

Security Features

Microsoft also took a similar approach to Apple when it came to the security architecture by utilizing a secure boot process supported by a Trusted Platform Module responsible for encryption of data and applications as well as maintaining device and firmware integrity. Importantly, Windows 8.1 also deployed the sandbox model to ensure application isolation from the system resources and other applications during runtime.

These security measures greatly enhanced the mobile OS's security especially in comparison to Microsoft's existing PC Operating Systems. Nonetheless, Windows 8.1 OS's best security feature was, unfortunately, its relative obscurity because it did not gain sufficient market share—less than 2% of the market. This was because it only ran on a handful of vendors' devices and this simply did not make it a viable target for malware developers. This is often referred to as "security by obscurity."

Windows 8.1 was made obsolete in 2016 with the introduction of Microsoft Windows 10 Mobile. However, it was to prove no more fruitful for Microsoft or attractive to third-party vendors in the mobile marketplace and was itself discontinued in 2020 leaving the field to Android and iOS.

iOS and Android Evolution

The introduction of Windows 10 Mobile possibly came too late for it to have any effect on the market. This was simply because Google's Android and Apple's iOS had the advantage of years of evolution and feature improvements via adaptations of their very popular mobile Operating Systems. Indeed, Android's deployment on a vast array of diverse vendors, devices, and platforms, from smartphones to tablets to IoT devices made it ubiquitous across the technology industry. This diversity fueled the need for innovation and feature-rich versions of the OS. In contrast, Apple's hugely popular albeit highly limited range of products allowed Apple to specialize and develop iOS to meet their customers' specific requirements. As such, both Android and iOS had matured into highly evolved mobile operating systems. This did not occur overnight and required sometimes major compromises and reevaluation of the security architecture in order to placate developers.

What we can see is that over the years, both Android and iOS were forced to shift away from their rigid security architectures with regards to sandboxing and application isolation. This was primarily due to developer and market pressure. The problem was that application isolation worked great as a security mechanism back in 2010 when the idea of applications being stand-alone code was prevalent. However, the advent of the microservices approach to programming as well as the availability of vendor APIs changed all that. Now, what was needed was applications that could interface with the outside world. This would require a security architecture that allowed application integration and easy coding of mash-ups—mixing and matching vendor APIs with some link code to make fully functional applications—hence, the concept of isolation became a serious roadblock for developers.

In response to this changing environment, both Google and Apple were encouraged to relax some of their security principles, as we will see in their version updates over the years. However, although both Android and iOS use the concept of a sandbox to isolate an application from the file-system and other apps, the techniques that Android and iOS use are quite different.

On one hand, with Android, each app runs as a specific "user," with its own user identity (UID), and the Linux kernel guarantees that different "users" are by default isolated and therefore unable to interfere with each other.

On the other hand, with iOS, all the apps run as the same user ("mobile"). This means that each app is chrooted, which gives it access to only its own specific part of the filesystem.

However, isolation as a means of limiting what apps can do is only one side of the sandboxing issue. This is because not every application is an island and there will often be a requirement for apps to securely and safely access things like APIs, contacts, location, and so forth. In order to achieve these secure access requirements, both Android and iOS use granular app permissions and these were introduced and refined over time as they evolved.

Android Version Evolution

Android 1.0 was launched in 2008 and is today barely recognizable as the same product. To see how Android evolved from version 1.0 to today's version 11, we need to consider its development timeline.

- 2008—Version 1.0 was a basic OS with integrated Google apps.
- April 2009—Version 1.5 (Cup Cake) introduced a framework for third-party apps.
- September 2009—Version 1.6 (Donut) had support for CDMA and different screen sizes.
- October 2009—Version 2.0–2.1 (Eclair) had navigation and speech-to-text function amongst others.
- May 2010—Version 2.2 (Froyo) had flash support and several under-the-hood performance upgrades.

- December 2010—Version 2.3 (Gingerbread) had more aesthetic changes to the screen layout and color scheme.
- February 2011—Version 3.0–3.3 (Honeycomb) was a tablet specific release of Android with a different UI.
- October 2011—Version 4.0 (Ice Cream Sandwich) brought many of the UI experiences from Gingerbread to the smartphone.
- 2012–2013—Version 4.1–4.3 (Jelly Bean) introduced Google Now and the Quick Settings panel.
- October 2014—Version 4.4 (KitKat) introduced OK, Google, and reimagined the screen layout and UI.
- November 2014—Version 5.0 (Lollipop) introduced the Material Design Concept, which provided material themes for third-party app developers to use in their mobile apps.
- 2015—Version 6.0 (Marshmallow) introduced granular third-party apps permissions.
- 2016—Version 7.0 (Nougat) introduced Google Assistant and third-party application notifications by batch.
- 2017—Version 8.0–8.1 (Oreo) introduced an enhanced notification channel for third-party apps and ways to make it easier for vendors to roll out software updates.
- 2018—Version 9.0 (Pie) prevented applications running in the background accessing the camera, microphone, and sensors and provided clear notification to the user if any device is being accessed by an application.
- 2019—Version 10 (no more dessert nicknames) updated the permissions system that gives the user more control over how apps access location data as well as an expanded system for protecting unique device identifiers.
- 2020—Version 11.0 introduces expanded and more granular permissions that allow access to devices (location, camera, microphone) on a single user, single use basis.

In addition to these releases, Google is also introducing stricter privacy and security measures to prevent unscrupulous developers from gaining access to location data without explicit permission.

Android sandboxed applications are by default isolated from each other. Therefore, applications must explicitly share resources and data. To do this, the applications use granular permissions introduced in Android version 6 (Marshmallow). This means declaring the required permissions that the application needs to function or for access to additional capabilities that are not provided by the basic sandbox. These granular permissions permit an application to gain appropriate access to another application or a device's features such as the camera, location, or microphone.

Apple iOS

The iOS version are as follows:

- 2008—Version 1.0 launched with the introduction of the iPhone, described as an OS version of Apple's version X Desktop OS for mobile devices.
- 2008—Version 2 introduced the App Store, Push email, and GPS.
- 2010—Version 3 introduced with the iPhone 3GS with 3G carrier support.

- 2011—Version 4 introduced CDMA support.
- 2011—Version 5 introduced SIRI and iCloud and Twitter integration.
- 2012—Version 6 introduced LTE support and Facebook Integration.
- 2013—Version 7 introduced Airdrop for iOS, and Notifications Control Center.
- 2014—Version 8 introduced Apple Pay, iCloud Drive, and Handoff.
- 2015—Version 9 introduced Android migration.
- 2016—Version 10 introduced access to Apple Apps for third-party developers.
- 2017—Version 11 introduced Apple Pay Cash, updated Siri, wallet, and Maps as well as a new App Store.
- 2018—Version 12 improved Safari privacy protection, which prevents browser activity tracking.
- 2019—Version 13 features quick unlock with face recognition and improved Siri voice.

Looking at the timeline, we can see that the major milestone from a security perspective was introduced in 2016, when Apple deliberately relaxed the sandbox concept of application isolation. In 2016, the release of iOS 10 allowed for applications to access other third-party apps and even Apple's own apps including Siri.

Over time and via incremental updates, both Android and iOS have addressed the problem of securely isolating applications from the filesystem and other system resources while at the same time allowing specific access via granular permissions. They have relaxed strict isolation to allow interconnectivity, integration, and collaboration without lessening their security posture.

Security Challenges of Handoff-Type Features

In 2014, Apple released a new feature called **Handoff**, with Windows and Google working on similar technology. In order to use Handoff, each device must be connected to the same Wi-Fi network and be signed into iCloud with the same Apple ID. Handoff allows users to move seamlessly between devices, continuing right where they left off when they switch devices. So if, for example, a user is interrupted while composing an email on a Mac, that user can switch to another device—say, an iPhone—and continue where they left off. The same would apply to browsing webpages on an iPhone. A user could switch devices, and the browser would continue on the alternative device where they left off.

When connected to the same local ad hoc wireless network, Handoff also allows a user to make calls from a Mac or an iPad. This of course requires more than just some nifty shuffling of files and work profiles in iCloud storage; it also requires interdevice continuity, using both Bluetooth 4 and Wi-Fi direct connections. This provides communication, data, and workflow between devices. From a security standpoint, however, this could lead to data being handed off to unmanaged devices or even into unauthorized apps.

With Microsoft and Google pursuing their own versions of this technology, Handoff could fundamentally change the way people use devices in the future. What is more, developers can add Handoff capability to their applications with the iOS 8 app extensions. This would enable apps to communicate with each other, thereby relaxing the sandbox concept. It is feasible that data could be passed through applications via Handoff, allowing, for example, data and information from corporate apps to be used and communicated to social media apps.

BYOD and Security

The emergence of consumer smartphones such as the iPhone brought about a shift in corporate strategy with regard to employees' personal devices. It soon became clear that employees would much prefer to access their business email accounts from their own smartphones.

This initially wasn't a big problem for IT departments, because applicable secure technologies already existed, such as Outlook Web Access and Exchange Active Sync (EAS). Soon, however, gaining email access wasn't enough. Workers found that receiving an email late in the evening requesting urgent information was pretty stressful if they couldn't access their business files and applications. The obvious solution was to provide virtual private networks (VPNs) and web-based applications, either in-house or as Software as a Service (SaaS), and allow employees to use their own devices to connect remotely. Eventually, there was talk of increased productivity and better work/life balance and the like. Before long, employees wanted to use their personal devices within the network boundaries. Soon, personal laptops, tablets, and smartphones began invading the workplace. For IT, this became a major challenge. The tide was irreversible, and solutions to the ensuing major security dilemmas were required.

The solution, as far as security was concerned, was to implement some measure of control and management on devices that companies did not own. To do that, however, they needed a BYOD policy that was acceptable to both employees and IT management. The policy was needed to help IT with the complex task of enforcing company governance policy to ensure the network's security was not compromised and to protect company assets (data). Data leakage—a term used to describe company data leaving the traditional network boundaries—was and remains a major security concern.

BYOD exacerbates the risk of data leakage because thousands of uncontrolled mobile devices have access to or hold company data. Compounding the problem, there soon followed a variant of BYOD: Bring Your Own Applications (BYOA). This added into the already-confusing equation nonstandard and perhaps unlicensed software as well as personal subscriptions to Google Apps and SaaS applications. A second variant of BYOA soon followed. Suddenly, employees were uploading data to iCloud, Google Drive, and Dropbox. The prospect of controlling data leakage seemed bleak.

From the perspective of IT, Windows laptops were not a problem. There had been ways to lock down and secure the Windows platform for years. The problem was with the new technologies—the smartphones and tablets that were now prevalent in the office. Fortunately, the main vendors of the most popular devices, Apple and Google, were already addressing some issues by introducing security and management features into their operating systems. These features would enable IT to set a policy template that could restrict undesirable and unsecure features and configure on the devices a uniform security policy complying with the company's IT governance.

The Apple iPhone had by far the largest presence in the corporate workplace. It was the ideal BYOD client, because there was just one OS and only one or two models of each version of the device. In addition, Apple iOS was a closed system. If the device remained secure (that is, it wasn't jailbroken), then many of the potential threats could be mitigated by the device's own operating system and antimalware. Consequently, creating a security policy was straightforward. Apple iPhones and iPads had strong security and management feature sets built into the OS. IT could devise and configure standard, uniform, and granular security policies on iOS devices. Perhaps Apple's natural fit with BYOD was what made it so successful in the corporate market. It soon ousted BlackBerry as the business device of choice.

Microsoft, on the other hand—the undisputed market-share leader when it comes to corporate desktop PCs and laptops—found itself being brushed aside by the iPhone and iPad and lagging behind. Windows Phone devices simply were not competitive. To address this, Microsoft concluded an exclusive deal with Nokia to use the new Windows Phone 7 operating system on its range of handsets. Unfortunately, Windows Phone 7 was introduced with no security or management features, making the configuration of a security policy for BYOD a nonstarter. It wasn't until Windows Phone 8.1 was released in 2014 that Windows finally addressed those vital business requirements in its OS.

Android OS phones were and still are a BYOD nightmare. They are quite the reverse of Apple. The OS, which is severely fragmented, is hosted on hundreds of diverse devices from a multitude of manufacturers. Unfortunately for IT departments, however, Android devices have proven very popular with consumers, and have become a major presence in the BYOD workplace. Creating security and management policies that were standard and uniform across the Android platform was complex, despite Android having the largest security and management feature set of any of the other operating systems. This was purely due to the diversity and range of features and applications on the devices—a side effect of Android being an open system and using open source code.

Clearly, if IT departments were to successfully manage BYOD, there had to be a system and policy in place that could provide, manage, and remotely deploy security policies to each device. The system would need to be able to identify and authenticate the device and check its configuration for compliance with company policy before allowing it to enter the network. The system that IT uses to provide, configure, and manage mobile devices in the workplace is mobile device management (MDM). Whereas MDM is device oriented, mobile application management (MAM) is used to manage application software on mobile devices. MDM and MAM are considered subsets of enterprise mobility management (EMM).

Security Using Enterprise Mobility Management

Enterprise mobility management (EMM) is a framework that consists of sets of people, processes, and technologies required to manage mobile IT within the enterprise. Mobile IT has come at a considerable cost, as significant security risks were introduced with the acceptance of BYOD. To combat these threats, it became necessary to find a turnkey solution that could help IT departments secure and manage a broad range of operating systems and devices. EMM has a wide scope that includes disciplines such as security, application management, and financial management.

With regard to security, system administrators' lack of access to devices, combined with the problems of OS and device diversity, make configuring and installing applications on BYOD devices problematic. EMM solutions typically deploy middleware MDM to automate the management of a wide range of mobile devices. With the introduction of BYOD, there is a need to automate the processes involved with securing and managing mobile devices. The large numbers of diverse devices mean that a physical, hands-on approach would be feasible only in small companies. For larger companies, an automated, over-the-air provisioning, deployment, and configuration method is required.

Mobile Device Management

Mobile device management (MDM) is typically built on a client-server model whereby the MDM server controls and manages the client agent that is installed on the mobile device. MDM software can automatically detect a device on the network and can send and collect information, send updates, and configure the device over the air. MDM also allows for remote locking and remote wiping of company data from a lost or stolen device. Furthermore, updates can be sent to one or a group of devices—all iOS 8, all Android 4.4, or all Windows Phone 8.1—simultaneously. This greatly reduces the burden of administration and management of large numbers of company-owned or BYOD devices. In addition, some MDM systems have a user web portal that hosts device-specific apps, drivers, and updates. Employees can access this portal on a self-serve basis. On the administration side, reports can be run on just about any device's activity. These reports include call logs, messages sent or received, apps installed or removed, configuration changes, antimalware and antivirus versions, and last run date, all run from the central MDM server web console.

Of all these features, the most useful is automatically configuring a registered device with a security template over the air as it attempts to join the network, by placing the device initially into a secure quarantine zone. The MDM software then interrogates the mobile device agent to retrieve security and management settings from the phone. Finally, it compares the settings to the device's security template. The MDM can then automatically download to the device the latest versions of the security and management feature template, switching certain features on or off. (This is entirely dependent on the company policy.) Subsequently, all registered mobile phones and tablets can be kept up to date with the latest security policy automatically.

MDM servers can be hosted on-site or in the cloud and delivered as a SaaS. Most MDM systems should be able to handle the tasks outlined in **TABLE 12-1**.

As can be seen from the list of MDM functions, there are some general application and software management features. However, in enterprise ecosystems, software and application management is normally handled by the MAM system.

Mobile Application Management

The main difference between MDM and mobile application management (MAM) is that MDM handles the device activation, enrollment, and provisioning, whereas MAM assists in delivering software. MAM provides a distribution platform for apps with application lifestyle management and also handles software licensing, configuration, and usage tracking.

In a real-world situation, MAM and MDM work together to provide parts of the overall EMM solution. The MDM system authenticates the user and inventories the device before handing control over to the MAM system to continue the process. MAM may then try to verify the presence of mandatory applications and any optional apps that are installed on the device. It might then push a catalog or individual apps to the device in accordance with the user or group profile. It also records results for software maintenance, audits, and compliance reports.

MAM has another very important role when dealing with Apple iOS or Windows Phone devices. Both of these are closed systems that place major emphasis on application provenance. As such, both of these operating systems are configured to allow only digitally signed

TABLE 12-1 Common MDM tasks.

TASK CATEGORY	EXAMPLE OF TASK	DESCRIPTION OF TASK
Inventory management	Maintaining a device inventory	This should include the device model and ID, firmware version and OS level, network adapters, and removable memory cards.
	Inventory maintenance	MDM software periodically polls each device to update the inventory records.
	Physical tracking of mobile devices	This should be done if it is necessary to know both who carries a company mobile device and where it is located. Note that this is a controversial feature for BYOD.
Device provisioning	Device management	This depends on supporting many device characteristics such as OS, vendor, and platform.
	Device registration	MDM software can register user devices directly, or users can use an enrollment portal.
	Device activation	Some devices, such as those by Apple, ship with native MDM client software, so can be activated directly. Others may need a client download to be sent via an SMS with a web address where the user can download the client app.
	Device configuration	MDM software can reconfigure the device to suit company policy.
Software distribution	Patch updates	MDM software can automatically push patch updates to registered clients.
	Mobile optimization	MDM software can be used to manage bandwidth and offer compression or incremental updates over known poor wide area network (WAN) links.
	Software packages	MDM software can bundle related updates and applications for delivery to clients as packages to a group.
Security device management	User authentication	This can be integrated with Active Directory or other single sign-on solutions.
	Password enforcement	MDM software can ensure secure passwords are enabled in compliance with security policy.
	Remote device wipe	This is the ability to wipe data from a lost or stolen device.
	White lists/black lists	These can be used to allow or block, respectively, certain unsecure features and applications.
Data protection	Data encryption	This refers to the ability to enforce a policy of hardware encryption. Most smartphones can support this feature.
	Backup/restore	MDM software can provide scheduled, over-the-air backup of data to a cloud-backed repository.

TASK CATEGORY	EXAMPLE OF TASK	DESCRIPTION OF TASK
	Data tracking	MDM software can maintain an audit of corporate data on the device and control and report on sensitive files transferred between devices during over-the-air sync or onto removable media.
Remote technical support	Self-help portals	MDM software often has the facility to provide client self-help portals for technical support, FAQs, and known issues.
	Diagnostics	MDM software can show present settings, convey recommended settings, and provide real-time health checks on memory, battery, and network connectivity.
	Remote control	MDM software often provides remote control to allow an administrator to take control of the phone.
	Audit and compliance	MDM software can provide reports for assessment, remediation, and compliance reporting.

applications to be downloaded and installed on the devices. This is a major problem if the enterprise has its own in-house developed applications that it needs to run on iPhones, iPads, or Windows Phone 8 systems. The solution is to use the vendor's enterprise developer certificate to sign the applications. But they still need to be placed into an in-house enterprise distribution platform from which the devices can register and download the required applications. MAM provides that enterprise distribution platform and can manage the entire life cycle of the application from release to retirement. It follows, then, that MAM is a very important part of the EMM solution.

CHAPTER SUMMARY

The use of application provenance and sandboxing is common among the three main mobile platforms and has proven to be a sound foundation for securing mobile devices. Application provenance helps to ensure that developers and applications are vetted before applications are released. Sandboxing helps ensure that applications are self-contained so that issues with one application do not affect others.

The efficacy of the application provenance model is essentially proven by the fact that the one platform that does *not* use this model is the one that has the most issues with malware. Also, most issues on those platforms that do use this model are the result of users circumventing provenance control.

Beyond the device-specific security models, IT teams also require the ability to provision, control, and secure devices on a large scale to meet the dual requirements of BYOD support and organizational risk management. In response, each of the main platform providers has developed mechanisms to support the needs of large organizations.

KEY CONCEPTS AND TERMS

Always on Listening
Android sandbox
Application provenance
Auto Content Update
Crypto primitives
Enterprise mobility management (EMM)
FreeBSD
Handoff

CHAPTER 12 ASSESSMENT

1. Android permissions are the key to ensuring that one application cannot access other applications or application resources.
   A. True
   B. False

2. In what way does Android take advantage of Linux's multiuser environment?
   A. By growing its social media applications
   B. By sharing resources between applications
   C. By adapting the multiuser-based protection of isolating Android applications
   D. By developing open source code

3. Which of the following best describes application provenance?
   A. It helps ensure that users can trust the authenticity and integrity of the applications in the App Store.
   B. It prevents applications from accessing common resources.
   C. It guarantees that applications are safe and are free from malware.
   D. All of the above.

4. In what way does Apple's sandbox method differ from Android's?
   A. Apple's App Store is more secure.
   B. Apple requires the owner's permission before granting access to an application or system feature.
   C. Apple allows applications free access to system resources and features.
   D. There is no difference.

5. Most security issues with Apple iOS are the result of jailbreaking, which circumvents Apple's ability to secure applications.
   A. True
   B. False

6. Windows Secure Boot and application integrity ensure the authenticity and integrity of apps, making it very difficult for cybercriminals to target Windows Phone devices with effective malware.
   A. True
   B. False

7. In what way is the Windows Phone 8 policy control different from Android's?
   A. Android's Google Play allows for better policy control.
   B. The Windows Phone 8 policy control can be enforced on any Windows Phone device, whereas Android is fragmented across many devices.
   C. Android security policy and control can be enforced on any Android mobile device, whereas Windows is fragmented across many devices.
   D. There is no difference.

8. Although the Apple Handoff feature can increase productivity, it may also do which of the following?
   A. It could potentially introduce malware.
   B. It could undermine the benefits of application provenance.
   C. It could undermine the benefits of sandboxing.
   D. It could change permissions control.

9. What is the main difference between MDM and MAM?
   A. MDM is used on Apple phones, and MAM is used on Android phones.
   B. MDM handles the device activation, enrollment, and provisioning, whereas MAM assists in the delivery of software.
   C. MAM handles the device activation, enrollment, and provisioning, whereas MDM assists in the delivery of software.
   D. MDM can perform integrity checks on applications.

10. MAM plays an important role in providing application provenance for custom enterprise apps on Apple iOS or Windows Phone devices.
    A. True
    B. False

CHAPTER 13

Mobile Wireless Attacks and Remediation

THERE ARE MANY DIFFERENT RISKS ASSOCIATED WITH SMART DEVICES, including lost data from lost or stolen devices, access and device control issues, and malware, to name a few. Another risk is the fact that smart devices are also wireless client devices, and as such present risks to wireless networks. Whether used in unintentional acts by unknowing users or deliberate attempts to gain unlawful access or otherwise sabotage a wireless network, smart devices present an entirely new arsenal of potential attacks for which IT security teams must account. This chapter looks at the risks that mobile clients present to corporate networks, as well as the tools and techniques used to mitigate these risks.

Chapter 13 Topics

This chapter covers the following concepts and topics:

- How to scan the corporate network for mobile attacks
- What client and infrastructure exploits exist
- What network security protocol exploits exist
- What browser application and phishing exploits exist
- What mobile software exploits exist and how to remediate them

Chapter 13 Goals

When you complete this chapter, you will be able to:

- List the most common mobile software exploits
- List the remediation methods for common mobile software exploits
- Explain why there is a need for particular focus on mobile clients when scanning wireless networks
- Describe ways in which weak infrastructure security can be used to exploit smart devices
- Describe how network protocol exploits can be used to target mobile devices
- Understand why and how browser and phishing exploits are particular concerns on mobile devices

© Cherezoff/Shutterstock

Scanning the Corporate Network for Mobile Attacks

The advent of mobile devices, including smartphones and tablets, and their subsequent prevalence in the enterprise due to the Bring Your Own Device (BYOD) trend, have created a huge security issue. Data leakage is but one concern because these devices open up other attack routes for cybercriminals: backdoors. Employees who bring jailbroken, rooted, or otherwise compromised devices into the network create a gateway for cybercriminals to exploit not only the devices but also the networks to which they are attached.

Security professionals must regularly scan the network for unauthorized smart devices. Ironically, that's just what cybercriminals do when gathering information prior to an attack. Therefore, scanning a wireless network is both an essential administrative function and a hacker activity. Gathering information on the network is an essential tool for administrators and hackers alike, and they use many of the same tools and techniques. Both are looking for weaknesses in the infrastructure; only their intentions differ.

Complicating the matter for IT teams is the fact that they must scan not only known networks but also unauthorized ad hoc networks. These temporary networks have become a major source of threats and have great potential for abuse due to the ease with which smart devices can establish them.

The good news is that mobile phones and Wi-Fi equipment manufacturers have worked hard to reduce the *attack surface* (the sum of all potential attack routes) for potential cybercriminal attacks through their devices. In doing so, they have greatly reduced the number of security vulnerabilities available to an attacker—assuming, of course, that users have not intentionally or unintentionally undermined these efforts. This leads to a key aspect of mobile security: security awareness.

Security Awareness

One problem with technologies aimed at consumers is that developers often overestimate the general public's technical literacy. For example, when it comes to Wi-Fi security, it's a good bet that few members of the public understand the difference between Wired Equivalent Privacy (WEP), Wi-Fi Protected Access 2 (WPA2), and Lightweight Extensible Authentication Protocol (LEAP)—let alone understand the implications of not understanding. As a result, despite warnings about potential hacks on unprotected wireless systems, many access points remain unprotected or use outdated security protocols.

This same problem exists with smartphones and tablets. Indeed, it may be worse, because users may understand the need for security on their Wi-Fi router but remain oblivious to the need for securing their mobile phone or their tablet's Wi-Fi connection. This makes exploiting mobile devices very attractive to cybercriminals.

From a security perspective, this lack of awareness on the part of the public is worrying. After all, how can a network be secure if authorized users—now equipped with intelligent networked accessories—have little or no concept of security, or are willing to circumvent security measures in a cavalier fashion? For this reason, the IT department, with the full support of management, must deliver security awareness training to employees. In addition, employees must understand the security policies, such as the BYOD security policy, before bringing their devices onto the company network.

CHAPTER 13 | Mobile Wireless Attacks and Remediation

This problem is compounded by today's prevalence of smartphones, tablets, and wearable devices on the network. Gone are the days when there was a roughly 1:1 ratio of employees and devices. With BYOD, employees might bring to work a laptop, one or two smartphones, a tablet, and a smart, Wi-Fi–enabled watch, of which only one or two devices may be authorized.

The real problem is that those unauthorized devices regularly leave and reenter the premises as the employee comes and goes. If a device is unauthorized, it likely uses default wireless settings and attempts to establish a wireless connection with every Wi-Fi network it encounters as the employee moves around. The employee's device might promiscuously connect to any of these networks, whether it's a free hotspot from a legitimate source (which can still be compromised) or a hacker-deployed device. Even if the device does not promiscuously connect, the employee who owns it may seek out and connect to a network, unaware of the dangers. This is normally not the concern of IT, but when the employee returns to the company, bringing their newly infected—and perhaps highly contagious—smartphone, it becomes IT's problem.

Keeping track of authorized smartphones and tablets on the network is hard enough without having the additional burden of tracking down unauthorized devices connecting through mobile hotspots or ad hoc connections, or piggybacking on Bluetooth connections. Therefore, even with strict enterprise mobility management (EMM) in place, administrators must still regularly scan the wireless network and assess it for vulnerabilities, noncompliance with security policies, and vulnerability to malicious attacks.

Scanning the Network: What to Look For

The first step in assessing the wireless network is to conduct a network search, or scan, to discover which entities are communicating on the 20 MHz and 40 MHz channels in the 2.4 GHz and 5 GHz unlicensed bands. Typically, when performing a wireless network scan, the information that the administrator or attacker will collect from discovered access point entities is as follows:

- Media access control (MAC) address
- Extended service set identifier (ESSID)
- Channel
- Average/peak signal-to-noise ratio
- Power levels
- Network type (802.11a/b/g/n)
- Beacon security parameters (WEP, Temporal Key Integrity Protocol [TKIP], or Advanced Encryption Standard-Counter Mode Cipher Block Chaining Message Authentication Protocol [AES-CCMP])
- Beaconed quality of service (QoS) parameters
- Location

Additionally, the administrator or attacker will document the following from discovered wireless stations:

- Associated access points or peer stations
- 802.1X identity

NOTE

Scanning a network is not illegal, but a network administrator is likely to take a dim view of an employee running an unauthorized network scan on the corporate network. At present, there is no U.S. Federal law that makes scanning a network illegal; however, in many countries, your employment contract and acceptable usage of the network could forbid such activities leading to termination of employment. Hence, should you even consider performing a legitimate scan you should obtain written permission beforehand.

The administrator can then compare the results of the network scan against the inventory and the as-built network designs to uncover any discrepancies. Alternatively, an attacker can identify weak points in the network to exploit.

After identifying unknown devices, the administrator can set about investigating them. Discovering the presence of such rogue devices is a relatively straightforward task, but actually locating and eliminating them is quite difficult. This is because they may not always be malicious devices. Rather, they could be devices owned by neighbors, vendors, guests, or employees. To compound the problem, it takes far too long to investigate and evaluate these devices using tools such as NetStumbler or Kismet. An alternative is the commercial inSSIDer Office tool, which enables you to scan and visualize the network more quickly.

Ideally, an enterprise should have a wireless intrusion protection system (WIPS) architecture with central control, a reporting console, and distributed agents, which protect each wireless segment. As an effective automated solution, a WIPS agent should be running 24/7 on each wireless network segment, scanning for and identifying probe requests by unauthorized devices.

Obviously, before an administrator or attacker can uncover a rogue entity, there must be a baseline inventory of authorized devices. Unfortunately, this is not always the case. This is especially true in some large companies, where authorized access points and BYOD devices were rolled out indiscriminately to meet urgent employee requirements.

Scanning for Vulnerabilities

Wireless scanning is not just about finding rogue devices. It is also about verifying that security measures are in place on authorized access points. All access points should be subjected to the same vigorous security testing as routers and firewalls because they are subject to association with both trusted and untrusted devices. Specifically, access points should be checked for the following:

- The latest firmware and security patches
- The correct security credentials
- The appropriate security protocols
- Encrypted administration interfaces
- Appropriate protocol filters
- Vulnerability to common attacks such as denial of service (DoS) attacks via authentication floods

Authorized smartphones and tablets are also a concern. They must be checked for the latest operating system (OS) and security patches using a vulnerability scanner such as OpenVAS. They should also be checked for antimalware and antispyware software and for information regarding their usage—for example, when this software last scanned the device. Other points of interest would be the existence of a firewall and of open ports that may be targets for exploit.

A typical weakness in smartphones is the default configuration to automatically reassociate with previously associated access points' service set identifiers (SSIDs). This is done to aid in connection. However, many Android and iOS devices can connect to any neighboring access point if it has a strong signal. For this reason, it is important to be selective about which networks roaming devices will be allowed to associate with.

Service providers often ignore this, because convenience tends to trump security (for users at least). As a result, they often configure their smart devices to automatically connect to any wireless network that is nearby. The result is that the device, such as an iPhone or iPad, will always poll and try to access these networks. Although this is helpful for quick association and connectivity, it is also a vulnerability. Any rogue device can impersonate an access point and therefore capture client station connections.

Another issue is that a smartphone may simultaneously connect to both a mobile and a Wi-Fi network. In doing so, it can bypass the company network's routers and firewalls by transferring data between the phone's interfaces and bridging the two independent networks. This is an inherent vulnerability in smartphones and is a major concern with regard to data leakage.

These are just a few of the wireless security issues that apply to unauthorized smartphones from a security perspective. The prevalence of smartphones as data-communication devices has made securing wireless networks very challenging. At issue is the fact that you can only protect (or attack) what you know is there. Therefore, scanning the wireless network topology is the key first step for any audit or attack.

In general, the wireless network should be subject to at least the same level of diligence and penetration testing as all other external components, such as firewalls and routers. Most of these tests are not specific to wireless, and will be part of the fixed and Internet-facing network infrastructure program. However, given the nature of wireless and the much broader attack surface, special attention should be paid to wireless networks and Wi-Fi–enabled smart devices.

> **FYI**
>
> One of the most famous examples of a simultaneous Wi-Fi and mobile connection was the AT&T Wi-Fi connection called ATTWIFI. AT&T and Apple phones *should* have connected to the wireless access point using the device's unique MAC address. However, if the phone had connected to another AT&T network within the previous 24 hours, it simply ignored the MAC address verification check and connected to any network claiming to be ATTWIFI. This made hijacking Wi-Fi connections not just possible but very easy. Although this occurred almost a decade ago, user awareness has not really improved since then. It is still common to see smartphone users indiscriminately connecting to any free Wi-Fi network they come across.

The Kali Linux Security Platform

To scan and assess vulnerabilities in an enterprise, a professional security administrator or serious attacker requires a security toolkit based on a secure and mobile platform. The Linux OS supports an open source security platform called **Kali Linux**. This is a restricted version of Linux with single-user root access, but it hosts a multitude of security and penetration testing tools.

Kali Linux is a security platform containing tools that enable an administrator to check or verify security measures in categories such as the following:

- Information gathering
- Vulnerability analysis
- Wireless attacks
- Web applications
- Exploitation tools
- Forensics
- Stress testing
- Sniffing and spoofing
- Password and hardware hacking

Kali Linux also includes Metasploit and Aircrack-ng as well as more than 300 other penetration-testing tools, along with recipes for validating and penetration testing the security of the network.

The Kali Linux security platform, which can be downloaded as open source and loaded on a laptop, a USB drive, or even an Android phone, is based on Debian Linux. This makes it a very dangerous tool in the wrong hands. It is not necessary, but is advisable, to have some Linux knowledge before trying to use this tool. However, Metasploit includes tools that enable users to target virtual hosts to improve their skills before attempting a security assessment on a live network.

Some of the tools included with Kali Linux are used for scanning wireless networks. For example, Fern Wi-Fi Cracker and Aircrack-ng make the task of scanning a network much easier. After all, the goal in scanning a wireless network is to uncover all the wireless devices on the network, seek out vulnerabilities, hunt down rogue and unauthorized devices (that may be open to attack), test the authorized access points, and update the inventory. This is before considering taking action and eliminating or exploiting vulnerabilities.

Scanning with Airodump-ng

An example of a wireless network scanning tool is Airodump-ng, which is part of the Aircrack-ng suite bundled with Kali Linux. With Airodump-ng, the audit results returned from a wireless scan for each discovered device are as follows:

- **Basic service set identification (BSSID)**—An association with a station
- **Extended service set identification (ESSID)**—The network name
- **MAC address**—The unique ID for each access point and station
- **Probe**—The network that the station is trying to reach
- **Power level**—The level the entity is broadcasting on (the higher the power, the better the chance to crack the password)
- **Cypher**—WEP, TKIP, or CCMP
- **Authentication**—Preshared keys (PSK), Remote Authentication Dial-In User Service (RADIUS), or none

The administrator then collates and compares this data to the inventory to detect discrepancies. This is when the investigation really begins.

Client and Infrastructure Exploits

Cybercriminals tend to seek out weak points that they discover in a network infrastructure or in a device's hardware or OS. The weak points in a wireless network relate to the association and authentication of devices. In a wired network, much of this can be taken for granted, because a physical connection is required via a cable and switch. With a wireless network, this is not the case. Any device listening on the same 2.4 GHz or 5 GHz band can listen and perhaps negotiate an association and join the network.

In the early days of 802.11, wireless networks were easy to penetrate because encryption was weak and people didn't understand best practices. Today, IT departments have good standards and better tools with which to secure wireless networks. However, weak points in the network still exist. One of the main problems with wireless networks is that clients broadcast to any device that cares to listen. It is therefore imperative that wireless communications be protected against unauthorized access or eavesdropping.

Mobile devices, such as smartphones and tablets, have Wi-Fi capabilities that enable them to listen and communicate over wireless networks, both licensed (Telecom, 3G, LTE) and unlicensed (802.11). The fact that most devices can do this simultaneously creates a backdoor into corporate networks through which data (leakage) can flow undetected.

Other potential security vulnerabilities are mobile devices configured to automatically back up data to cloud data storage services such as iCloud, Google Cloud, One Drive, and Azure. This off-device storage of data may be okay for personal data, but it might not be the right choice for corporate data. Worse, in many cases, the company may not even be aware that it's being done. At issue is the fact that data stored on these sites is also available to other devices and applications that operate on the same user account. This leaves serious security questions unanswered, such as where does this data reside and under what secure conditions?

Client-Side Exploits

Interestingly, Wi-Fi communication has become far more secure in recent years. The same is true of authentication with Extensible Authentication Protocol–Transport Layer Security (EAP-TLS) and encryption techniques detailed in 802.11i. Consequently, attackers have started to shift to client-side exploits via malware through attachments, downloads, website browsing (drive-by attacks), and direct USB connections.

Client-side attacks have always been part of the attacker's arsenal, infiltrating clients with Trojans and malware through a variety of techniques. Smartphones are inherently more secure than PCs, however. This creates a bit of a dilemma for cybercriminals. One "solution" is to attack a smartphone via the PC's USB port.

For example, a cybercriminal could begin an attack by using social engineering to gather email addresses and other information. After finding email addresses of trusted partners (for example, the company's mobile phone supplier), an attacker could use the company's email address as the source of a forged email. Typically, all it takes to get someone to open an attachment is a convincing email (and subject line) offering an invitation to a special event or a change in billing that must be approved.

The attacker can then construct their client-side payload using Metasploit exploits such as the Adobe util.printf, which targets a vulnerability in Adobe Reader. The goal of the attack is to inject a payload disguised as a file with a .pdf extension (attached in an email, as noted). The file executes a reverse TCP/IP connection back to the attacker's machine to load a USB exploit. When loaded, the program can gain remote control of the USB ports. The malicious PDF file will pass an antivirus check and will be activated when opened, thereby initiating a connection back to the attacker's machine and simultaneously capturing control of the USB ports. Gaining remote control of the USB port gives the attacker the opportunity to take control of a smartphone when it is attached to the port. The attacker can then jailbreak or root the device, steal data, or infect it with a Trojan or other malware, depending on their intent.

This type of client-side exploit is difficult to remediate. This is because the attacker uses social engineering to gather information and then uses sophisticated tools to embed the malicious payload into an attachment sent from what appears to be a trusted source. Once the payload is activated, however, a WIPS should be able to detect traffic anomalies and raise an alarm.

Other USB Exploits

The smart device's vulnerability lies in the handshake between the USB port and the device when it is connected. Among other things, this handshake determines what the plugged-in device is. For example, a rogue USB device might claim to be a human interface device such as a keyboard, at which point all keystroke information typed on the infected PC will be forwarded to the connected USB device. Called a *keylogger*, this exploit is common on both fixed and wireless peripherals. This attack is often deployed in hotel business lounges, where an attacker plugs a USB key into a PC used by dozens of people to check various accounts. This works because very few people think to look at what is physically connected to the machine they are using. The hacker can come back weeks later to harvest a potential windfall of account name and password combinations.

Of specific concern for smart devices is that USB standards allow a promiscuous association that makes the USB connection between a vulnerable PC and a tethered smartphone very dubious. Therefore, many cybercriminals target weak PC clients (lower-priority targets). By infecting them, they have direct control that enables them to infect smartphones and tablets (high-priority targets) when using the USB connection. Kali Linux features several tools designed to reprogram USB controllers and allow exploits on the hardware vulnerability that affects all devices. Furthermore, to the delight of cybercriminals, the USB vulnerability has no remediation other than to disable the USB controller.

 **NOTE**
One implementation of Kali Linux, NetHunter, is designed to run on the Google Android Nexus range of smartphones, making it a very lightweight and portable auditing or attack tool.

Network Impersonation

Another common network attack is *network impersonation*. With this method, the attacker impersonates an access point by broadcasting an identical SSID (evil twin), but on a network interface on which they have boosted the power setting. By having a stronger radio power signal than the genuine access point, the attacker's rogue access point can capture clients trying to connect. For this attack to be effective, though, the original access point must drop all its associations with clients by sending out deauthorization (deauth) packets. Using tools such as hostapd software and a couple of wireless adapters, an attacker can easily set up a rogue access point on an Android smartphone or just about any other wireless device capable of running Kali Linux.

When the attacker sends the deauth command to the original (valid) access point, that access point will drop all connections. At that point, the clients will attempt to reassociate. Always favoring the stronger signal, the clients will associate with the rogue access point, and all communications will flow through the Kali Linux box. From there, an attacker can run any man-in-the-middle (MITM) exploit, such as Evilgrade (used to inject fake updates) or SSLsplit (to take over an encrypted session). Hackers can also spoof Domain Name System (DNS) queries by running the Dnsmasq tool to redirect the victim's smartphone's browser requests to a phishing site.

Creating an evil twin access point is pretty straightforward. Indeed, there are Kali Linux recipes for constructing these rogue access points. Unfortunately, regular audits will not mitigate the risks, as these rogue access points might be on a smartphone that will be difficult to locate due to its size and portability. Using a smartphone-based rogue access point, the hacker can move around or operate intermittently, lying dormant for long periods to avoid detection. A WIPS is a critical tool for listening for and detecting rogue devices on the network's spectrum. A WIPS will not stop them from occurring, but it will make it much easier to detect them, which may prove to be a deterrent.

Network Security Protocol Exploits

In addition to exploiting clients and infrastructures, cybercriminals can (and do) exploit network protocols and services. This is especially effective when network security protocols and services can be spoofed. In that scenario, the very foundation of an organization's security is used against itself.

RADIUS Impersonation

The preceding section outlined how a hacker could use a rogue access point to direct traffic to a false DNS server. Along those same lines, FreeRADIUS is a tool that enables a rogue access point to intercept and capture clients' logon credentials by passing the RADIUS authentication requests to a rogue host running FreeRADIUS. Packages such as PwnSTAR and easy-creds can automate the setup of the fake environment, with Karmetasploit (another Metasploit tool bundled with the Kali Linux platform) impersonating access points, capturing passwords, and harvesting data.

In the RADIUS server exploit, the attacker first sets up a fake access point and RADIUS server and waits for (or more likely forces) clients to authenticate. The attacker can then capture the challenge-and-response traffic between the RADIUS server and the client. These are encrypted, but the cipher can be broken using a variety of password-cracking tools. When this is complete, the attacker has a user account and password, which, depending on the user's credentials, could allow the attacker access to sensitive data. These credentials can work on a variety of access methods, from remote access to virtual private networks (VPN) to Outlook Web Access.

This attack works because self-signed certificates are often used in 802.11 environments, and these are installed on the RADIUS server. Unfortunately, Windows clients are forced by default to accept any certificate the server presents to them. (Protected EAP [PEAP] configuration does not require clients to validate the RADIUS server certificate.) Furthermore, clients are often configured to support external **certificate authority (CA)** certificates. (A certificate authority is a company or agency that issues digital certificates.) If an external certificate is enabled, an attacker sending a forged certificate will prompt the client for confirmation. The client will almost always confirm the false certificate, causing the forged external certificate to become trusted. This authenticates the attacker's device, providing the attacker access to many network resources.

To remediate this type of attack, administrators should use EAP-TLS. It is not vulnerable to this type of attack because it uses mutual authentication. Unlike with PEAP, where only the client authenticates to the server, with EAP-TLS, the client authenticates to the server and the server also authenticates to the client. An alternative is to install an internal CA certificate and disable external or public signed certificates, which can be forged. By forcing clients to validate RADIUS certificates and by manually configuring the internal server certificate (and deselecting all other certificates), administrators can prevent this exploit. This remediation is best accomplished through Active Directory, which allows centralized configuration and ensures distribution to all wireless clients.

Public Certificate Authority Exploits

Certificates are used for authentication in communication with servers in banking and retail websites to assure clients that the sites are indeed who they claim to be. This Secure Sockets Layer (SSL) protocol uses certificates from trusted CAs to prove the authenticity of the server and ensure the confidentiality of the data. Most browsers strictly enforce a policy of checking the validity of the SSL certificate and immediately throw a warning if the certificate is out of date or invalid. However, many mobile applications are deployed without requiring a check of the validity of SSL certificates.

An ongoing issue with SSL certificates was the typically long validity period of often 3 years. This was favorable to administrators as no one wanted to be going through the renewal process any sooner than necessary. However, the vendors of the popular browsers, Google and Apple, with Chrome and Safari, respectively, were actively campaigning for shorter certificate validation periods. The logic behind this was that a shorter validity period forced the regeneration of encryption keys. If this was done in a regular fashion then it would enhance security and mitigate much of the risk associated with the loss of existing encryption keys or passphrases. Indeed, in Sept 2020, Apple's Safari browser no longer trusted SSL/TLS leaf certificates with validity of more than 398 days. It would of course continue to support SSL/TLS leaf certificates issued prior to that date.

Compounding the issue is the growing number of self-signed but fraudulent certificates. These include fraudulent certificates for Facebook, Google, Apple, and a Russian bank. Although few up-to-date browsers would pass these fraudulent, self-signed certificates, many mobile applications do. This is a problem because if an application does not validate the authenticity of an SSL certificate, an attacker can use a forged certificate to launch MITM attacks on banking and e-commerce sites.

The **Heartbleed** security vulnerability discovered in OpenSSL in early 2014 was a staggering blow to what was believed to be secure communications and reliable authentication processes. This shook peoples' confidence in online banking and e-commerce platforms. Moreover, OpenSSL was used in many open source and commercial software programs, including the Android OS. In fact, the Android OS had vulnerable versions of OpenSSL in all releases from Jelly Bean to KitKat. Fortunately, these releases had heartbeats disabled, so only early versions of Android OS (4.1.1) were vulnerable to this exploit.

> **FYI**
>
> SSL keeps connections alive by sending periodic keep-alives, or *heartbeats*. The problem was that OpenSSL used recently dumped data in the memory stack to send back as the heartbeat. This posed a security risk because that data could be sensitive. To compound the problem, the client could request a heartbeat of, say, 64 KB. This was greatly in excess of what was required for the keep-alive function and potentially contained the decrypted security keys. Indeed, it was proved that encryption keys could be recovered through the Heartbleed vulnerability. Most frightening was the realization that the OpenSSL vulnerability had been susceptible to this exploit for years.

Developer Digital Certificates

In addition to communication and authentication, certificates are also used for application validation. In this case, an encrypted digital signature in the code, signed by a trusted source, is used to ensure that an app has not been tampered with. For example, the trusted source could be Apple, Microsoft, or Google within its own application portal—the App Store, the Windows Store, or Google Play, respectively. In corporate settings, where an enterprise may need to side-load or download in-house applications to BYOD devices, applications can be verified with enterprise developer certificates. Developer certificates can be stolen, however, and sold on the black market. Or, they can be forged and then used to sign malicious applications, which can then be downloaded from a captive portal or side-loaded using a USB cable.

Browser Application and Phishing Exploits

Another class of indirect attacks on mobile devices is browser and phishing exploits. These types of exploits have existed for years on PCs and are essentially unchanged on mobile devices. What is different is the small size of mobile clients, which makes it more difficult to spot some of the telltale signs of these exploits, and a more nonchalant approach to clicking on links and opening emails on mobile devices.

Captive Portals

Captive portals play an important role in IT security. They are often used to validate guest users on a per-guest basis and to communicate terms of use for Wi-Fi networks. However, like many tools and technologies, captive portals can and are also used by hackers and cybercriminals. In addition to being used for MITM attacks, captive portals can be used to steal user credentials. Usually, this occurs on a website that allows users to access the Internet or a private intranet.

The security tool PwnSTAR can be used to front-end a rogue access point for this purpose. Client mobile devices are directed to the captive portal through DNS spoofing. The client then autoconnects without the owner even being aware that they have been redirected. Once the client is on the captive portal, the hacker can steal Wi-Fi Protected Access (WPA) handshakes and email credentials, serve up phishing webpages, and launch assorted exploits such as Aireplay-ng and Airdrop-ng to deauthenticate stations.

> **NOTE**
> Captive portal–based hacks have become very common with the rise in popularity of smartphones and tablets and a new breed of mobile consumers who are less aware of security threats. It's easy to fool people with captive portals because they serve a legitimate purpose. As a result, users tend to not look too closely at them.

As an example, a cybercriminal in a shopping mall could use routing tables to route clients to a captive portal. After users enter their credentials, the captive portal could act as a real hotspot by allowing users to access the Internet. Similarly, the portal could allow Internet access but only after the user downloads a client-side PDF attack that exploits the Adobe Reader vulnerability in the client-side attack detailed earlier. In addition, the attacker could launch several browser exploits to track usage and to log keystrokes.

Drive-By Browser Exploits

Captive portals and phishing sites are not the only dangers of which mobile users must be wary. **Drive-by browser exploits** are becoming more common. Drive-by browser exploits target web browser plug-ins on mobile devices for Java, Adobe Reader, and Flash. These attacks are most often launched via legitimate but compromised websites that infect browser software running on mobile clients. The fact that they can do this without any user interaction is worrying—just browsing the site is enough to be contaminated.

An example of a drive-by exploit on Kali Linux is the Browser Exploitation Framework (BeEF). This tool can be easily installed and launched to assess or exploit mobile devices through their web browsers. BeEF uses a JavaScript hook, which runs on the client's browser. The script is embedded into the code of the webpage, requiring that the site be compromised. Once this is done, setting up the exploit is easy. Any number of means can be used to get the victim to the website, including, but not limited to, social engineering and DNS spoofing. When the victim visits the page, the code injects a Trojan payload into the browser.

Keeping browsers and plug-ins up to date and hardening websites are the best defenses against this exploit for users and site owners, respectively. Users can also disable JavaScript plug-ins, but this can seriously degrade the web-browsing experience.

Mobile Software Exploits and Remediation

Like all network-based devices, mobile devices are subject to attacks that seek to exploit holes, backdoors, or other weakness in the software responsible for running the basic functions, the communication protocols, and the applications that reside on the device. Given the combination of varied usage patterns, broad adoption, relative newness, and complex software, it's no surprise that mobile devices are attractive targets for hackers and cybercriminals.

CHAPTER 13 | Mobile Wireless Attacks and Remediation

> **FYI**
>
> These mobile risks and the required remediation are primarily the concern of developers rather than IT security specialists. However, it helps for IT security specialists to understand these issues. These personnel are "downstream"; as such, if these problems manifest themselves on a network, they're the ones who will act as first responders. An understanding of these issues may help with postevent actions, which may include disallowing the offending app and communication with its developers.

The **Open Web Application Security Project (OWASP)** is an international foundation and open community dedicated to enabling organizations to develop secure applications and to raising the visibility of software security so that individuals and organizations can make informed decisions about security risks. Recognizing mobile threats as their own class of security issues, the OWASP lists its top 10 mobile risks as follows:

- Weak server-side security
- Unsecure data storage
- Insufficient Transport Layer protection
- Unintended data leakage
- Poor authorization and authentication
- Broken cryptography
- Client-side injection
- Security decisions via untrusted inputs
- Improper session handling
- Lack of binary protections

Each of these is detailed in the sections that follow.

Weak Server-Side Security

Weak server-side security can allow the injection of malicious code to infect the mobile browser using drive-by exploits. This vulnerability exploits the web interface or web application programming interfaces (APIs), and the impact can be severe. The solution to this problem is to use applications from trusted sources that have a reputation for developing secure application software.

Unsecure Data Storage

Typically the result of developers who assume that users or malware will not be able to access their device file systems, this vulnerability occurs when sensitive data is stored in locations with no or inadequate security. If the device is breached or compromised, the result could be data loss, fraud, or even stolen credentials. Attackers commonly exploit this vulnerability through malware, and the impact can be severe. Large numbers of users can be affected in a single breach.

Remediation methods vary by platform. For iOS developers, best practices include the following:

- Never store credentials on the device's file system.
- If the storage of credentials on the device is necessary, use a secure iOS encryption library.
- For items stored in the password keychain, use the most secure API designations.
- Consider a layered approach to encryption, and don't just rely on the device's hardware encryption.
- For a database stored on the device, use SQLCipher for SQLite data encryption.

Best practices for Android devices include the following:

- For secure storage of data on the device, consider using the enterprise Android Device Administration API. This can force encryption to the local file store.
- Secure all plaintext data with a master password and AES-128 encryption.
- Consider a layered approach to encryption, and don't just rely on the device's hardware encryption.
- Ensure that the shared preferences properties are not set to Mode_World_Readable.

NOTE
From an IT perspective, an incident traced back to one of these issues should result in disallowing the offending app and communicating with the developers.

Insufficient Transport Layer Protection

Mobile applications often do not protect network traffic that uses secure SSL/TLS for authentication and then reverts to cleartext HTTP after secure logon. This makes them vulnerable to MITM attacks and eavesdropping. The impact is moderate, because it typically affects only one person/session at a time.

To protect against this, best practices for all developers are as follows:

- Ensure that digital certificates supplied by the server for authentication are valid.
- Use the secure transport API in all cases.
- Do not allow self-signed certificates from external websites and servers.

In addition, Android developers should remove all code that refers to applications that accept all certificates by default, such as Apache's AllowAllHostnameVerifier. This is the same as trusting all certificates, regardless of origin.

Data Leakage

Data leakage is often the result of developers inadvertently placing sensitive data in easily accessible storage locations on the mobile device. This is another common and easy-to-exploit vulnerability. In mobile devices, it typically manifests itself in the way the OS caches data and images and handles logging and data buffers.

To protect against this, best practices for application developers working on iOS and Android devices are as follows:

- Check for URL, copy and paste, and keyboard caching.
- Check HTML5 data storage.
- Check browser cookies.
- Check analytics and other information being sent to third parties.

Poor Authorization and Authentication

Poor or insufficient authentication mechanisms enable an attacker to gain access to and execute functions on mobile devices. The authentication failures on mobile devices are usually the result of input forms on mobile screens, which tend to use simple authentication schemes. This vulnerability is common and simple to exploit, but the impact can be severe. Remediation in this case falls on developers, who should ensure that server-side authentication mechanisms are enforced rather than relying on device authentication on the user's part.

Broken Cryptography

Poor or weak encryption used for authentication or data communications can be easily captured and cracked by an attacker. In iOS applications, code is encrypted to prevent reverse engineering or tampering. However, upon startup, the iOS loader must decrypt the application in memory before it can execute the code. That means a rogue developer can load an application on a jailbroken iOS device and take a snapshot of the decrypted code resident in memory. Attackers can use tools such as ClutchMod and GBD to accomplish this task just before the application starts to execute. This vulnerability is common, and can have a severe impact. With the right tools, this vulnerability is easy to exploit.

Remediation methods for all developers are as follows:

- Do not rely on built-in code encryption processes.
- Manage keys effectively and securely.
- Do not use deprecated algorithms such as RC2 (short for Rivest Cipher 2, after Ron Rivest, the algorithm's developer), MD4 (short for Message Digest), MD5, or SHA-1 (short for Secure Hash Algorithm).

Client-Side Injection

Client-side injection leads to the execution of malicious code on a mobile application. This can generate malformed data that is intended to be malicious and can result in Structured Query Language (SQL) injection, JavaScript injection, or gaining access to application interfaces or functions. Client-side injection vulnerabilities are common and easy to exploit, although their impact is considered moderate. This is because the client-side code typically runs with the same security permissions as the user, which can restrict the impact on other services or servers.

Best practices for mobile application developers include checking applications for the following:

- SQL injection (via input validation)
- JavaScript injection (via input validation)
- Local file inclusion
- Extensible Markup Language (XML) injection
- Classic C attacks, such as C functions prone to injection, including strcat, strcopy, strncat, sprint, and gets

Security Decisions via Untrusted Inputs

A mobile application can accept data from many sources via an **inter-process communication (IPC)**, which allows different phone processes and subdevices to share information. To prevent malicious code from infecting other applications, a sandbox approach should be used in development. Only applications that are explicitly required to communicate with an app should be permitted to do so. This occurs through the use of a **white list**, a list of permitted applications or users. The prevalence of open communication between processes is a common security weakness, and the impact of exploitation is severe—especially given that it is also considered an easy vulnerability to exploit.

To mitigate this, best practices for developers are as follows:

- Only allow applications to interact if there is a pressing business requirement.
- Actions that may access sensitive data or functions must require user intervention.
- All input received from external sources should undergo strict validation testing.
- Do not use the Apple Pasteboard for IPC communications because it can be read by all applications.

Improper Session Handling

Sessions must be handled for users during interactions with web applications due to the stateless nature of HTTP. For a user to follow a seamless set of transactions with a web server without having to authenticate each time, the session must be maintained so that the server can uniquely identify the user. This is done through tokens or cookies, which identify and maintain the session between the server and client user. The vulnerability here lies in the fact that an attacker can gain access to the cookie and use it to hijack a session by impersonating the user (or server). Attack routes tend to be through physical access to the device, over-the-air capture of a communication stream, or malware. The vulnerability is common, and the impact can range from moderate to severe. This exploit is easy to initiate.

The best option for remediation is for developers to understand that they must invalidate sessions on the server side *and* on the mobile device. Otherwise, there's a window of opportunity for an HTTP exploit tool to capture the existing session. Developers should also place adequate timeout limits on applications based on the application's sensitivity—for example, 3 minutes for a secure app and 1 hour for an app that's less secure.

Another common mistake is failing to rotate cookies during session management. Cookies should be rotated and existing sessions torn down and re-created during events such as the following:

- A user switching from anonymous to logged on
- A change in logged-on users on the device
- A switch from a regular user to an elevated privileged account (admin or root)
- After each session timeout

Lack of Binary Protections

Many mobile developers provide little in the way of binary protection to their code. **Binary protection**, also referred to as *binary hardening*, is a software security technique in which binary files are analyzed and modified to protect against common exploits. Binary protection prevents rogue developers from reverse engineering or tampering with application code. Without it, cybercriminals can alter code and inject malicious code in the form of malware into the application to perform some hidden functionality. The vulnerability is common, and the impact can be severe. Reverse engineering the code is not a trivial task, however.

To mitigate this, best practices for developers include ensuring the application code adheres to secure coding techniques, especially with regard to jailbreak detection, checksum controls, certificate pinning controls, and debugger detection. Additionally, the mobile app must be able to determine at the time of execution if there is a code-integrity violation.

CHAPTER SUMMARY

The prevalence of mobile devices that pass freely between public and corporate networks and a general lack of security awareness among many mobile users have made smart devices a preferred target for cybercriminals. Not only do many of the common PC exploits also work on these devices but cybercriminals now use some traditional PC exploits to leapfrog onto mobile clients, which are considered higher-value targets. To combat these threats, IT teams must expand their efforts to include new mobile-based attacks, perform regular scanning of the corporate network, and increase their emphasis on user security training and awareness.

KEY CONCEPTS AND TERMS

Binary protection
Certificate authority (CA)
Client-side injection
Drive-by browser exploits
Heartbleed
Inter-process communication (IPC)
Kali Linux
Open Web Application Security Project (OWASP)
White list

CHAPTER 13 ASSESSMENT

1. Weak server-side security does not pose a direct threat to mobile clients.
   A. True
   B. False

2. Which of the following is not a risk from client-side injection attacks?
   A. SQL injection
   B. Data leakage
   C. JavaScript injection
   D. Hackers gaining access to application interfaces or functions

3. Which of the following describes binary protection?
   A. It is a two-way handshake for authentication.
   B. It prevents rogue developers from reverse engineering or tampering with application code.
   C. It is a form of session security.
   D. It prevents Bluetooth-based attacks.

4. In addition to searching for rogue devices, scanning wireless networks is also used to do which of the following?
   A. To see which websites employees are using
   B. To prevent data leakage
   C. To frequency-jam unauthorized access points
   D. To verify that security measures are in place on authorized access points

5. Given the improvements developers have made on their mobile platforms, user security awareness is no longer an issue.
   A. True
   B. False

6. The ability of many smart devices to access mobile and Wi-Fi networks simultaneously creates potential issues with which of the following?
   A. Corporate governance
   B. Data leakage
   C. Security patches
   D. Sandboxing

7. PC-based USB exploits are not a concern for mobile devices because most of these devices do not have USB ports.
   A. True
   B. False

8. Why are certificate authority (CA) exploits an issue on smart devices?
   A. Mobile browsers do not enforce the validation of certificate authority certificates.
   B. Many mobile applications are deployed without requiring a check of the validity of SSL certificates.
   C. Mobile devices are small, which makes the certificates hard to read.
   D. All of the above.

9. Which of the following is *not* true about captive portals?
   A. They can be used to steal user credentials.
   B. The can be used to front-end rogue access points.
   C. They are used only for hacking purposes and should be avoided.
   D. They can be used to launch attacks.

10. Drive-by browser exploits target which of the following?
    A. Access points with no encryption
    B. Mobile devices on highways
    C. Near-field communication–based applications
    D. Web browser plug-ins for Java, Adobe Reader, and Flash on mobile devices

CHAPTER 14

Fingerprinting Mobile Devices

THE PROCESS OF IDENTIFYING A DEVICE on a network (or the user behind the device) is called fingerprinting. As the name suggests, the process involves identifying some set of characteristics that uniquely identifies one device or user.

This chapter looks at the fingerprinting of mobile devices both directly (that is, devices connecting to a network) and indirectly (that is, devices connecting to one or more websites). Given the nature of mobile devices, you'll find that the methods used to fingerprint stationary PCs do not work so well on mobile devices. However, some characteristics of mobile devices lend themselves to new and somewhat daunting fingerprinting capabilities.

Chapter 14 Topics

This chapter covers the following concepts and topics:

- What the nature of device fingerprinting is
- What the types of fingerprinting are
- What fingerprinting methods exist
- How unique device identifiers work
- How spyware for mobile devices works

Chapter 14 Goals

When you complete this chapter, you will be able to:

- Define fingerprinting
- Describe the nature of device fingerprinting
- List the two types of fingerprinting
- Discuss the various fingerprinting methods
- Understand how unique identifiers are collected
- Describe how sensors and components are used to fingerprint devices
- Discuss the difference between device fingerprinting, spyware, and spy software

© Cherezoff/Shutterstock

Is Fingerprinting a Bad or a Good Thing?

Fingerprinting is the process of identifying a device on a network or the user behind the device. There is a common point of confusion regarding the nature of fingerprinting and whether it is good or bad. The broad assumption among the general public (and the media) is that it is both new and bad—a sinister plot cooked up on the Internet to violate the privacy of unsuspecting victims. The reality is that while there are new methods of doing so, fingerprinting of devices and users has been around for about as long as networks. Similarly, while there are ways in which fingerprinting can be (and is) abused, a lot of good can come from it, too. Like many technologies, whether fingerprinting is "good" or "evil" depends on users and their intent rather on the nature of the technology itself.

On the good side of the ledger, fingerprinting can help network administrators understand what (and who) is on the network. This can improve both performance and security. Fingerprinting also makes for a richer user experience online, allowing sites to offer customized content and more convenient consumer transactions.

On the bad side, fingerprinting can be used as a means to aggressively target users with unwanted advertisements and pop-ups. This is problematic for users who have visited sites that they would prefer were kept private, because ads associated with those sites can pop up any time the browser is active. Fingerprinting can also be used by cyberstalkers, who can literally track a person's every move. (Of course, this is a positive for law enforcement.)

The increased potential for wrongdoing that results from the fingerprinting of mobile devices is a legitimate concern. While standard computers might contain financial or potentially embarrassing information about a user, mobile devices—a constant companion for many people—hold deeply private details about people's lives, making them even more attractive for fingerprinting. Imagine if you were asked to take part in a program where someone could have access to the following:

- Your location at all times
- Private photographs (which are stamped with time and location)
- Private correspondence (emails and texts)
- Online browsing and shopping history
- Audio and video at any given time
- Banking and credit card details
- Security and convenience systems in your home
- Names, pictures, and contact information for your spouse, children, friends, and employer

It's doubtful that anyone would ever subject themselves to that level of tracking, but this is pretty much what's available on most people's phones today.

This is not a problem in and of itself. Apart from some rare cases (the tracking of suspected terrorists, for example), people are not typically subject to this extreme level of tracking. However, the mere existence of this type of access creates the potential for some very serious privacy problems. Therefore, while fingerprinting is not inherently bad, the significant consequences of fingerprinting when used for nefarious aims require increased attention and caution on the part of both security professionals and individual users.

Types of Fingerprinting

There are two general categories of fingerprinting. These categories pertain more to *where* fingerprinting is done than to *how* it is done. (There are various methods.) They are as follows:

- **Proximity fingerprinting**—Fingerprinting that occurs on a network is called **proximity fingerprinting**.
- **Remote fingerprinting**—Fingerprinting that is done online is called **remote fingerprinting**.

> **NOTE**
> The distinction between proximity and remote fingerprinting is more pronounced with wireless-enabled mobile devices than with fixed equipment. This is because the unique characteristics of radio transmitters are one method by which a mobile device can be uniquely identified.

Network Access Control and Endpoint Fingerprinting

Network access control (NAC) is a security paradigm that aims to restrict access to who or what (people and/or endpoint devices) can access resources on a corporate network. The goal of NAC is to positively authenticate all users that join a network and to determine whether their devices meet the minimum network security policy regulations.

NAC ensures that insecure devices cannot inadvertently access sensitive network resources or introduce things that could potentially infect the network. However, to be able to identify noncompany endpoint devices requires a technique called endpoint fingerprinting.

The purpose of endpoint fingerprinting is to discover, classify, and monitor unknown noncompany devices and endpoints. This type of fingerprinting is done by discovering the endpoints' MAC addresses and then validating them against a company's authentication, authorization, and accounting (AAA) register.

Endpoint fingerprinting helps network and security administrators automate security profiling of attached endpoints like Internet Protocol (IP) phones and Internet of Things (IoT) devices that may be sprawling throughout a network by using passive scanning techniques. NAC is typically used in larger networks but this is not the only type of network scanning and fingerprinting that goes on and some other techniques and tools are suitable even for small business networks.

> **TIP**
> Visualizing and fingerprinting network topology are helpful from a security and administrative perspective. Doing so lets the administrator see everything on the wireless network, which can then be compared against a baseline of known and authorized devices.

Network Scanning and Proximity Fingerprinting

Proximity fingerprinting is typically performed by administrators to identify devices found during network scans. By scanning the wireless network using tools such as inSSIDer, the administrator can scan, discover, and visualize the network topology. Visualizing Wi-Fi networks and their devices can reveal ad hoc networks and unauthorized connections. In this case, the purpose of fingerprinting a device is to reveal the type, manufacturer, operating system (OS) version, and any other information that the scanning application will retrieve through the discovery process.

Wireless Anonymity

Any device connecting to a wireless network, whether authorized or not, must communicate and associate with an access point. As a result of this process, the access point will have a record of the media access control (MAC) address of the device with which it is communicating. That means all devices can easily be detected. However, detecting a device and identifying it are two different things. This is because most hackers strive for wireless anonymity.

For example, suppose a hacker occasionally initiates a malicious attack from within the wireless network, such as a denial of service (DoS) attack against an external target. Suppose, too, that this is done for a brief period of time, and that the hacker wants to remain anonymous. When the attack begins, the security administrator is alerted via the WIPS and attempts to identify the anonymous device that appears and disappears in bursts. To identify the attacker's device, the administrator must first fingerprint it. Meanwhile, the attacker tries to avoid being fingerprinted so that they can remain anonymous and continue with the attack.

To remain anonymous, the attacker's device must constantly change its MAC and IP addresses. It must also clear any **cookies** on the device. (Cookies are files that contain a unique user identifier that a website creates when a user first visits.) Clearing cookies prevents a captive portal on the access point from retrieving a cookie that would identify the device. Alternatively, the attacker might masquerade as another legitimate user by forging that user's MAC and IP addresses and using their cookie as a false identifier.

Whichever strategy the attacker adopts, it is clear that they can easily change the device's identity by altering three of the most common identifiers: the MAC address, the IP address, and a cookie. This makes it necessary to find other characteristics or indicators of identity—at least in the cases where there is evidence or suspicion of unauthorized anonymous access.

In the case of authorized users and devices, this is a pretty straightforward exercise. It's not as simple to detect and locate attackers, however. Rogue access points in particular are notoriously difficult to locate because they are typically not on all the time. (Hackers prefer to lay dormant for long periods, becoming active only for short spells.) A wireless intrusion prevention system (WIPS) can detect these intruders as they work using the same principles as fingerprinting, but it does not always help in pinpointing a location. There is also an issue with anonymity, because it's in an attacker's interest to remain anonymous by avoiding fingerprinting.

Online or Remote Fingerprinting

The type of fingerprinting that most people are familiar with (and have been subjected to) is remote or online fingerprinting. This type of fingerprinting plays a significant role in e-commerce and search analytics. In this case, remote fingerprinting is beneficial to users, but there are controversial uses as well. These tend to garner headlines, given the concern for online privacy.

One of these controversial uses involves Internet marketing firms who collect personal information via both fixed and mobile web browsers, then collate and sell that information to other marketing firms. Some of those third-party firms then use hyperaggressive tactics aimed at classes of users based on the users' web-browsing habits.

It's important to note that this is not spyware. Spyware is malicious software that enables a user to covertly obtain information about another user's computer activities. In remote fingerprinting, no software is loaded onto the device. Rather, remote fingerprinting is the act of identifying a device based on a collection of common characteristics shared by all devices of a certain type. For example, when fingerprinting an Apple iPhone, the goal for the administrator or hacker would be to find a unique characteristic that could be used to identify the device when compared to all other smartphones. To achieve this, the fingerprint technique must work across web browsers.

> **NOTE**
> Fingerprinting PCs as they browse the Internet is nothing new. Indeed, it is viewed as a necessary part of the service that many websites provide. Without fingerprinting, it would be much more difficult to navigate the Internet. In addition, e-commerce would be much less convenient.

Websites need some mechanism to identify and remember users from one visit to the next. This is necessary because Hypertext Transfer Protocol (HTTP) is stateless—that is, it has no way to "remember" users from one transaction to the next, because multiple users access the site simultaneously. Session IDs are therefore essential to maintain state and assist websites with identifying users. With session IDs, websites can track and collate each unique user's actions into seamless transaction streams on a per-user basis. Session IDs last only as long as the user remains on the site, however. Sessions are deleted when the user ends the session or after some predetermined time of inactivity. Therefore, while session IDs are unique fingerprints, they are of no use for remembering the user's device from one visit to another. The solution was, and remains, the humble cookie.

Cookies

As noted, a website creates a cookie when a user first visits. When a device returns to the website, it passes along the cookie for that site. The website then recognizes the cookie's unique identifier, which identifies the user.

The cookie itself is a simple file containing information stored as sets of name-value pairs—for example, userID/A9CF87546ABC, where the name of the pair is userID and the value is A9CF87546ABC. Most websites create cookies that store this information and nothing else. This is not because they care about the user's privacy, but because it isn't practical to store the user's information on the user's device. That's mostly because websites store a lot more about the user than just the name-value pair. Some common information includes the user's browsing history, what pages or advertisements the user clicked on, what preferences the user has configured, their shopping cart status, and their sales history. The list is endless.

This critical information is stored in the website's own database and referenced when the user visits the website. When the user clicks a link or types a URL to visit the site, the cookie is passed to the site and is used to look up this key user data. This is how websites such as Amazon.com remember their customers and are able to provide useful shopping hints based on previous sales or browsing activity. They can also remember and maintain the customer's shopping cart over multiple visits, because this information is stored on their databases. This is a simple solution to providing state information over the stateless Internet, and it's worked well for years.

Unfortunately, this system has its flaws. The biggest one, from the website's perspective, is that a user can delete the cookie file when clearing their temporary Internet files folder. (Incidentally, this happens to be the first step technical support asks the user to take when troubleshooting browser issues.) This is not a major problem, but it will be inconvenient when the user returns to their favorite website. Because the cookie file has been deleted, the site will no longer recognize the user. Instead, the website will consider them a new user and create a new cookie. Fortunately, the user's original information will still be in the website's database. The user can use their logon credentials to retrieve that information and reassociate their new cookie with the previously collated information stored in the website's database. This is a primary reason websites store only the name-value pair on the customer's device.

Despite the benefits that cookies provide, they've developed a bad reputation. Some say they invade users' privacy. As a result, many users now disable cookies as the default setting, enabling them only for specific reputable sites on a per-use basis. This is not a bad strategy, because a website should only store information relevant to its specific domain, not from third-party sites. Nevertheless, the real issue is not with cookies, but with browsers that allow cross-site profiling.

Cross-Site Profiling

Cross-site profiling occurs when a browser accepts cookie requests from third-party advertisers on the website the user is visiting. Most browsers allow third-party cookies by default, although Safari on the iPhone browser does not. (It must be specified in the Preferences screen.) Cross-site profiling is a problem because it allows advertisers and giants like Google to track users across many sites. Google, for example, advertises on many sites, and can track the customer's movements by collecting third-party cookies via JavaScript that runs in the ads on the host website. Although the information is anonymous, over time, with enough data, advertisers can profile the customer's interests and preferences, and through cookies uniquely identify the customer's device. Once that has been done, they can focus their advertising based on the user's web-browsing habits. That's why, when you visit a website you've never been to before, you might see ads for products that are similar to ones you've purchased on other websites.

For advertisers, this is of course a wonderfully rich set of data. Advertising networks sell advertising to their clients that can specifically target users who have browsed similar products or clicked similar ads. By using third-party cookies, they can fingerprint the user's device. This is neither good nor bad in and of itself. Supporters claim that it enhances the browsing experience and drives commerce. Detractors, however, say it is an unethical breach of privacy. This is a tough case to make unless the advertising network (or some other entity) makes a connection between the user name, address, and browsing history, and then chooses to exploit that information.

Developers, advertisers, and cybercriminals are searching for new ways to fingerprint mobile devices. They seek an alternative to cookies, which users can disable. As always, it is vital to educate mobile users to be cautious about which information they enter into untrusted websites, especially with a device that is also used for work purposes. As mentioned, even if a technology is developed for positive aims, it can still be twisted into something sinister.

> **FYI**
>
> Fortunately, the big three mobile device providers—Apple, Google, and Microsoft—are moving toward device diversity. With device diversity, a user can work on any device with their data stored in the cloud, using workflows that follow them from one device to another. This greatly mitigates fingerprinting performed via browsers because the user can easily switch back and forth between devices. This makes the task of fingerprinting more difficult.

Fingerprinting Methods

The problem with fingerprinting a mobile device is that to be effective, fingerprinting characteristics must be diverse enough to be unique. Uniqueness based on diversity is the best way to ensure that no two devices have the same fingerprint. This is not always an easy thing to achieve, however—especially with iPhones, which are closed systems with factory-set configurations.

A mobile device must also be stable. That is, its fingerprint must remain unique over time. Some sets of identifiers may be unique at any given time, but if the characteristics can be easily changed by a user, then they cannot be considered stable. If a fingerprint is not stable, then it cannot be used to track a user over time, which is the whole point of fingerprinting. An example of this is the Apple iOS IdentifierForVendor tag. A collection of these may be unique, but the tag persists only as long as the application remains on the phone. Therefore, it is not stable.

> **NOTE**
>
> Diversity and stability tend to work against each other. As diversity increases (through the use of multiple parameters), fingerprint stability decreases. This is because there is a higher chance that one or more characteristics will be changed.

Many characteristics can be collected. Generally speaking, however, the more distinctive a characteristic is, the harder it is to obtain. The two general categories of fingerprinting—proximity fingerprinting and remote fingerprinting—relate to *where* fingerprinting is done. Beyond that, there are two types, or methods, of fingerprinting: passive and active.

Passive Fingerprinting

Passive fingerprinting occurs without querying the client device. Instead, it analyzes information supplied by the device itself. On mobile devices, passive analysis typically focuses on certain protocols, including HTTP headers, TCP/IP headers, 802.11 Wi-Fi settings, and OS parameters.

Depending on the purpose of the fingerprinting, some of these protocols will be more useful than others. For example, if a network administrator wanted to fingerprint devices on the network, they might set up a laptop as an access point and use Wireshark to intercept the traffic on the Wi-Fi interface in promiscuous/monitor mode. Using Wireshark to decode the packets, the administrator would be able to determine the MAC and IP addresses of all the devices communicating on the segment. From the headers, device type, and OS, the administrator could create a reasonable set of device fingerprints on the network.

Examining TCP/IP Headers

Another approach to passive fingerprinting is to examine the TCP/IP headers. These headers contain very important data. For example, a device's IP and MAC addresses could help identify it as a smartphone. However, using IP and MAC addresses is not a great way to identify a device because they are not stable. IP addresses are easily changed with Dynamic Host Configuration Protocol (DHCP) and MAC addresses are easily altered through MAC spoofing.

A better option is to analyze the HTTP traffic that the device submits to a website, because it contains a lot of diverse options. This is how Panopticlick determines the distinctiveness of a browser. This website remotely checks for distinctiveness in browser characteristics by comparing the fingerprint of one browser against its database of more than 4 million samples (at time of this writing). It looks for characteristics such as the following:

- User agent
- HTTP_ACCEPT headers
- Browser plug-in details
- Time zone
- Screen size and color depth
- System fonts
- Whether cookies are enabled

Panopticlick shows that there are sufficient differences between PC browsers that PCs can be uniquely identified. Fortunately (or unfortunately, depending on one's perspective), the same is not true for mobile devices, especially Apple iPhones. This is because PCs are readily customized with different browsers, plug-ins, and system fonts, which greatly aids the chances of a browser being identifiable. Apple iPhones, however, do not have that diversity. They are typically all configured alike.

Application Identification

One method for fingerprinting smartphones is to examine not just the browser fields but also the applications residing on the device. This can be done through a port scan, using a tool such as Nmap, or passively, through application fingerprinting. **Application fingerprinting** looks more closely at the HTTP user agent headers for application-specific information. Many web and cloud apps work in the background on smartphones even when they are not active, occasionally connecting and synchronizing or looking for updates. What is more, each mobile application sets its own user agent request header. Therefore, it is possible to tell which application on which mobile device originated the HTTP request. This is a passive way to determine the application set on a smartphone or tablet without using obtrusive queries or violating the user's privacy. Additionally, by observing synchronization of applications between devices—for example, an iPhone, an iPad, and a PC—it can be inferred that they belong to the same owner. Of course, this will only work if the combination of applications is sufficiently diverse and if they remain installed.

Active Fingerprinting

Active fingerprinting differs from passive fingerprinting in that it queries the device in an invasive manner to access characteristics not readily obtained otherwise. For example, active queries attempt to find serial numbers and other unique characteristics that are both diverse and stable. An example of active scanning is using Simple Network Management Protocol (SNMP) to discover, map, and visualize a network. SNMP interrogates each device and retrieves information from agents to map the network.

The preceding section discussed the passive analysis of HTTP, which yields a number of parameters that could be used to fingerprint a browser. Active scanning can do the same, but it can dig deeper, interrogating the browser to obtain much more information. A typical technique used by website designers is to run JavaScript on an e-commerce webpage. When the browser loads the page, the JavaScript runs on the browser and requests additional information about the machine's identity. This is how Google Analytics works.

Active scanning can also involve probing the network and recording the responses from devices. Vulnerability scanners such as OpenVAS can also be used to actively scan the network. This will return OS versions and patch profiles, which could be useful in fingerprinting network devices.

Unique Device Identification

The Holy Grail of fingerprinting is a unique identifier that is everywhere, accessible, and permanent. Ideally, when scanning a device, one would interrogate the device for a unique identifier such as its serial number. Application developers often try to obtain one of these hardware identifiers to use as a unique identifier in their apps. Other solutions are for developers to give the device their own identifier, like a cookie, and download a unique serial number with the application. This is how a lot of cloud-based mobile apps work. However, users can circumvent this tactic simply by deleting the application and downloading a new copy with a new identity. While these are certainly ways for developers to solve the identity problem from their application-centric perspective, none of them is a widely employed method of device identification and fingerprinting.

Apple iOS

Every device has its own unique identifier. For example, all Global System for Mobile Communications (GSM) phones have an International Mobile Station Equipment Identity (IMEI), while Apple and Android phones have a Unique Device Identifier (UDI). Access to these identifiers is tightly controlled. In the wrong hands, this information could lead to abuse such as fraud and identity theft.

Apple phones used to have two unique identifiers: the universal device identifier (UDID) and the IMEI, which all GSM phones have. Not surprisingly, Apple took a dim view of applications that tried to access these parameters, banning them from its App Store. Any potential vulnerability created from gaining access to the UDID was mitigated when iOS 5 was deprecated, and UDID is no longer obtainable on iPhones.

In the place of the UDID, Apple introduced another identifier in iOS 6 aimed at advertisers: advertisingIdentifier. This identifier is unique to each device but can be removed if the device is wiped by the user. Another potential identifier is IdentifierForVendor, which is also available in iOS 6 and later. This identifier's purpose is to link the device to the appropriate application vendor for analytics. There is a different identifier for each vendor with an installed application. This identifier is present only as long as an application from the vendor remains on the phone.

Application developers could use these identifiers to fingerprint devices, but it only works for devices on which the user has installed that vendor's application. This is not a problem for the application developer, who only wants to be able to identify the device in their own application. However, for general fingerprinting, it has a major weakness—by design. If the user deletes the application, or simply does not run it, the identifier will not be available. Because of Apple's sandbox functionality, the identifier is not available to other applications, so the fingerprint only works when the specific application is running—and then, only for the application's developer.

Android

With Android phones, the IMEI can be used as a unique hardware identifier and is accessible through the getDeviceID method. Each device also has a device serial number and an Android ID. Both are unique numbers; the latter is a 64-bit number that is randomly generated on the device's first boot and that remains constant for the device's lifetime.

These numbers provide excellent means to identify an Android device in a specific application. For example, these identifiers could be used to fingerprint a device in a cloud service provider's mobile application. However, just as with the iPhone, it is not feasible to create an application that can be loaded on all Android phones for the purpose of fingerprinting beyond a specific application. And just as with the iPhone, once the application is removed, the identifier is lost.

HTTP Headers

An alternative way to identify a device when the UDID and IEMI are not available to the application is to create data sets and then use the information contained in each set to create a unique fingerprint for the device. For this to work, the browser must be actively scanned and queried for the following information:

- Country code
- Device brand
- Device model
- Device carrier
- IP address
- Language
- OS name
- OS version
- User agent
- Timestamp

This information returned can then be used to create a data set by aggregating all the attributes and applying weighting to fingerprint the device. Some of these characteristics are not unique and can be collected passively, but the aggregation of all the attributes will provide a fingerprint that is approximately 94 percent unique. Unfortunately, this data set is considered stable for only 24 hours. In other words, this solution has diversity-based uniqueness but lacks stability.

New Methods of Mobile Fingerprinting

In addition to the fingerprinting methods mentioned earlier—which, for mobile devices, are less than ideal—researchers have developed new methods that do a better job of finding characteristics that are both unique and stable. It turns out that with mobile devices, each device's physical features may yield sufficient, if not ideal, characteristics for fingerprinting.

An example of this type of research, carried out and published by Stanford University, involves detecting tiny differences in the manufacturing tolerance of the subcomponents embedded in mobile devices. Specifically, the research focused on finding unique identifiers by measuring the performance of the microphone and the accelerometer (a sensor that detects and measures motion, used to recognize screen tilt for displays and games). By using JavaScript running on a webpage, researchers interacting with the browser could accurately measure the tiniest defects in these components. What they discovered was that each accelerometer was predictably different, and that readings could be used as a fingerprint to uniquely identify devices.

This method of analyzing the flaws in sensors in a device to create a unique digital fingerprint has a lot of potential. No active software needs to be loaded on the device. Moreover, every smartphone, regardless of make or state (jailbreak or rooted or not), can be scanned to produce a fingerprint with a unique ID. Vendors, marketers, and even law enforcement could use that ID to positively identify individual users. For advertisers, this could be a windfall.

JavaScript

Running JavaScript on a website is an excellent way to actively fingerprint devices. By running JavaScript, the website can probe and retrieve many more attributes from the device than can be obtained by passively scanning the HTTP headers. For example, the Augur.js JavaScript library can identify and fingerprint devices using unique identifiers that provide a stable fingerprint of devices browsing the website. Augur.js can do this for both anonymous and registered users. Moreover, this active method is transparent to the device because JavaScript interacts with the device's browser. This is one of the reasons some smartphone manufacturers do not support JavaScript in their default browsers.

From the perspective of those looking to fingerprint and target users, this is a great method. Even if users were to detect the intrusion, they would not be able to adjust application privacy settings to mitigate the risk or delete the fingerprint ID. That is not to say users could not block the intrusion, however; simply disabling JavaScript would solve the problem. Unfortunately, this would have a negative impact on the overall browsing experience. The broader issue, however, is that a lack of user awareness will likely mean that the vast majority of users will be subjected to this type of fingerprinting when it becomes more popular with advertisers.

Similar to the research done at Stanford, a research team at the Technical University of Dresden, Germany, discovered that they could track smartphones using a variation in the radio signals emitted. Radio components such as amplifiers, mixers, and oscillators—due again to manufacturing tolerances and physical variations—can produce a predictable signature with which to identify the device. Similarly, researchers have found that the M7 coprocessor chip may also provide a durable fingerprint for mobile devices. The M7 coprocessor handles all the Quantified Self (QS) tracking (the voluntary tracking of movement and location use in many fitness and habit applications) of the motion sensor, including the device's accelerometer, gyroscope, and compass. This processor frees up the main processor and takes up less battery. Unfortunately for those who want to protect their privacy, this coprocessor in the iPhone stores 7 days' worth of data. Not only does it provide a unique fingerprint, it can also tell an accurate story of the user's location at various times.

The main flaw in these high-tech solutions (from the fingerprinter's perspective) is that to gather the fingerprinting information, the attacker must first get the user to visit a website running JavaScript. With user education on the dangers of fingerprinting and drive-by mobile malware, the chances of being subjected to this type of fingerprinting can be greatly reduced. Unfortunately, history suggests that even if privacy advocates push awareness, many users will unknowingly subject themselves to this type of fingerprinting and tracking.

Fingerprinting Users

Smartphones are set to become a useful device for law enforcement. Here, the goal is not to fingerprint the device, but to use the device to fingerprint a person. This will be done by providing police officers with a mobile system—a smartphone and a direct broadband connection—for scanning and processing fingerprints on the street. This allows police officers to be more productive by fingerprinting suspects in the field. The NYPD invested around $160 million in such a system, which it paid for using funds received in bank settlements.

Using the fingerprint sensor on the smartphone, officers will be able to scan and fingerprint suspects at the scene. The smartphone will process the fingerprint scan and send the data via high-speed broadband to the police and FBI databases to check for a match. Worryingly, the fingerprint sensors installed on smartphones and top-end laptops have been less than stellar to date, but as the technology improves, the field prints should stand up to legal scrutiny.

This is good news for law enforcement, but it's worrisome from a privacy perspective. It's not unlikely that a fingerprinting form of malware could be used to constantly scan the phone for fingerprints. Worse, users would likely not be aware that their fingerprints were being scanned, shared, and possibly misused.

Nonetheless, despite privacy concerns, user biometrics has become an area of intense interest in security because it potentially has the answer to the password conundrum.

Fingerprinting Users via Biometrics

The use of passwords as a method of authentication has been the bane of security for generations. Even the advent of two-factor authentication does little to mitigate the fundamental issue that not everyone in control of logon credentials are who they claim to be. Biometrics has long been the answer to this problem in physical access control via iris scans or fingerprints. However, much research has gone into biometrics over the last decade, as a means for authentication of online users. The principle behind this is that everyone differs in the way they interact with the user interface. The way a user types, moves a mouse, or navigates around a form or page is very telling.

Determining the unique characteristics of how a specific user types via speed, cadence, and even the time between inserting a @ or a period (.) within an email address forms a unique pattern that can be used to identify them. Combined with the recent advances in Machine Learning, biometrics has become the potential answer to dispensing with passwords to positively determine a user.

Similarly, voice recognition, which is another form of remote biometrics, is now deployed in many call centers, especially in the financial services industry, to authenticate a caller. As a result, biometrics has huge potential in fingerprinting and in positive authentication that could provide irrevocable proof of a user's identity.

Spyware for Mobile Devices

Spyware is different from fingerprinting. Fingerprinting identifies a unique device, while spyware tracks specific (and private) information about what a user is doing—sites visited, location, and so on. Spyware comes in many forms. The most prevalent (and annoying) is *potentially unwanted applications (PUAs)*. Not only do PUAs consume resources, including battery and bandwidth, but these applications also track information, including the user's web-browsing history, Global Positioning System (GPS) location, and contacts. Many developers fund their work by sharing the data they collect using these unwanted applications with third-party advertisers. PUAs can also change browser settings, opening the victim to further abuse.

> **NOTE**
> PUAs are a nuisance but are rarely malicious. In that respect, they're not that much different from cross-site profiling, except that they pull information directly from the device rather than from what is available through the browser interaction.

It may seem a bit shocking that this goes on, but it's perhaps even more shocking that developers are upfront about it— which is why it works. All this activity is actually spelled out in the application's terms and conditions, which makes it difficult for vendors such as Apple and Microsoft to block it. The problem, of course, is that very few people (even security experts) actually read these terms and conditions—typically consisting of pages of dense text filled with legalese. Essentially, these PUA developers are hiding in plain sight, taking advantage of the large population of mobile users who have yet to figure out that there is no such thing as a "free app."

A much more insidious and potentially damaging variety of spyware comes in the guise of legitimate applications. With this type of spyware, developers have again exploited the fact that few people read or care about the permissions they give smartphone applications. An example of this is a smartphone flashlight application, which, for no apparent reason, requests access to read and delete files on USB devices; switch on the microphone; access the camera, stored photos, and videos; and track the user via GPS. Again, the developers of this application may well have covered themselves by stating all this in the terms and conditions, but this type of intrusion is not so easily explained. This type of embedded spyware has potential for real spying, with malicious intent. Therefore, before you download any application onto a smartphone, you should vet its permissions.

Spy Software

One type of spyware that makes no attempt to hide its true nature (at least for the user) is spy software. Spy software is typically used by parents, employers, or other phone owners who wish to track a phone user's activities. Questions of legality often arise, but this type of software is legal under certain conditions. These include the following:

- The person or entity installing the software and viewing the information must own or have legal authority over the target phone.
- The person or entity installing the software and viewing the information must inform any adult user of the target phone that they are being monitored.

The key point here is that it is legal to monitor a company-owned smartphone as long as the employee is informed that their activity will be monitored. This should be clearly stated and appear in the acceptable use policy. Typically, no such notification is required for minors under the care of an adult.

> **NOTE**
> These legal considerations are generalities. Laws differ from place to place, so it's always best to verify the laws in effect in your location if you are asked to install this type of software in either a work or a private setting.

Despite these rules, it should be pretty obvious that loading spy software onto a smartphone has a real potential for misuse. Most spy software packages include the ability to monitor text messages, emails, web history, call logs, and GPS location. Other more advanced functions can include monitoring chat services such as WhatsApp and Skype, recording calls, recording background sounds, and remotely controlling the smartphone's features, such as the camera and microphone, without the user's knowledge. That is, the camera would be active, but the "camera on" indicator would not light up.

> **NOTE**
> There is a potential for irony if the IT department jailbreaks smartphones to enable tracking software (ostensibly to prevent employee misuse). Jailbreaking exposes the phone to myriad forms of malware, which could have far more damaging effects than employees taking extra-long lunches or checking Facebook throughout the workday.

Fortunately, there are some constraints. Spy software simply will not run on a secure or nonrooted iPhone. That is, the phone must be jailbroken. The same is true of Windows Phone 8.1. This is not an issue if the phone is rooted by a knowledgeable IT professional as part of a policy for company-owned phones. It is a significant problem, however, for average users who open themselves up to a great deal of abuse by jailbreaking their phone.

It's a bit different on Android smartphones. These phones support third-party apps, which can be side-loaded or downloaded from a website. Therefore, rooting is not essential, but may be required for some more-advanced spy software features.

An alternative approach for Apple devices—which does not involve jailbreaking the device—is to use an application on a PC, such as PhoneSheriff Investigator. This software, loaded on company-owned PCs, monitors the Apple iCloud backup rather than the phone itself. When the iPhone syncs with iCloud, it will back up everything. The spy software can then pull the data from the iCloud account without ever having to connect to the phone itself. The list of items that can be monitored includes text messages, iMessages, call history, GPS location, photos, contacts, Safari bookmarks, notes, and account details. It may not be as impressive as the spy software features loaded directly onto the mobile phone, but it is an elegant solution for iPhones that doesn't compromise a phone's security. That said, there is great potential for abuse of this software as well.

Spy Cells: Stingray

One method of spying on a phone is to use a *spy cell*—that is, to imitate a cell tower or base station, in much the same way an evil twin imitates a legitimate wireless access point. One well-known program that allows this is Stingray, which is essentially a fake tower used to intercept and perform man-in-the-middle (MITM) attacks on mobile networks. These spying base stations—sometimes called **International Mobile Subscriber Identity (IMSI)** catchers because they capture the unique IMSI number from a phone—are often used by law-enforcement and intelligence services to spy on and track mobile users in a given cell. (An IMSI is a unique identification associated with all GSM and Universal Mobile Telecommunications System (UMTS) network mobile phone users.)

Much like Wi-Fi clients, mobile phones are configured to search for an optimized signal—the more powerful, the better. With this in mind, the IMSI catcher boosts its power higher than the real mobile operator's signal. By masquerading as a base station, it captures the mobile-phone connections in that cell (or some other predetermined area of interest). The IMSI catcher also has a connection to the real base station through an intermediate device and can relay captured calls through itself to the real base station. The caller can detect none of this.

The clever part of this attack is the way the IMSI catcher masquerades as a base station to capture the signals in the vicinity. Because a base station always sets the encryption type during call setup with the mobile phone, it can force the mobile phone to use no encryption. This is completely transparent to the caller because it is an automated function. The user has no idea that the initial leg of the call, from the handset to the fake base station, is without encryption, and that all of their data (and voice) will be in cleartext. (The second leg, from fake base station to the real base station, uses standard encryption because it is at that point a normal call.)

An earlier weakness with this MITM attack was that call initiation was unidirectional, so the tapped phones couldn't receive calls (which would be a giveaway). To remedy this, newer equipment has patch-through technology that allows incoming calls to be passed to the device. Another weakness was that the technology worked only on GSM base stations, not on the later UMTS node-B stations. However, because practically every network supports older 2G phones over 3G networks, this permits base stations and node-B to coexist on the same network, resolving that issue.

NOTE

Stingray and other fake base stations can only work on 2G GSM. Hence, the requirement to force modern 3G/4G capable phones to switch to 2G mode. This weakens the effectiveness as a good indicator that something may be amiss is if a 3G or 4G phone suddenly drops down to using 2G, especially in areas where there should be more advanced networks. Interestingly, even the latest 5G networks are vulnerable to this type of attack, as we will discuss later in this chapter.

NOTE

Traditionally, the United States has had very strict laws regarding phone tapping, but the Patriot Act loosened some of those requirements—much to the dismay of privacy advocates.

The Stingray mobile phone tracker was initially developed for the military, but local and state law-enforcement agencies in the United States have widely adopted it. Stingray has both an active and passive mode. In active mode, it works as a base-station simulator. In passive mode, it works as a digital analyzer. When in active mode, Stingray can force all nearby mobile phones to connect with it because it simulates a real base station. It can then extract data such as the IMSI numbers and electronic serial number (ESN). This is often a necessary step because the Stingray will capture many mobile devices, and the operator will have to identify the target device by extracting data from the device's internal storage. Once the IMSI has been identified for the target device, surveillance can continue on that phone.

Perhaps more troubling than law enforcement's encroachment on privacy was a survey carried out by ESD America (a provider of defense and law-enforcement technology in the United States) in the mid-2000s. It uncovered 19 fake base stations around the country—some in big cities such as New York, Chicago, Denver, Dallas, Los Angeles, Seattle, Houston, and Miami. These fake base stations, which had previously been unidentified, were potentially installed by criminals in the United States copying a Chinese business model that involved spamming users' phones to obtain banking details via Short Message Service (SMS). By intercepting the SMS in cleartext, they could harvest the data without anyone noticing.

Fingerprinting on Modern Cellular Networks

The world of mobile/cellular communication networks is not as homogeneous as you might think. After all, 2G and 3G networks are still commonplace in some parts of the developing world. However, in more developed countries, 4G and even 5G IP data networks have replaced the earlier Time Division Multiplexing (TDM) technologies. This has brought tremendous improvements in bandwidth, speed, and lower latency, as well as user density per cell but also their own unique security issues.

In today's modern mobile/cellular networks, attackers can take advantage of known weaknesses in existing 4G technologies such as LTE. This is due to a flaw whereby the initial connect process where a device connects with the network the user identity and device capabilities transference occurs unencrypted. What this means is that the smartphone, IoT device, or smart vehicle will send its device capabilities and requested functionality to the core network in clear text. This occurs between the user's device and the eNodeB (4G network access point) during the initial attach process. However, because this initial conversation occurs unencrypted then this allows an eavesdropper to either fingerprint the user device or launch a MITM attack.

To understand how this flaw occurs we need to understand that the 4G and LTE architecture is made up of base stations, which cover a specific area. These base stations run on IP so they connect to the core network via an edge cloud or IP backhaul. However, a device, a smartphone, IoT device, or smart vehicle will connect over the air to the base station (eNodeB) in order to establish a suitable connection to the carrier's network. But because devices vary in their capabilities and use cases, the devices must pass a list of their own capabilities and requirements to the base station. These will typically be things such as the type of encryption primitives it can support, whether it needs voice support, SMS, vehicle to vehicle communication (V2V) support, as well as the device category that relates to the power of the device's processor, which is critical for IoT devices. As a result, an eavesdropper can ascertain with a high level of probability the nature of the devices making the connection requests. Furthermore, in the case of smartphones or modems, they will be able to detect what frequencies are being used and whether the device requests the radio access network (RAN) to use MIMO (multiple input/multi output) antennas for high bandwidth throughput. Armed with this fingerprinting data, they can then tailor exploits to match the device's use case.

In late 2019 to early 2020, there was an accelerated shift from LTE networks to full scale 5G commercial network roll outs. These 5G networks bring the promise of superfast high-bandwidth networks with extremely low latency. Nonetheless, they also require a rethink in the way that they are secured. Unfortunately, in the mad rush to be first to market with a commercial 5G network, many of the security protocols and algorithms for 5G were inherited from the 4G standard, which still had not mitigated the flaw that allowed device fingerprinting via unencrypted connection data for targeted attacks as well as for MITM attacks.

MNmap

Security researchers in Europe and the United States were able to sniff the information sent by 4G/5G devices in plaintext. Using this information, they were able to create a map of devices connected to a given network. What is more is that they could fingerprint all the devices and with high probability tell if a given device was running iOS and Android operating systems, if it was an IoT device, a smartphone, a car modem, or even a router connected to a vending machine.

A lot of information can be deduced from this type of fingerprinting because there are only five main baseband manufacturers—Huawei, Intel, Mediatek, Qualcomm, and Samsung. Moreover, they do not use the same set of mandatory capabilities, so this can aid in fingerprinting a device.

Man-in-the-Middle Attack

In addition to MNmap, MITM attacks allow for more sophisticated attacks on the user device. For example, if you can use a MITM relay to hijack the device information before security is applied, you could take this data and modify the device's capabilities. For instance, in 4G and 5G networks, it is the RAN that defines the speed at which a device receives data. Hence, it would be feasible to suppress the speed at which data is sent to the device by altering the device's capabilities. If the RAN is led to believe that the category of the device is something other than a smartphone, such as an IoT device, it could send data at a much lower rate. This type of attack is called bidding down, as the attacker deliberately understates the capabilities of high-end devices. But bidding down is also open to the imagination of the attackers because they can do many other things, such as prevent handovers or roaming, prevent the use of MIMO, or they could disable voice over LTE, which makes a phone fall back to 3G/2G.

Thankfully, a fix was implemented in Release 14 of the 5G standard. This was done to ensure that encryption is applied before the device sends its capability information to the network and thus prevent eavesdropping on the capability exchange upon connection. However, at time of writing, no 5G commercial networks had implemented this upgrade.

CHAPTER SUMMARY

Although fingerprinting is not necessarily a bad thing, it can and often does lead to abuses of privacy or worse. This is especially true with mobile devices. Because these devices are such a big part of people's lives, fingerprinting them can enable others to capture a very accurate profile of an individual and their behaviors.

Thus far, finding unique characteristics to fingerprint a device over time has been difficult. Many features provide a unique profile, but few are stable. Unfortunately, methods that appear to identify unique characteristics, down to very specific devices and their users, appear to be on the horizon. This new way of fingerprinting is in many ways much like the real thing—unique, stable, and very much a part of each person's identity. And therein lies the problem, from both an advertising and a criminal perspective. It is a huge advantage to be able to track someone as they browse the web or wander around out in the real world.

Not surprisingly, the groups that want this information are investing heavily in better ways to get it. The good news is that the big three mobile device providers—Apple, Google, and Microsoft—are moving in the opposite direction, toward device diversity. With device diversity, a user can work on any device with their data stored in the cloud, using workflows that follow them from one device to another. This greatly mitigates fingerprinting performed via browsers because the user can easily switch back and forth between devices. This makes the task of fingerprinting more difficult.

The bad news is that there are newer, more detailed, and more permanent methods of fingerprinting, not just browsers, but actual devices and users. Because of this, it's more important than ever for IT security to help educate users. This is not only for the sake of the users, but also for the IT teams that will have to deal with the resulting breaches that will inevitably leak into the corporate space.

KEY CONCEPTS AND TERMS

Active fingerprinting
Application fingerprinting
Cookies
Cross-site profiling
Fingerprinting
International Mobile Subscriber Identity (IMSI)
Passive fingerprinting
Proximity fingerprinting
Remote fingerprinting

CHAPTER 14 ASSESSMENT

1. Device fingerprinting is a relatively new networking phenomenon and is always malicious.
   A. True
   B. False

2. Which of the following describes fingerprinting?
   A. It makes for a richer user experience online, enabling sites to offer customized content and more convenient consumer transactions.
   B. It can lead to abuse from aggressive advertisers.
   C. It can lead to serious security issues on jailbroken phones.
   D. All of the above.

3. Which of the following describes proximity fingerprinting?
   A. It works only on wired networks.
   B. It is a standard practice for managing networks.
   C. It relies on the use of JavaScript.
   D. All of the above.

4. Remote fingerprinting is primarily accomplished with spyware.
   A. True
   B. False

5. Which of the following are *not* used in passive fingerprinting?
   A. HTTP headers
   B. TCP/IP headers
   C. GPS locations
   D. 802.11 Wi-Fi settings

6. On mobile devices, it is difficult to find unique characteristics for fingerprinting. Once found, however, they tend to be stable.
   A. True
   B. False

7. The advertisingIdentifier tag in Apple-based applications does which of the following?
   A. It is commonly used for cross-site selling.
   B. It remains active even if the application is deleted.
   C. It enables application vendors to collect analytics on their applications only while they are installed.
   D. It protects against jailbreaking.

8. What is one of the simplest ways to avoid mobile device fingerprinting?
   A. Disable JavaScript.
   B. Delete the cookie file after each use.
   C. Regularly reset the phone to its factory settings.
   D. Lie about your preferences on e-commerce sites.

9. Which of the following describes spy software?
   A. It is easy to install.
   B. It is always illegal and unethical.
   C. It can open up a phone to much greater abuse because it typically requires jailbreaking.
   D. It does not require advance notice when employers own the phones.

10. Spy cells enable law enforcement to set up rogue access points on local area networks.
    A. True
    B. False

Mobile Malware and Application-Based Threats

CHAPTER 15

WITH THE POPULARITY OF SMARTPHONES, it's no surprise that cybercriminals have shifted their focus to them, creating a new battlefield for IT security. Mobile malware presents some unique challenges, however, that make this more than just another front in an ongoing war.

One troubling aspect of the smartphone and smart-device culture is that users trust links and download apps without a second thought. This greatly increases the likelihood of users being subject to malware and other attacks. Combine this with the fact that there is now a free flow of devices between work and public networks and suddenly, the IT department is faced with a whole new set of challenges, the scale of which dwarfs the malware issue on company-owned PCs and laptops.

This chapter examines mobile malware, exploring not just where it is most likely to exist but also what forms it takes and how it finds its way onto devices. It also looks at mitigation strategies that will help prevent malware from finding its way to corporate resources.

Chapter 15 Topics

This chapter covers the following concepts and topics:

- What the malware landscape is for Android devices
- What the malware landscape is for Apple iOS devices
- How mobile malware is delivered
- How mobile malware is executed
- How to defend against mobile malware
- How mobile device management can be used to protect mobile devices
- How to use penetration testing with mobile devices

© Cherezoff/Shutterstock

Chapter 15 Goals

When you complete this chapter, you will be able to:

- Explain the reason behind the disproportionate distribution of malware on Android phones
- Describe the nature of malware and the goals of cybercriminals
- List the techniques used to deliver malware
- Discuss what madware is and how it differs from malware
- Understand the implications of and issues with excessive permission requests within applications
- Describe the role that social engineering plays in mobile malware
- List the two main categories of mobile application exploits and provide examples of each
- Discuss the role that mobile device management plays in protecting companies from mobile malware

Malware on Android Devices

When it comes to mobile malware, cybercriminals have focused their efforts on the market leader: Android. In fact, 90 percent of all malware targets the Android operating system (OS). Android competitors, such as Apple iOS, barely register on the scale. Despite this obvious threat, Android has built up huge market share in the consumer sector, capturing more than 85.6 percent of the global market at the time of this writing.

Although Android's dominance in the market is a factor in why cybercriminals have targeted that OS so vigorously, it is not the only one. The main driver is that the Android OS is an open source system. This is in contrast with Apple iOS, which is an infamously closed system.

Apple locks down and secures their products in what is termed a closed-system. That is, they require their customers to obtain applications from their own proprietary marketplace—the App Store. In contrast, for Android, Google opted for an open system, whereby customers could obtain applications from any number of outlets—secure or not. This decision resulted in more choices and greater flexibility for users, but it also opened up the Android OS to malware.

> **FYI**
>
> If there is no inherent vulnerability in the Android OS, why is it targeted even beyond what its proportion of market share suggests? The answer is simple: because that's where the money is. Remember, the primary driver behind cybercrime is financial gain. Because cybercriminals are in it for the money, it makes sense that they will direct their efforts where there is a greater opportunity for profit. Because of Android's dominant market share and lack of application provenance, an opportunity for profit is exactly what the Android OS represents.

This is not to say that the Android OS is inherently insecure. Although it does have a few known vulnerabilities (fewer than Apple's iOS), the Android OS is actually very secure and is not inherently vulnerable to malware. The real reason for Android's susceptibility to malware is more a function of software fragmentation among vendors; poor vendor OS-update-management practices; and importantly due to their vast customer base, poor customer vulnerability awareness, or user security hygiene.

All OS software is very complex. Due to this complexity, vendors, users, and hackers regularly find security vulnerabilities, even after the software is released. When these vulnerabilities are found, the race is on to uncover the root cause, code a fix, test it, and distribute it. Then it's up to users to update their devices—which they may or may not do. The lag between the time a vulnerability is found and when it's patched is a time of increased risk, because cybercriminals look to exploit the vulnerability. This is hard enough for developers in closed systems to mitigate. For Android, it's especially difficult.

Consider an example. Suppose Google becomes aware of a vulnerability in its Android software and issues a patch. In a closed system, the patch would be delivered directly to users, who would then adopt it at some normal rate. However, because Android is an open source OS, Google must first release the patch to the many hardware vendors whose devices run the operating system. These vendors must then determine if and how the patch applies to their software. For example, they may have modified the base OS, so implementing the patch may require additional development on their end, not to mention all the necessary testing. Only after that occurs can these various patches be released (at varying rates) so users can adopt the fix.

Software Fragmentation

Because the Android OS is open source, it can be (and has been) adopted by many different mobile hardware vendors—hence its dominant position over the proprietary Apple iOS and Windows Phone in the marketplace. But although the many vendors who use Android OS all begin with the same core software code, each one modifies it to suit its own requirements. As a result, the code becomes very fragmented, increasing the overall complexity of the code as well as the security vulnerability.

Of course, Android works not only on smartphones but also on tablets and **phablets** (devices that are somewhere between a phone and a tablet), each with its own diverse software features and hardware configurations. This is wonderful from an innovation and production perspective, but it creates even more security vulnerabilities due to even more complexity. In contrast, Apple can make global updates available to its user base via download because the code will be identical on every phone, tablet, and phablet. This is a huge advantage in terms of closing vulnerabilities.

Google does not have that option. Instead, as mentioned, Google must identify vulnerabilities and engineer security patches, which it makes available to the many device manufacturers that host Android. The manufacturers must then study the update for compatibility with their versions of the Android code and, if necessary, engineer the patch to fit. Compounding the problem is the fact that hundreds of manufacturers use Android, some of whom support only limited versions of their modified code—in some cases, only those models still in production or devices less than 2 years old. As a result, there are many devices for which

> **NOTE**
> The fact that some manufacturers support only limited versions of their code may help to explain why, according to Google, less than 20 percent of Android users run the latest version of the OS. Compare this to Apple, who claims that 91 percent of iPhone users run the latest version of iOS.

patches may not be available, meaning that many devices remain susceptible to the vulnerability. For those vendors that do roll out upgrades, there is the problem of distributing them through their partner network. Partners must perform regression testing and confirm that the updates are not harmful to their customized code before releasing them. It takes time to make these patches available for end users to download, even on the latest phone models.

For the bad guys, this is ideal. This lag provides them with a window of opportunity to exploit the vulnerability. The longer the window is open, the more lucrative the vulnerability may be.

This explains why there is a disproportionate focus on Android. It's not because the core OS is less secure than its competitors. It's because the open system through which it is distributed, modified, and supported results in greater complexity, which makes it harder to mitigate those vulnerabilities that do exist. This, along with the fact that it takes more time to distribute fixes, gives cybercriminals more opportunities to exploit vulnerabilities.

Still, this is only part of the story. The rest is that unlike the walled garden approach taken by Apple, Android vendors allow developers to distribute and users to sideload software through a wide variety of channels. Outside of the Play Store, Google cannot police these applications. This is ideal for cybercriminals. Vulnerabilities that remain active on devices and a lack of application policing (at least compared to Apple) have provided a profitable environment in which cybercriminals can ply their trade.

Criminal and Developer Collaboration

The large number of active smartphones operating on older, unsecure versions of Android is a major temptation for both criminals and rogue developers. Consequently, security experts suspect that there is broad collaboration between cybercriminals and dubious developers who focus their attention on Android malware—potentially unwanted applications (PUAs) and commercial mobile adware (madware). Not surprisingly, the top types of Android malware, as reported by the security company Sophos, are financially focused. As of 2020, these malware categories, or families, included the following:

- **Remote Access Tools (RATs)**—These offer a means of backdoor access to infected victim devices and are often used for intelligence collection. RATs are typically deployed to get access to installed applications, or to mine personal information such as a user's call history, address book, web browsing history, and texting data. However, mobile RATs are also used to send SMS messages, enable device cameras, and capture location data.
- **Bank Trojans**—This type of malware aims to steal the user's logon credentials for predetermined mobile apps. These are typically overlay attacks whereby a malicious app catalogues all of the installed apps on a device and waits for an app of interest to be opened. If that app is one of its several hundred target apps it will produce a screen overlay that matches the target app's logon screen so that it is disguised as a legitimate application to compromise users who conduct their banking or retails business from their mobile devices.

- **Ransomware**— This is used to lock out a user from their device often by encrypting the data, making it useless. The attackers will then demand a "ransom" payment. Access will only be granted to the device once a ransom is paid.
- **Cryptomining malware**—This is specifically designed to allow attackers to covertly mine cryptocurrency on the victim's mobile device, allowing them to generate cryptocurrency. The malware will execute calculations on a victim's device overloading the CPU and draining the battery. Cryptomining malware is often hidden in legitimate-looking apps.
- **Advertising Click Fraud**—This is a hugely popular type of malware that is designed to let an attacker gain access to a device in order to generate income through click fraud.
- **Stalkerware**—There are two types: trackers and fully fledged stalkers. The former generally focus on stealing the victims' GPS coordinates and perhaps intercepting text messages. These were common in many apps on Google Play until 2018 when a change in policy meant that most of them were removed from the Play store. The latter are commercial apps designed to steal almost any type of data from a compromised device, not just the location data, such as photos, texts, contacts, screen taps (keylogging), and so on. Fully fledged stalkers have in recent times gained notoriety in advanced persistent threat (APT) attacks against specified targets. These tailormade attacks are typically launched by state sponsored or sophisticated criminal gangs against an individual of interest for political or financial gain. In a bid to stay on the right side of the law, many vendors that develop these fully fledged stalker toolkits only sell to reputable (white-listed) governments, but that doesn't necessarily ensure their ethical use.

> **FYI**
>
> **Premium SMS** is a pay-as-you-go service available on smartphones. Charges can be linked directly to a credit card or tacked onto the monthly bill. Hackers often take advantage of this service, using malware to initiate premium SMS messages from hacked phones back to services that they own or control or from which they can otherwise extract payment. Cybercriminals love premium SMS because it allows them to quickly monetize their efforts.

Bear in mind that the Android ecosystem is very fluid. The list of malware categories is meant to illustrate the focus of high-level cybercrime at a specific purpose and time rather than to serve as an enduring list. There are thousands of variants of these categories, which shift on a weekly basis. For example, during the early first quarter of 2014, the number of families of malware had risen to 369—not 369 *instances*, but 369 *families*— each with dozens and even hundreds of variants. Today, the number of identified malware families is 1,267.

More worrying is that over the past few years, cybercrime gang structure has morphed, taking on more of a hierarchical pyramid structure. This is in great contrast to the fabled perception of the socially inept hacker working alone in a dark basement. The new hierarchy consists of an army of enablers or distributors of software payload or malicious code, comprising the core of these dynamic "organizations." Atop them are divisions of technical analysts and financiers of various projects. These organizations are formidable in their scale, expertise, and ability to execute.

For obvious reasons, the exact makeup and strategies of these groups are not entirely known. What is known is that the groups tend to focus on the following goals:

- Claiming personal or confidential information by gaining access to a phone
- Gaining control of phones for other attacks, such as launching distributed denial of service (DDoS) attacks to cause business disruptions or sending premium SMS messages for financial gain
- Gaining access to GPS and other location information to sell to third-party advertisers
- Gaining control of phone file systems to steal data, photos, and the like, or to lock out the user for ransom
- Gaining control of features, such as the camera and microphone, for surveillance or location tracking—prevalent in commercial and business espionage and cyberstalking

In some cases, these groups focus on a single target over time in an APT. These continuous, coordinated attacks usually target organizations and/or nations for business or political motives.

Cybercriminals go about achieving these goals through a variety of techniques, which target the inherent risks that apply to mobile smartphones. These risks include the following:

- Insecure data storage
- Weak server-side controls
- Insufficient Transport Layer protection
- Poor authorization and authentication
- Improper session handling
- Data leakage
- Poorly implemented encryption
- Sensitive data disclosure

Sensitive data disclosure is a common aim of malware, PUAs, and many "genuine" applications that are not considered malicious but still harvest information, such as location, contacts, photos, and other sensitive data. Furthermore, many benign mobile apps are very talkative, commonly exchanging data with third-party application programming interfaces (APIs)—especially advertising and marketing networks. In short, there is a near constant flow of information to known and unknown recipients from smart devices.

Interestingly, the growth in malware evolution is fueled by an ever-changing target landscape. For example, back in 2014, the main mobile malware threat was premium SMS hacks rather than credit cards. At first, this may be surprising, given what makes the headlines. Nevertheless, it does make sense when one looks at the numbers.

For example, a single credit card can be sold on the black market for approximately 60¢. Compare that to a premium SMS Trojan. Cybercriminals often deliver their malware through an affiliate program, paying people a commission to deliver the payload. Some affiliates make up to $10 per successful SMS Trojan activation. Additionally, malware that sends premium SMS text messages can provide an ongoing revenue stream—at least until the unfortunate victim notices the charges on their mobile phone bill or credit card statement.

SMS later fell out of favor because means were found to mitigate many of the attacks; therefore, cybercriminals switched tracks to focus on ransomware. In fact, premium SMS and ransomware attacks were so lucrative that entire ecosystems were formed to support them, complete with contact centers, supported by agents, developers, distributors, partners, and affiliates. These ecosystems closely parallel legitimate software ecosystems, the primary distinction being that they support a criminal enterprise.

To illustrate the constantly shifting mobile threat ecosystem by 2019, both SMS and ransomware were no longer the lucrative targets they once were and the focus had changed to bank malware, which was designed to steal banking app logon credentials from mobile devices and was hugely lucrative for a time. Today, the most common threat is still targeting banking apps through advertising and fake review malware. Indeed, 4 of the top 10 malware families today are from the adware category. An emerging category of mobile threats are targeted at consumers via stalkerware, with this new malware becoming very popular with cybercriminals since 2019.

Madware

As mentioned, madware is a form of very aggressive adware that is prevalent on mobile devices. Due to the way the "free application" business model works on the Internet, developers consider madware to be not only harmless but a legitimate aspect of application behavior. Although some free applications that use advertisements as a revenue source are legitimate and benign, others run background processes that access GPS information, scan address books, and send out stolen data via HTTP to third-party APIs. Some also track and share location details, browsing histories, and contact lists, often with the phone owner's unknowing permission.

Of course, developers maintain that the terms of use have been explicitly stated and that the user's permission has been granted. However, developers also know that very few users read the entire terms and conditions, which are typically very long and filled with complicated language. Indeed, it's likely that few users ever give these agreements more than a glance as they scroll through to the Accept button. As a result, application vendors have begun to take more and more liberties with application permissions. Indeed, in recent studies of Android applications, researchers found that many applications—even some Top 20 Google Play applications—request far greater access to the phone's features and data than is justified.

> **NOTE**
> Of course, it is not always clear to the end user which access to features or data an application genuinely requires.

One could argue that users have a responsibility to read and understand the terms and conditions before agreeing to an application, and that in a "buyer beware" market, perhaps they deserve what they get if they don't. This is debatable. Regardless, when users allow these apps on devices that hold company data and access company resources, it becomes the IT department's problem. Knowing users' tendencies, it's in IT's best interest to monitor app stores and make users aware of applications with excessive or unwarranted permissions.

Excessive Application Permissions

A recent study by SnoopWall on the threat assessment of the Top 10 Android flashlight apps revealed that all the flashlight applications they tested obtained access and information well beyond their needs. One naturally wonders why developers would ask for permission that their applications don't require to function. It's especially suspicious when refusal of these permissions seems to have no effect on the execution of the application, which would indicate that access to these resources is unnecessary.

Consider the scale and scope of permissions and access that developers request to run these flashlight applications. Here's a list from one app:

- **Retrieve list of running apps**—For a flashlight application, this is a dubious requirement.
- **Modify or delete contents of your USB storage**—Why would a flashlight application want to delete files on a USB device? Other applications, such as a camera interface or music system software, may have a legitimate reason to delete uploaded or downloaded files, but it's hard to make a case for a flashlight doing this.
- **Test access to protected storage**—This is dubious. Why would a flashlight application need to read or write to protected memory outside its container?
- **Take pictures and videos**—This is a rather frightening ability for a flashlight if it's done without user notification.
- **View Wi-Fi connections**—This is another unnecessary permission for a flashlight.
- **Read phone status and identity**—A request to read the phone status is a yellow flag, but this one does have a genuine use. Developers need to retrieve this information for analytics, to identify the phone model, and to identify the OS on which end users are using their product.
- **Receive data from the Internet**—Most applications request this access because they are client/server models with the main application residing on a remote server. The danger here is that client-side injection could cause security issues. For a flashlight app, this may be needed to receive advertisements.
- **Control flashlight**—This permission seems legit.
- **Change system display settings**—This seems acceptable, because the flashlight application needs to change screen display parameters.
- **Prevent device from sleeping**—This may be a necessary function for a flashlight to prevent it from shutting off while in use.
- **View network connections**—This is yet another unnecessary request for permission.
- **Gain full network access**—This is likely a request for unrestricted Internet access, which hints at uploading personal information.
- **Approximate location (network based)**—This is an obvious requirement for adware.
- **Precise location (GPS)**—This is totally unnecessary.

As you can see, this flashlight app requests 14 permissions. Of the 14, only 4 have anything resembling a legitimate need to operate the application. In general, a long permission list like this should raise a red flag—whether in Google Play, the App Store, or the Windows Store and on the phone itself when the user reads the terms and conditions of use. Unfortunately, this is not possible with a jailbroken or rooted device, because the security permissions will likely have been circumvented.

Objections to requests for excessive permissions do not always stop malware developers. They use a variety of tricks to fool the OS security. One way to ask for a long list of permissions without raising an alarm is to split the application into modules, each requesting only one or two permissions each. The first module typically asks only for a couple of innocuous permissions but also requests permission to download updates. As each subsequent module is constructed, it is sent as an update. These often pass undetected by the OS security. Of course, the phone user must permit each download, which might raise concern (if the user is even paying attention).

> **NOTE**
> Even if users are vigilant, there are ways around that, as proved by the Android Jmshider threat. In that case, developers signed the application with an Android Open Source Project (ASOP) certificate, which allowed installation and further downloads of updates to take place without any user interaction.

Malware on Apple iOS Devices

Unlike Android, there appears to be very little iOS malware out in the wild at the time of this writing. Perhaps this is testimony to Apple's very secure closed system and focus on application provenance. Surveys suggest, however, that roughly 50 percent of iPhones in China have been jailbroken. If that figure is correct, it is surprising there are not more Apple-based malware apps going around.

That said, just because malware is rare, it doesn't mean it doesn't exist. One example is the flashlight app noted in the previous section, which requested excessive permissions, including access to sensitive features. While there is no evidence to suggest this flashlight is indeed malware, it certainly looks to be a PUA that contains unwanted and perhaps aggressive advertising. It also violates the user's privacy by seemingly sending location details to an advertising network.

Although malware targeting Apple iOS is not common, there are proven cases of it. Recent research by Kaspersky Lab revealed a spy Trojan called Xsser mRAT that targeted iOS 7 in China and Hong Kong. Kaspersky discovered that the infection was delivered through contaminated PCs or Mac OS malware. Typically, the malware first targeted the PC or Mac, where it lay dormant, waiting for an iPhone to be attached via USB. The Trojan was then activated via the remote control system (RCS), at which point it attempted to silently jailbreak the device and infect the iPhone. This sophisticated iOS Trojan could then discreetly conduct its spying activity with little impact on battery performance. It could perform many functions, including tracking location via GPS coordinates, stealing personal data, taking photos and video, and spying on SMS and other messages.

However, like other forms of malware, Xsser mRAT had limitations. One was that the phone had to be jailbroken. Typically, if a user keeps their device up to date with the latest (approved) version of iOS, the iPhone is far more likely to be protected. Of course, this assumes that the user (or another form of malware) does not jailbreak the phone. As noted, roughly 50 percent of iPhones in China have been jailbroken—many of those with sponsored malware from APT groups or syndicates.

Other, older iOS malware exists, but some are threats only on jailbroken devices. These are outlined in **TABLE 15-1**.

At the time of this writing, only 11 known iOS malware instances are found in the wild, and only a few of these were found inside the App Store. This is negligible when compared to the thousands of variants of Android malware, but it does show that while Apple malware is rare, it's not a myth—it's more a narwhal than a unicorn.

TABLE 15-1 Older iOS malware.

MALWARE	YEAR DISCOVERED	PHONES AFFECTED	DESCRIPTION
iOS/TrapSMS	2009	Jailbroken	SMS forwarder
SpyMobileSpy! PhoneiOS	2009	Jailbroken	Spyware
iOS/Eeki.A!worm	2009	Jailbroken	Worm
iOS/Eeki.B!worm	2009	Jailbroken	Worm
iOS/Torres A!tr spy	2009	Jailbroken	Rogue application
Adware/LBT MiOS	2010	Any	Call premium phone number
Spy/KeyGuardiPhoneOS	2011	Jailbroken	Keylogger
iOS/FindCall A!tr spy	2012	Any	Privacy Trojan
Riskware/Killmob!iOS	2013	Jailbroken	Spyware
iOS/AdThief A!r	2014	Jailbroken	Ad
iOS/SSLCred A!tr pws	2014	Jailbroken	Password stealer

Mobile Malware Delivery Methods

When it comes to mobile malware, it's one thing for hackers to be able to use available tools to construct a mobile exploit package. It is quite another to deliver it successfully to the target. This is why cybercriminals find value in organized groups. With enough teams attempting to deliver various payloads, it doesn't matter what information they collect through the spyware. It seems the more the merrier. Once the payloads are delivered, it is up to the analysts to sort through all the data and find a way to monetize the information collected. The job of the distributor or enabler is simply to deliver the exploit payload to infect as many devices as possible and then harvest the maximum amount of data.

Cybercriminals often deliver their malware through an affiliate program. That is, they pay people a commission to deliver the payload, just as with any other legitimate affiliate program. In fact, commercial software companies use the same model. The tactics used by affiliates vary considerably, but they tend to rely on social engineering to send targets to infected websites for drive-by attacks. A web attack on an Android or jailbroken Apple iPhone is one of the easiest ways to infect a smartphone.

Other mobile malware delivery methods include the following:

- **Binding an application to a genuine popular free app available in Google Play and the App Store**—Unless done well, this effort is likely to be short lived. This is especially true with the App Store, which is manually curated. That is, Apple employees check the authenticity and integrity of all applications in the store.
- **Loading malware to the many third-party stores**—These are plentiful for Android phones as well as for jailbroken iPhones.
- **Creating a webpage infected with a drive-by download**—Cybercriminals then use a variety of means to get users to visit and click on the page (for example, offering "free" merchandise).

- **Using a hybrid attack with a dual payload to target both PCs and smartphones**—If the victim device is a PC, it will activate the PC version of the code and wait for a smartphone to be connected via the USB port. When the smartphone is connected, the malicious code on the PC will try to jailbreak the phone and inject a Trojan payload.
- **Using social engineering**—The idea here is to trick targets by building up a rapport via social media into downloading a Trojan dropper or using premium SMS to connect to infected sites, where they may download a Trojan payload.
- **Creating malicious QR codes for malware distribution**—A QR code can represent as many as 7,089 numeric or 4,296 alphanumeric characters. These are very popular with marketers and poster advertisers and can also be used to distribute malware.
- **Using a Trojan dropper**—The benefits that droppers bring are that they do not exhibit any malicious intent because they are simply code used to perform mundane tasks such as to download, decrypt, and execute other programs. This is a good way to circumvent App Store and Play Store curation. The second issue is to get people to download them so they are often disguised as lightweight, useful, and harmless apps such as currency convertors.

Mobile Malware and Social Engineering

Not a day goes by in which cybercriminals don't collaborate to discuss how to attack what is most often the weak link in corporate network security: the end user. These are typically employees with no idea of security awareness, who also happen to be equipped with powerful and intelligent smartphones. Compounding this issue is the fact that cybercriminals have proven to be highly collaborative. Despite being a loose collection of syndicates and individuals, new exploits are communicated throughout these informal networks much more quickly than information is spread in many corporations with well-defined information-dissemination processes.

 NOTE

Thankfully, most of the vulnerabilities targeted by these operations get patched. Users should remain vigilant, however, because new ones are always being developed. Also, as mentioned, patches work only if users keep their software up to date. Shockingly, a Google survey revealed that roughly **80 percent** of all Android phones run on deprecated OS code.

Captive Portals

The most common tactic cybercriminals use to take advantage of users involves the use of a captive portal. That is, the cybercriminal launches and advertises an attractive free Wi-Fi portal, realizing that many people will join any hotspot—for example, in a mall or supermarket—if it is free. Many of these same users will also click on ads that offer too-good-to-be-true deals or giveaways. This takes them from the captive portal to a phishing website, which uses JavaScript (or other means) to perform a drive-by browser attack. This JavaScript will run on the user's browser to implant a Trojan, which will then load malicious code.

Drive-By Attacks

Drive-by attacks involve injecting malicious code into websites to exploit vulnerabilities in browsers. The result: A user's device can be infected simply by visiting a website. Drive-by attacks can occur not only on hacker websites specifically created to launch an attack but also on legitimate sites that have been hacked and infected with malicious code.

Drive-by attacks are launched when a user visits an infected website. In cases where a site has been specifically set up for an attack, there is typically an effort made to entice a victim to visit the page (for example, by offering free items or services). Remediation methods include disabling JavaScript (when feasible) and setting preferences to allow Java and Flash to run only on trusted sites.

Clickjacking

Even if users manage to avoid a drive-by attack, they might still fall for **clickjacking**. The purpose of this exploit is to confuse the victim by creating a webpage with a background of invisible frames. The attacker then superimposes another set of frames or buttons that the victim can see. If the victim clicks on a frame button once or twice and nothing happens, they will tend to swipe the mouse around, clicking indiscriminately, hence triggering the exploit.

Likejacking

The **likejacking** exploit is usually very successful on smartphones due to their smaller screens, which are more difficult to see. In this case, an invisible frame or button is placed directly on top of a Facebook Like button. Clicking the button works as normal but also infects the device.

Plug-and-Play Scripts

The small screen on a smartphone also makes it difficult to use the navigation controls. Malware developers have found this to be to their advantage. By using JavaScript (a hacker favorite), they can cause any mouse click to activate the target button, which will infect the device. If the user inadvertently clicks in the wrong place (easy to do on a smartphone), it causes a malware script to run.

Browsers are not the only weak link, however. There are far more vulnerabilities in plug-ins and scripts. Users should be advised that if they do not need plug-ins, they should switch them off. This helps to reduce the attack surface area. The main takeaway, though, is to switch off scripting in the browser.

Mitigating Mobile Browser Attacks

Website attacks on mobile browsers are commonplace, and many are rather sophisticated. As a result, you must make sure any browser on a smartphone is hardened to protect against client-side threats. Best practices include the following:

- **Practicing due diligence**—This is especially true when downloading Android apps because Android phones are susceptible to malware.
- **Using HTTPS when entering credentials**—If a site doesn't offer HTTPS, users should think twice before keying in private or sensitive data. HTTPS is indicated by the site's URL, which should begin with HTTPS:// or sometimes by a security lock icon.

- **Maintaining robust antimalware software**—Ensure it's always up to date and scans often.
- **Blocking pop-ups (via the device's Preferences screen)**—This prevents malicious pop-ups, which are attack vectors for many forms of malware, from appearing.
- **Checking app permissions**—Always check an application's permissions before downloading and installing it to make sure they are relevant.
- **Removing unwanted apps**—Surveys have revealed that 81 percent of downloaded applications were only used once. If you don't use an app on a regular basis, it's best to delete it.
- **Switching off autofill dialog boxes, JavaScript, and HTML5**—Some advanced features, such as autofill, are meant as time savers, but can be exploited because they store personal information in the browser caches. Features such as JavaScript and HTML5 can be used in drive-by attacks, so it's best to disable them when possible, turning them on only as needed for trusted sites.
- **Enabling fraud warnings**—These prevent users from visiting phishing sites.
- **Clearing the cookies, history, and cache**—Regularly deleting your browsing history can prevent the theft of sensitive data by spyware and adware.

Mobile Application Attacks

The various mobile application security exploits generally result from one of two main causes:

- **Malicious unnecessary functionality**—This refers to mobile code that is both unwanted and dangerous. With malicious unnecessary functionality, users think they are downloading an application or game for a specific purpose, unaware that it comes loaded with spyware, adware, links to phishing sites, or the ability to secretly handle premium SMS messages. Malicious unnecessary functionality might include monitoring activity on the device, data theft, the unauthorized sending of premium SMS text messages or emails, the unauthorized dialing of the phone, unauthorized network connectivity, and system modifications such as jailbreaks, the installation of rootkits, or tampering with the file system.
  Think back to the example of the flashlight app, in which a seemingly legitimate app hid various features that the consumer would have been tricked into installing. The consumer doesn't know that these transparent features—which are most likely not working with the user's best interests in mind—will operate in the background.
- **Errors in design and implementation**—These typically occur because of the complexity (and relative immaturity) of both iOS and Android code and the hundreds of thousands of third-party applications, many of which are coded by amateurs and hobbyists. Vulnerabilities include sensitive data leakage, unsafe storage of sensitive information, unsecure communication and data transmission, and hard-coded passwords and keys.

The problem—again, as illustrated by the flashlight app example—is that many applications are not designed and developed with security or consumer privacy in mind. Because many applications are offered free of charge, developers monetize their efforts by selling users' locations and browsing history to ad networks. It is understandable that with applications being rushed to market and often offered free, developers sacrifice security, diligence, and user privacy. This is how they make a quick financial return for their efforts. After all, Google Play and the App Store, are marketplaces for independent developers.

Mobile Malware Defense

Malware and PUAs are not easily distinguishable. Therefore, the common advice to users is to run antimalware and antivirus software on their computers. So why aren't these services standard on mobile devices?

The problem with antivirus and antimalware tools is that they have the same restrictions as any other application—they are restricted to their sandbox. Unfortunately, unless these services have privileged access, their usefulness is limited. Testing on some products found antivirus software for Android devices had a detection rate of less than 10 percent, and that versions for iOS and Windows Phone were merely placebos. The same holds true for antimalware apps—the strict application isolation that exists across all OSs makes them ineffective. To work, an antivirus application would have to break out of the sandbox. This would require the application to run as root, which would of course open the device to many new forms of malware—hardly worth the trade-off.

If antimalware and antivirus software are not the solution, what is? The answer lies in many of the best practices already discussed. In review, they are as follows:

- Providing security-awareness training for end users
- Prohibiting applications from noncertified developers and third-party marketplaces
- Restricting users to vetted and authorized marketplaces
- Prohibiting the unlocking or jailbreaking of devices and the sideloading of apps
- Policing third-party downloads
- Adhering to policy using mobile device management

Mobile Device Management

A key strategy for protecting mobile devices in the workplace is to have a Bring Your Own Device (BYOD) policy and a mobile device management (MDM) system. Following MDM best practices for supporting BYOD smartphones and tablets in the enterprise will go a long way toward mitigating the risks.

Initially, it's important to address the basics: controlling passwords, encryption, remote wipe, and data containers. You must address these essentials by setting a policy that includes requirements for strong passwords, auto-locking after 5 minutes of inactivity, auto-wiping after 10 consecutive failed logon attempts, and segregating business data in its own isolated container.

Also, you must develop a lightweight inventory system and allow administrative access to those who require it. This includes members of your human resources team, who will need to reclaim mobile devices during exit interviews. This practice also alleviates the burden on the IT department. For example, giving help-desk employees access to the

> **NOTE**
> Segregating business data on a BYOD phone is a must. Doing so enables you to perform a wipe of the business data without risk to the employee's personal content. If you don't segregate this data, employees may be slow or reticent to report a lost phone for fear of their personal data being wiped.

inventory enables them to access devices via remote control to troubleshoot and repair them. Another great timesaver is to empower users to enroll their own devices, lock and wipe devices if they have been stolen, reset their own pass codes, and locate lost devices.

Penetration Testing and Smartphones

When it comes to mobile malware defense, it's a best practice to set realistic policy and security templates for each type of authorized device, and to become familiar with the devices used and their vulnerabilities. The best way to do this is to test the security measures in place via penetration testing (pentesting) on sample phones or emulators.

Pentesting smartphones involves the following stages:

- **Recon**—During this stage, the tester gathers information by identifying the types of mobile devices on the target network.
- **Scanning**—After the types of phones have been identified, the scanning phase can proceed. When scanning for mobile devices, the tester looks to identify networks sought out by mobile devices.
- **Exploitation**—In the exploitation stage, the tester actually captures and gains control of the mobile device.
- **Post-exploitation**—During the post-exploitation stage, the tester inspects the phone for sensitive data in common areas such as notes as well as SMS and browser history databases. The tester might also search the device for stored passwords in the keychain or payloads such as a backdoor, depending on the scope of the project.

As always, when pentesting on corporate networks, it is necessary to obtain permission in writing from senior management. Equally important is having the right tools for the job. The community version of Shevirah Inc.'s Dagah tool, formerly known as the Smartphone Pentest Framework (SPF), is a good option because it runs on the Kali Linux platform. The Dagah software testing tool can identify and port-scan a victim smartphone and look for vulnerabilities such as the default Secure Shell (SSH) password for jailbroken iPhones. Additionally, the Dagah Pentest framework features a range of remote, client-side, and social-engineering exploits.

When pentesting a smartphone, the tester first uses basic social-engineering techniques to construct a convincing SMS with an embedded link. All that is required is to get the victim to select the link, which will take them to a webpage. The level of success is proportional to the diligence of the social-engineering and information-gathering process. The user must trust both the source and the link. If the SMS is well crafted and the user falls for the trap, their device's browser will be directed to a Dagah-controlled webpage running a client-side attack.

The exploits used will depend on the goals. A typical pentest objective would be to download to the phone via the compromised browser a Dagah agent with a variety of payload options. The tester could then interrogate the phone and receive data back in HTTP or SMS replies. There are also remote-control functions to demonstrate Dagah's control of the device, such as taking a picture or sending a text message.

Another example of an Android smartphone pentest tool is Armitage. A graphical user interface (GUI) for Metasploit, Armitage is used to create an Android application package that carries a malicious payload. In this case, the payload is a simple reverse connection back to the tester's console. Getting the user to launch the application requires either a client-side exploit on the Android's browser or, more likely, some clever social engineering to entice the user to download and open the app. When the user is reverse-connected to the tester's console, the tester establishes a connection to the compromised Android phone. This only creates a reverse connection, however. Another exploit needs to accompany it in the payload—perhaps a remote-control agent. However, even if these methods are not taken to their logical conclusions, if nothing else they can prove to be a good assessment of users' level of security awareness.

CHAPTER SUMMARY

The combination of a large, often naïve population of mobile device users and the prevalence of BYOD presents a difficult problem for IT security teams: mobile malware. With the free flow of devices and data between work and nonwork, criminals have a much easier pathway into very lucrative targets: corporations and their employees.

This problem is compounded by the fact that the tools that IT departments can bring to bear on the PC side of the network, such as antivirus software, do not apply on mobile devices. In addition, users are far more likely to download software or apps—often from questionable sources—onto a phone than onto a PC. In fact, being able to download an app in the heat of the moment is a fundamental aspect of the smartphone experience.

There are ways for IT to combat this issue, such as using MDM and compartmentalizing data on BYOD devices. But these mostly just help mitigate damage after a phone is infected with malware. They don't prevent the infection from occurring. In the end, the best tool for an IT department is one that cannot be stressed enough: user education. Only educating users to raise awareness and cataloging known good sources or known apps to avoid will give IT a fighting chance in BYOD environments.

KEY CONCEPTS AND TERMS

Clickjacking
Likejacking

Phablets

Premium SMS

CHAPTER 15 ASSESSMENT

1. Malware tends to be evenly distributed across Android and Apple iOS devices.
   A. True
   B. False

2. Which of the following describes Android-targeted malware?
   A. It exists because of a weakness in Android code.
   B. It is easy to get through Google Play.
   C. It mostly exists due to software fragmentation.
   D. It is not that big a problem.

3. In what way are the top categories of mobile device malware financially focused?
   A. They target financial and banking apps.
   B. They target banks directly.
   C. They extract data or services that are easily monetized.
   D. All of the above.

4. Madware often tracks people illegally by pulling data from a device without the owner's knowledge or consent.
   A. True
   B. False

5. Mobile malware tends to focus on which of the following?
   A. Gaining control of phones to launch DDoS attacks or send premium SMS messages
   B. Gaining access to the ports that control GPS and other location information to sell to third-party advertisers
   C. Gaining control of phone file systems to steal data, photos, and so on, or to lock out the user for ransom
   D. All of the above

6. Any application that harvests data from applications or storage is considered to be either malware or madware.
   A. True
   B. False

7. Which of the following describes Apple iOS malware?
   A. It is proportionate to the company's market share.
   B. It primarily targets iPhones that have been jailbroken.
   C. It is a growing problem in the App Store.
   D. All of the above

8. Social engineering does not play a prominent role in mobile malware.
   A. True
   B. False

9. Which of the following is an example of a malware-delivery technique?
   A. Captive portals
   B. USB exploits that jailbreak phones when they are connected to PCs
   C. Likejacking
   D. All of the above

APPENDIX A

Answer Key

CHAPTER 1	The Evolution of Data and Wireless Networks
	1. E 2. A 3. B 4. D 5. D 6. B 7. B 8. A 9. C 10. A 11. C
CHAPTER 2	The Mobile Revolution
	1. C 2. B 3. B 4. A 5. C 6. B 7. D 8. B 9. B 10. D 11. D
CHAPTER 3	Anywhere, Anytime, on Anything: "There's an App for That!"
	1. C 2. A 3. A 4. D 5. B 6. E 7. B 8. A 9. D 10. B 11. B
CHAPTER 4	Security Threats Overview: Wired, Wireless, and Mobile
	1. A 2. B 3. A 4. A 5. B 6. A 7. A 8. E 9. B 10. D
CHAPTER 5	How Do WLANs Work?
	1. A 2. E 3. C 4. C 5. A 6. C 7. B 8. B 9. E 10. C
CHAPTER 6	WLAN and IP Networking Threat and Vulnerability Analysis
	1. A 2. D 3. A 4. C 5. B 6. D 7. A 8. A 9. E 10. B
CHAPTER 7	Basic WLAN Security Measures
	1. B 2. C 3. D 4. C 5. A 6. A 7. D 8. B 9. A 10. C
CHAPTER 8	Advanced WLAN Security Measures
	1. A 2. D 3. E 4. D 5. A 6. B 7. B 8. A 9. A 10. A, D, E
CHAPTER 9	WLAN Auditing Tools
	1. B 2. E 3. C 4. B 5. D 6. A 7. C 8. B 9. A 10. B
CHAPTER 10	WLAN and IP Network Risk Assessment
	1. A 2. B 3. F 4. C 5. B 6. A 7. A 8. A 9. C 10. C
CHAPTER 11	Mobile Communication Security Challenges
	1. B 2. B 3. B 4. D 5. A 6. A 7. B 8. A 9. A 10. C
CHAPTER 12	Mobile Device Security Models
	1. A 2. C 3. D 4. C 5. A 6. A 7. B 8. C 9. B 10. A

APPENDIX A | Answer Key

CHAPTER 13 Mobile Wireless Attacks and Remediation
1. A 2. B 3. B 4. D 5. B 6. B 7. B 8. D 9. C 10. A

CHAPTER 14 Fingerprinting Mobile Devices
1. B 2. D 3. B 4. B 5. C 6. A 7. A 8. A 9. C 10. A

CHAPTER 15 Mobile Malware and Application-Based Threats
1. B 2. C 3. D 4. A 5. D 6. B 7. B 8. B 9. D

APPENDIX B

Standard Acronyms

AES	Advanced Encryption Standard	IPS	intrusion prevention system
ALE	annual loss expectancy	IPSec	Internet Protocol Security
AP	access point	IPv4	Internet Protocol version 4
API	application programming interface	IPv6	Internet Protocol version 6
APT	advanced persistent threat	ISO	International Organization for Standardization
ARO	annual rate of occurrence		
ASDL	asymmetric digital subscriber line	ISP	Internet service provider
ATM	automatic teller machine	L2TP	Layer 2 Tunneling Protocol
BYOD	Bring Your Own Device	LAN	local area network
C-I-A	confidentiality, integrity, availability	MAC	mandatory access control
DDoS	distributed denial of service	modem	modulator demodulator
DES	Data Encryption Standard	NAC	Network Access Control
DHCP	Dynamic Host Configuration Protocol	NAT	network address translation
DMZ	demilitarized zone	NIST	National Institute of Standards and Technology
DNS	Domain Name System		
DoS	denial of service	NMS	network management system
DSL	digital subscriber line	OS	operating system
DSS	distribution system service	OSI	open system interconnection
EULA	End-User License Agreement	PBX	private branch exchange
FCC	Federal Communications Commission	PCI	Payment Card Industry
FTP	File Transfer Protocol	PCI DSS	Payment Card Industry Data Security Standard
GLBA	Gramm-Leach-Bliley Act		
HIPAA	Health Insurance Portability and Accountability Act	PII	personally identifiable information
		PIN	personal identification number
HTML	Hypertext Markup Language	PoE	power over Ethernet
HTTP	Hypertext Transfer Protocol	RAT	remote access Trojan
HTTPS	Hypertext Transfer Protocol Secure	SaaS	Software as a Service
IaaS	Infrastructure as a Service	SHA	secure hash algorithm
ICMP	Internet Control Message Protocol	SLA	service level agreement
IEEE	Institute of Electrical and Electronics Engineers	SLE	single loss expectancy
		SNMP	Simple Network Management Protocol
IP	Internet protocol	SOX	Sarbanes–Oxley Act of 2002 (also Sarbox)

SQL	Structured Query Language	**VPN**	virtual private network
SSID	service set identifier (name assigned to a Wi-Fi network)	**WaaS**	Wi-Fi as a Service
		WAN	wide area network
SSL	Secure Sockets Layer	**WAP**	wireless access point
SSL/VPN	Secure Sockets Layer virtual private network	**WEP**	Wired Equivalent Privacy
		Wi-Fi	wireless fidelity
SSO	single system sign-on	**WLAN**	wireless local area network
TCP/IP	Transmission Control Protocol/Internet Protocol	**WPA**	Wi-Fi Protected Access
		WPA2	Wi-Fi Protected Access 2
USB	universal serial bus	**XML**	Extensible Markup Language
VLAN	virtual local area network	**XSS**	cross-site scripting
VoIP	Voice over Internet Protocol		

Glossary of Key Terms

802.1Q tagging | The process of inserting a virtual local area network (VLAN) ID in the header of frames from clients assigned to a VLAN.

Access control lists (ACLs) | Lists of rules that a networking device will process to allow access rights to ports, applications, or services from clients or connections.

Acrylic Wi-Fi | A set of Wi-Fi Analysis and mapping tools.

Active fingerprinting | A method of fingerprinting in which requests are transmitted a remote device to gain information from the corresponding replies.

Active scanning | A process of wireless discovery in which the client proactively scans the network by sending out probe pulse requests.

Additional authentication data (AAD) | A method of ensuring tamper-proof authentication.

Address Resolution Protocol (ARP) poisoning | A type of attack in which a hacker sends a falsified ARP response, which tricks the client into sending information to the hacker's address.

Ad hoc mode | A wireless configuration in which a client forms a peer-to-peer connection with another client.

Advanced Encryption Standard (AES) | A U.S. government encryption standard used in IPSec, WPA2, and other data-protection schemes.

Advanced Mobile Phone System (AMPS) | The first, or 1G, standard for mobile communication. This is an analog-only technology.

Advanced persistent threats (APTs) | Multi-phased attacks used to break into a network to harvest valuable information while avoiding detection. These highly complex, long-term infiltration attacks present a significant risk to financial institutions and government agencies, among others.

AirMagnet WiFi Analyzer | A Wi-Fi troubleshooting tool that provides visibility into wireless network performance and security

Always on Listening | A phone feature that keeps a microphone active at all times to listen for key phrases. This is typically used to enable touchless voice commands, but other apps can take advantage of the feature as well.

Android sandbox | A security feature in the Android OS that isolates applications (and the resources they use) from each other.

Annualized loss expectancy (ALE) | The product of the annual rate of occurrence (ARO) and the single loss expectancy (SLE).

Annual rate of occurrence (ARO) | The probability that a risk will occur in a particular year.

Application fingerprinting | A type of fingerprinting that looks at the HTTP headers for application-specific information.

Application Layer | Layer 7 of the OSI Reference Model. It is the interface between the user-facing software and the layers that prepare and send or receive networked data.

Application programming interface (API) | This allows a mobile application to interact with an external application or service using a simple template based upon a set of commands.

Application provenance | The practice of requiring users to download applications only from an approved portal or source. This practice allows vendors to identify application developers (who must register their apps) and gives them the ability to test and check application code for malware.

AT command codes | Short for attention commands. The configuration commands used for modems. The AT command codes provide specific instructions to the device when set in configuration mode.

Authentication, authorization, accountability (AAA) | A system used in IP-based networking to track user activity and control user access to computer resources.

Authentication Header (AH) | Part of the Internet Protocol Security (IPSec) protocol suite, the AH confirms the source of a packet and ensures that its contents (header and payload) have not been changed since transmission.

Auto Content Update | A feature that enables information and content to be loaded to a phone app without having to request download permission from the user.

Autonomous access points | Access points with switch-like intelligence that can operate at the control and data functional layers.

Base controller station (BCS) | A device that handles signaling and communication between cellular towers and the mobile phone network or public switched telephone network (PSTN).

Base transceiver station (BTS) | A device that manages cellular traffic between mobile devices (via the cell tower antennas) and the mobile network (via the base controller station [BCS]).

Basic service set (BSS) | A wireless local area network that includes all the wireless devices.

Behavior analysis | A scanning technique that relies not on version data, but on how the system responds to requests, the aim being to find unexpected responses to queries.

Binary protection | A software security technique in which binary files are analyzed and modified to protect against common exploits. Also referred to as *binary hardening*.

BlackBerry Enterprise Server (BES) | The software and server that connect enterprise email and other services to BlackBerry devices.

BleedingBit | a set of vulnerabilities within Bluetooth Low Energy chips which, if exploited, allow attackers to gain access to networks without detection.

Block cipher | A cryptographic algorithm that operates on fixed-length groups of bits, called blocks.

BlueBorne | A Bluetooth vulnerability that exposes any affected device with Bluetooth enabled. The vulnerability is due to vendor implementation rather than with Bluetooth itself.

Bluebugging | A technique used to gain access to mobile phone commands by exploiting Bluetooth vulnerabilities.

Bluejacking | Misuse for advertising purposes of a Bluetooth feature whereby a mobile phone can exchange a "business card" or messages with another phone in the vicinity.

Bluesnarfing | The unauthorized access of information from a wireless device through a Bluetooth connection.

Bluetooth | A standard for short-range wireless interconnection.

Bots | Derived from the word *robot*, programs that perform automated tasks that would otherwise be conducted by a human being.

Bring Your Own Application (BYOA) | The practice of employees installing and using third-party applications for business purposes. Typically, these are cloud-based applications such as Google Docs and Dropbox.

Bring Your Own Cloud (BYOC) | Similar to BYOD (Bring Your Own Device) in which corporate employees utilize personal private or public clouds services from third-party providers.

Bring Your Own Device (BYOD) | The practice of allowing employees to use their own computers or smart devices for work purposes.

Broadcast domains | Logical partitions in a network in which all nodes behave as if they are on the same physical LAN segment.

Brute-force attacks | A method of attack, typically against passwords, in which every possible combination is tried until the right one is found.

Captive portal | A webpage to which users are routed prior to gaining access to the Internet. Captive portals are commonly used by hotspots for payment or for the acknowledgment of user agreements, but can also be used as a form of authentication or to check credentials.

Carriage return line feed (CRLF) | A common HTTP vulnerability in which an HTTP packet is split using a carriage return followed by a line feed. After splitting a packet in two, with one packet containing legitimate header and protocol information, the attacker can pack a malicious payload into the second packet.

Cell towers | The physical mount for cellular antennas. Typically constructed as tall poles but sometimes disguised as trees or other natural objects.

Cellular | A generic term for mobile phone systems or devices. Refers to the portioning of frequency coverage maps into cells.

Glossary of Key Terms

Certificate authority (CA) | An authority in a network that issues and manages security credentials and public keys for message encryption.

C-I-A triad | The three main components of information security: confidentiality, data integrity, and availability.

Circuit switching | A telecommunications method that uses a dedicated channel or circuit to connect two endpoints.

Clickjacking | An exploit that uses a webpage with a background of invisible frames. If the victim clicks the page, the exploit is triggered.

Client-side injection | An exploit that seeks to install malware through the injection of malicious content from a custom-built hostile service.

Code Division Multiple Access (CDMA) | Part of the 2G wave of technologies, CDMA allows multiple communication streams to share a common communications channel.

Collocation model | An access point deployment in which there is 100-percent overlap between the two access point service areas.

Command line interface (CLI) | An interface for issuing commands to a computer (or any computerized device) by way of typing. This was the primary means of interaction with most computer systems until the introduction of video display terminals and is still in use today as a shortcut method for interfacing with devices.

Completely Automated Public Turing Test to Tell Computers and Humans Apart (CAPTCHA) | A webpage feature that attempts to distinguish a human user from a machine user, typically by requiring the visual recognition of a phrase in the form of a picture (rather than text). This helps to prevent spam and automated logons.

Compliance | The adherence to regulations required by government and/or industry for IT security and data protection.

Cookies | Typically small text files that are stored on a computer to keep track of the user's movements on a website, resume interrupted sessions, and remember logon credentials.

Corporate owned personally enabled (COPE) | The practice of allowing an employee to choose, manage, and to some degree customize a device that is purchased by the company for which the individual works. This is considered to be the opposite of BYOD.

Counter Mode Cipher Block Chaining Message Authentication Code Protocol (CCMP) | An enhanced data cryptographic encapsulation mechanism created to address the vulnerabilities presented by WEP.

Cross-site profiling | A technique companies use to collect data from various websites to find and compile information about users, which the companies can then use to target ads.

Crypto primitives | Well-established, low-level cryptographic algorithms that are frequently used to build cryptographic protocols.

Customer resource management (CRM) | A system used to manage customer interactions—for example, with sales or support.

Cybercrime | Any crime (including trespassing) committed over the Internet or other computer network.

Data Link Layer | Layer 2 of the OSI Reference Model. It specifies how data is communicated over local area networks.

Dead spots | Areas without wireless coverage.

Deep packet inspection | A form of packet filtering that inspects the data portion of a packet for malicious code rather than looking only at the packet header.

Defense in depth | The practice of implementing several layers of network and data security.

Demilitarized zone (DMZ) | An intermediate area of a network that allows outside access to certain assets (such as a web server) but limits internal access to the back-end control servers.

Desktop virtualization | The reproduction of the user's desktop on an Internet-accessible server.

Diameter | An authentication protocol used for Mobile IP networks.

Dictionary password crackers | Password crackers that decode passwords by using a lookup table or database of large numbers of common and stolen passwords and their hashes for different encryption methods. Because most people don't use strong or varied passwords, these lookup tables are far more efficient than brute-force methods.

Digital Advanced Mobile Phone System (D-AMPS) | The standard for what was the second generation of mobile networks.

Direct sequence spread spectrum (DSSS) | A spread spectrum technique that uses data encoding to spread data across several channels for transmission.

Distribution medium | The physical medium to which the access points ports connect, typically an Ethernet LAN.

Distribution service (DS) | The ability of a wireless access point (WAP) to recognize, reframe, address, and deliver packets between two interfaces or mediums.

Distribution system service (DSS) | The internal software that controls the switch-like intelligence and manages client station association and disassociations.

Domain Name System (DNS) | A system for naming computers and network services so they can be located and communicated with by other networked devices.

Dotted decimal | A numerical format expressed as a string of numbers separated by periods.

Drive-by browser exploits | Exploits that target web browser plug-ins on mobile devices for Java, Adobe Reader, and Flash. These attacks are most often launched via legitimate but compromised websites that infect browser software running on mobile clients.

Dynamic Host Configuration Protocol (DHCP) | A network protocol that enables a server to automatically assign IP addresses.

Encapsulation Security Payload (ESP) | Part of the IPSec set of protocols. The ESP is inserted into an IP packet to provide confidentiality as well as data origin authentication and integrity.

End User License Agreement (EULA) | A legal agreement between a software developer and users.

Enterprise mobility management (EMM) | A set of people, processes, and technology focused on managing mobile devices, wireless networks, and related services in a business context.

Evil twin | A hacker-controlled access point set up with the same SSID as a legitimate access point.

Extended service set (ESS) | A wireless local area network that includes one or more basic service sets (BSSs) as well as their associated local area networks.

Federal Communications Commission (FCC) | An independent U.S. government agency charged with regulating communications over radio, television, wire, satellite, and cable.

File Transfer Protocol (FTP) | A network protocol used to transfer computer files from one host to another over a TCP-based network.

Fingerprinting | The process of identifying a device on a network or the user behind the device.

FreeBSD | An operating system for a variety of platforms derived from BSD, the version of UNIX developed at the University of California, Berkeley.

Free space optics (FSO) | Laser WAN links with up to 10 Gbps throughput and a line of site range about 3 miles. Typically used for data backhaul from remote locations or for remote IoT sensor networks.

Frequency Division Multiple Access (FDMA) | An access method that allows for multiple users through the assignment of frequency channels.

Frequency hopping spread spectrum (FHSS) | A method of transmitting radio signals by rapidly switching among many frequency channels.

Frequency reuse | The practice of assigning multiple users to the same frequency channel. Achieved by the physical separation and power management of the transmission streams.

Full duplex | Communication in both directions simultaneously.

General controls review (GCR) | An assessment of an organization's internal controls over information technology and information security to help align controls with industry best practices.

General Packet Radio Service (GPRS) | The first packet-switched technology used on mobile networks. Often referred to as 2G+, it allowed web-based access from mobile phones.

Global System for Mobile (GSM) | The primary technology used for 2G mobile systems. GSM was the dominant digital standard for mobile communications.

Gramm–Leach–Bliley Act (GLBA) | A U.S. federal law enacted to reorganize the financial services industry and control how financial institutions deal with individuals' private information.

Groupe Spécial Mobile (GSM) | The dominant 2G mobile phone system standard. GSM was the standard throughout all of Europe but saw competition from CDMA in the United States and parts of Asia. The name was later changed to Global System for Mobile.

Glossary of Key Terms

Hackers | The name generally applied to those who commit cybercrime, although in the early days of networking, hackers were more curious technologists than criminals.

Half duplex | Communication in both directions, one direction at a time.

Handoff | A feature that enables a user to begin working in an application on one device and then hand off, or continue work, on a second device.

Hash | A number generated from a string of text. The hash, also called a hash value or message digest, is substantially smaller than the text itself and is generated by a formula in such a way that it is extremely unlikely that some other text will produce the same hash value.

Health Information Technology for Economic and Clinical Health (HITECH) | Part of an economic stimulus package, HITECH was created to enable the use of electronic health records (EHRs)

Health Insurance Portability and Accountability Act (HIPAA) | A U.S. law aimed at making it easier for people to keep health information private.

Heartbleed | Discovered in 2014, a security bug in the OpenSSL cryptography library, which is a widely used implementation of the TLS protocol.

Highly directional antennas | Directional antennas that radiate greater power in a single direction. Also called *beam antennas*.

High Speed Downlink Packet Access (HSDPA) | An enhanced 3G communications protocol that allowed faster data rates than the original 3G systems. HSDPA is referred to as 3G+ and 3.5G.

Hypertext Transfer Protocol Secure (HTTPS) | A secure communication protocol in wide use on the Internet.

Identity and access management (IAM) | The management of an individual's authentication, authorization, and privileges within or across system and enterprise boundaries.

Independent basic service set (IBSS) | An ad hoc wireless network that does not have an access point. An IBSS cannot connect to other basic service sets.

Information security | The processes and practices that must be implemented to secure the digital assets you wish to protect from various threats.

Infrastructure mode | The most common topology for a WLAN. It uses an access point as a connection hub and portal to a distribution system.

Integration service (IS) | The process of recognizing, reframing, addressing, and delivering packets between wireless and wired mediums.

International Mobile Station Equipment Identity (IMEI) number | A serial number that uniquely identifies a mobile station internationally.

International Mobile Subscriber Identity (IMSI) | A unique identification associated with all GSM and UMTS network mobile phone users.

International Mobile Telecommunications-2000 (IMT-2000) | A set of 3G standards that define global roaming and network interoperability.

International Telecommunications Union (ITU) | A United Nations (UN) agency charged with coordinating telecommunications operations and services throughout the world.

Internet Control Message Protocol (ICMP) | A protocol used by network devices to send error messages.

Internet of Things (IoT) | The networking of electronics and other physical objects via embedded electronics.

Internet Protocol Security (IPSec) | An open standards encryption method of ensuring private, secure communications over Internet Protocol (IP) networks.

Internet Protocol version 4 (IPv4) | The addressing scheme that defines private networks. Version 4 has been used for most of the Internet age.

Internet Protocol version 6 (IPv6) | An updated addressing scheme that defines private networks. Version 6 offers many more addresses and more features than version 4.

Inter-process communication (IPC) | A set of programming interfaces that allow a programmer to coordinate activities among different program processes that can run concurrently in an operating system.

Intrusion detection systems (IDSs) | Devices or applications that monitor networks for malicious activities or policy violations.

Intrusion prevention systems (IPSs) | Network security appliances that monitor networks and systems for malicious activity.

IP addressing | A logical (nonpermanent) label assigned to networked devices to establish a location for transmitting and receiving data.

ISM unlicensed spectrum | The ISM radio bands are sections of the radio spectrum that are unlicensed (not for telecommunications) and purposefully set aside for commercial or public use.

Jailbreaking | The act of *rooting*, or hacking into a smart device to allow users to attain privileged control (known as *root access*) within the device's subsystem.

Jitter | The variation in the time between the arrival of packets from the same transmission, caused by network congestion, timing drift, or route changes.

Kali Linux | A variant of Linux designed for digital forensics and penetration testing.

Key performance indicators (KPIs) | A set of values against which to measure the quality or success of an operation or process.

Knowledge workers | Professionals whose job involves the use and manipulation of data.

Latency | A measurement of delay. Typically, the amount of time it takes for a packet to get from one point to another.

Least privilege | A policy whereby users are given access only to the systems and data they need to perform their jobs, and no more.

Likejacking | An exploit that uses an invisible frame over a Facebook Like button on a webpage. When a user clicks the button, the exploit is triggered. This exploit works especially well on mobile devices.

Local area network (LAN) | A collection of networked devices in a small location, such as a building, office, or home. LANs can be small home networks with a few users or an enterprise network with thousands of users and devices.

Long-Term Evolution (LTE) | A 4G mobile communications standard.

Machine-to-machine (M2M) communication | Direct communication between devices most often used for industrial instrumentation and sensor/meter communication.

Macrocells | Cells within a mobile system with a large coverage area.

MAC service data unit (MSDU) | A service data unit that is received from the logical link control (LLC) sublayer (a portion of the Layer 3 Data Link Layer).

Madware | An aggressive form of advertising that affects smartphones and tablets. Short for *mobile adware*.

Man-in-the-middle (MITM) attack | A form of network eavesdropping in which the attacker inserts themselves between two machines after making them believe they are talking directly to each other.

Masquerading | An attack in which the hacker attempts to impersonate an authorized user and gain that user's level of privileges.

Media access control (MAC) address | A unique identifier assigned by manufacturers to any network-connected device. The MAC address is used to establish the source or destination of data flows on a local area network.

Mesh basic service set (MBSS) | A basic service set that forms a mesh of stations.

Message information base (MIB) | A component of SNMP, an MIB is a local database containing information relevant to network management.

Message integrity code (MIC) | A short piece of information used to authenticate and provide messaging integrity.

Microcells | Cells within a mobile system with a small coverage area.

Mobile application management (MAM) | Software and services for controlling access to mobile apps used in business settings on both company-provided and BYOD smart devices.

Mobile cloud computing (MCC) | *A* combination of *cloud and mobile computing* to deliver resources to *mobile* users.

Mobile device management (MDM) | Best practices for deploying, securing, monitoring, integrating, and managing mobile devices in the workplace.

Mobile IP | A communication standard that allows users to maintain an IP address and web session as they roam between different networks or network segments.

Mobile Remote Access Trojans (mRATs) | Malware programs that give an attacker administrative control over a smartphone or tablet.

Modem | Short for *modulator/demodulator*. Modems prepare information for transmission over a network and reassemble the data on the receiving end.

Multipath | A phenomenon that results in radio signals reaching the receiving antenna by two or more paths.

Multiple input/multiple output (MIMO) antennas | A technology that allows multiple antennas to transmit and receive concurrently.

Glossary of Key Terms

Multi-user MIMO (MU-MIMO) | A *multiple*-input and *multiple*-output (*MIMO*) communication technology for multipath wireless communication, in which *several devices*, communicate with one another.

Narrowband | Transmitting or receiving signals over a narrow range of frequencies.

Near-field communication (NFC) | A communication standard for smartphones and other devices to establish radio communication with other devices by bringing them near each other.

Network address translation (NAT) | A method of masking or hiding a private address from a public network.

Network cloaking | The act of hiding a WLAN by not advertising the SSID in the beacon.

Network effect | The increased value of all devices whenever a new device is added. For example, the first fax machine was useless by itself, but adding a second fax machine made the first one usable. With each fax machine added, the usefulness of all fax machines increased.

Network Layer | Layer 3 of the OSI Reference Model, where routing protocols such as IP are defined.

Network management system (NMS) server | A combination of hardware and software used to monitor and administer a computer network or networks.

Nomadic roaming | A type of roaming in which, when the device moves from one wireless coverage area to another, the session is broken from one transmitter and then reestablished with another.

Nonce | A one-time randomly generated number used to create encryption keys.

Nonrepudiation | The act of providing undeniable evidence that an action was taken and by whom. Nonrepudiation is important in e-commerce and financial transactions such as online trading.

Omnidirectional antenna | An antenna that transmits and receives signals in all directions.

Open share | A method of sharing files directly between clients over an air interface.

Open System Authentication | A process by which a client gains access to a wireless network.

Open Systems Interconnection (OSI) Reference Model | A multi-layered, vendor-neutral description that defines the protocols and communication procedures for networks.

Open Web Application Security Project (OWASP) | An online community dedicated to web application security.

Orthogonal frequency-division multiple access (OFDMA) | A multiplexing scheme that allows access points to connect to multiple users at the same time.

OS fingerprinting | The process of analyzing TCP/IP packets to detect what operating system a machine is running.

Packet switching | A method of data communication that chops data into smaller parts for transmission and reassembles them on the receiving end. Packet switching greatly improves efficiency over dedicated circuit switching because packets from different communication streams can share a common circuit and packets from a single transmission stream can use different circuits during transmission.

Passive fingerprinting | The process of analyzing packets from a host on a network. In this case, the fingerprinter acts as a sniffer, and doesn't put any traffic on a network.

Passive scanning | With passive scanning, a client waits until it "hears" a beacon advertising an SSID from an access point. If the client hears a beacon with a matching SSID, it selects the access point with the strongest signal.

Password management system (PMS) | A software application or system that helps a user store and organize passwords, which are applied automatically when the user logs into different sites.

Payment Card Industry Data Security Standard (PCI DSS) | An industry-driven data privacy standard that describes best practices and certifications for the secure storage, processing, or transmitting of credit cardholder data.

PDCA cycle | A four-step problem-solving repetitive technique used to improve business processes. The four steps are *plan*, *do*, *check*, and *act*.

Penetration testing (pentesting) | A method of testing and evaluating network security by simulating an attack on the network. Pentesting requires permission in advance from the network owners.

Personally identifiable information (PII) | Information that can be used to identify, contact, or locate a single person, or to identify an individual in context.

Personal area network (PAN) | A computer network limited to an individual's workspace. PANs typically connect a person's computer, smartphones, and tablet. It is very useful for sharing files among devices.

Phablets | Mobile devices that are somewhere between a phone and a tablet.

Phishing | The act of defrauding a person by posing as a legitimate organization (typically a bank), most often in an email or website. The victim is tricked into giving away their logon information to criminals, who then obtain valuable personal information or take money from the victim's account.

Phreakers | Those who attempt to access and gain free use of telephone networks.

Physical Layer | Layer 1 of the OSI Reference Model. Defines the standards for the various signal paths over which data is transmitted.

Picocells | Small hotspot cells that offer wireless (Wi-Fi) connectivity via a mobile carrier.

Piconet | A network created using a Bluetooth connection.

Point-to-Point Protocol (PPP) networks | Data links between two locations (or clients) without the use of any intermediate devices.

Port-based Network Access Control (PNAC) | Part of the IEEE 802.1 group of networking protocols, PNAC provides an authentication mechanism through the use of an authenticator device, which passes logon credentials to an authentication database for approval prior to allowing access to the network.

Port mirroring | A method of monitoring network traffic in which the network sends a copy of all network packets seen on one or more ports to another port, where the packets can be viewed and analyzed.

Port scanning | The process of probing servers and hosts on a network for open ports. Port scanning is used by administrators to verify security policies and by attackers to identify services on a host.

Potentially unwanted applications (PUAs) | Programs that contain adware, install toolbars, or have other unclear objectives.

Premium SMS | A pay-as-you-go service available on smartphones. This service is often taken advantage of by hackers. They use malware to send premium SMS messages from hacked phones to services that they own, control, or can otherwise extract payment from.

Presentation Layer | Layer 6 of the OSI Reference Model. It defines the formatting of information sent to and from applications.

Proximity fingerprinting | Fingerprinting that takes places via data collected from nearby sources, typically via wireless sniffing.

Public switched telephone network (PSTN) | All of the world's telephone networks, which, while independently owned and operated, are connected and interoperate worldwide.

Quality of service (QoS) | A method of prioritizing time-sensitive traffic (usually voice or video) on a network.

Quarantining | The process of isolating a client that is not compliant or is out of date with antivirus and security patches. Usually, such a client is quarantined to a guest virtual local area network (VLAN), where it can access the Internet but not internal systems. It typically remains there until it is security compliant.

Rainbow table | A table that is populated over time and used for reversing cryptographic hash functions, usually for cracking password hashes.

Remote Access Trojans (RATs) | Malware programs that give an attacker administrative control over a computer.

Remote Authentication Dial-In User Service (RADIUS) | A network protocol that provides authentication, authorization, and accountability (AAA) services for devices or users connecting to a network.

Remote fingerprinting | Fingerprinting that takes place online.

Replay attack | An attack in which legitimate traffic is captured using a packet analyzer, modified, and then retransmitted.

Robust Security Network (RSN) | A protocol for establishing secure communications over an 802.11 wireless network. The Wi-Fi Alliance refers to its approved, interoperable implementation of the full 802.11i as RSN.

Rooting | The process of allowing users of smart devices to attain privileged control (known as *root access*) within the device's subsystem. Also called *jailbreaking*.

Sandbox | An approach to software development and mobile application management (MAM) that limits the environments in which certain code can execute.

Glossary of Key Terms

Sarbanes–Oxley Act (SOX) | A 2002 U.S. federal law that set new accounting standards for publicly traded companies and requires them to certify that controls are in place to protect their information. Also referred to as *SarbOx*.

Seamless roaming | A type of roaming in which the session is not disrupted when the device moves from one wireless coverage area to another.

Secure Simple Pairing (SSP) | A secure method of pairing or connecting Bluetooth devices.

Security Associations (SA) | The establishment of shared security attributes between two network entities to support secure communication.

Semi-directional antennas | Antennas designed to transmit and receive with greater effectiveness in a particular direction, most often at the expense of all other directions. This improves performance in the intended direction while limiting interference in the nontargeted areas.

Service level agreements (SLAs) | Contracts between a network service provider and a customer that specify, usually in measurable terms, what services the network service provider will furnish. An SLA can also be an internal agreement between internal service providers (such as the IT team) and their constituents.

Service-orientated architecture | An "as needed" software service where each key component of a software architecture is provided on a ad hoc basis over a network.

Service set identifier (SSID) | The name (in the form of a text string of up to 32 bytes) assigned to a wireless access point.

Session Layer | Layer 5 of the OSI Reference Model. It defines the communication setup and teardown between two networked devices.

Shared key authentication (SKA) | A type of authentication that assumes each station has received a secret shared key through a secure channel independent from the wireless network.

Short Message Service (SMS) | A standardized text message service for use over mobile phones and other devices.

Signaling System 7 (SS7) | A protocol used in the setup and teardown of telephone calls.

Simplex communication | Communication that flows in only one direction, such as broadcast radio.

Single loss expectancy (SLE) | The expected monetary cost from the occurrence of a risk on an asset.

Single sign-on (SSO) | A property of access control whereby a user logs on once and gains access to all systems without being prompted to log on again at each of them.

Single-user MIMO (SU-MIMO) | A *multiple*-input and *multiple*-output (*MIMO*) communication technology for multipath wireless communication, in which *a single device is communicated to* via multiple frequencies and/or antenna

Small office/home office (SOHO) | A office with just a handful of employees or an office in a person's home.

Small to medium business (SMB) | A business that, due to its size, has different IT requirements and often faces different IT challenges than do large enterprises, and whose IT budget and staff are often constrained.

Smartphone | A mobile phone with advanced computing and connectivity capabilities.

SNMP traps | Alerts generated by Simple Network Management Protocol (SNMP) agents on managed devices. SNMP traps are generated when certain activities are flagged or set to a threshold. When the conditions are met, an alert is sent to the network management system.

Social engineering | The practice of teasing out from people information that should not be shared to use it to one's advantage.

Software development kit (SDK) | A programming package that enables a programmer to develop applications for a specific platform. Typically, an SDK includes one or more APIs, programming tools, and documentation.

Spread spectrum | A data-transmission technique in which a signal is transmitted across the entire frequency space available.

Spyware | Malicious software that enables a user to covertly obtain information about another user's computer activities.

SSID cloaking | A relatively weak method of wireless security in which the network name (service set identifier) is prevented from being publicly broadcast.

Stack | In the context of the OSI Reference Model, the hierarchical relationship of the seven layers.

Station (STA) | A device that has the capability to use the 802.11 protocol.

Stream cipher | An encryption algorithm in which plaintext is combined with a random key stream. In a stream cipher, each plaintext digit is encrypted one at a time, creating a stream of ciphertext.

Subnets | A logical segment or partition within a private IP network.

Subscriber identity module (SIM) | A "smart card" that is either embedded in or attached to a mobile device to enable mobile network access. SIM cards often hold a user's personal information and can be switched from one phone to another.

SweynTooth | A group of 12 Bluetooth low energy (BLE) vulnerabilities identified by researchers.

T1/E1 | The standard digital carrier signals that transmit both voice and data. T1 rates are 1.544 Mbps and E1 operates at about 2 Mbps.

Telegraphy | A one-way message protocol invented by Samuel Morse that used start and stop signals of dots and dashes transmitted over copper wires and later radio signals.

Telephony | The design or use of telephones, limited to voice communication.

Telnet | A network protocol that allows one computer to log on to another computer on the same network.

Temporal Key Integrity Protocol (TKIP) | A security protocol used in the IEEE 802.11 wireless networking standard as an interim solution to replace WEP without requiring the replacement of legacy hardware.

Thin access points | Wireless access points with limited or no switch-like intelligence. Thin access points are configured and managed via a central controller, which provides the transition between the wireless and wired networks.

Time Division Multiple Access (TDMA) | A multiple user access scheme that partitions users into time slots on the same channel.

Transmission Control Protocol/Internet Protocol (TCP/IP) | The dominant suite of communication protocols used to communicate over the Internet. TCP/IP defines the communication setup as well as how packets are formatted, transmitted, routed, and received.

Transport Layer | Layer 4 of the OSI Reference Model. The bridge between the network and the application-processing software on devices. This is where data from applications is broken down into small chunks, or packets, and then reassembled on the receiving device.

Version analysis | A scanning technique whereby the scanner sends out requests to a target system and, upon receiving a response, analyzes the headers for the version details.

Very high throughput (VHT) | The 802.11ac standard, a faster version of 802.11n. It is referred to as giga wireless or VHT.

Virtual LAN (VLAN) | A logical partition on a local area network (LAN) that allows workstations within to communicate with each other as though they were on a single, isolated LAN.

Viruses | Malicious pieces of code that attach to a program. Once installed on a machine, a virus can replicate itself, often replacing or corrupting critical files in the process.

Voice over WLAN (VoWLAN) | A wireless-based Voice over Internet Protocol (VoIP) system.

Walled garden | An environment that controls the user's access to web content and services.

Wardriving | The act of searching for and using an unsecured or open wireless network.

White list | A security or access control technique whereby all options are denied except those specifically permitted. Those on the permitted list are said to be *white listed* (allowed).

Wi-Fi | An alternate term for wireless LAN (WLAN). Refers to a system that allows for wireless connectivity to the Internet or a private network.

Wi-Fi as a Service (WaaS) | A cloud-based Wi-Fi management system.

Wi-Fi Protected Access (WPA) | A technology used for security and data protection over wireless networks. More powerful than WEP, WPA uses 128-bit encryption.

Wi-Fi Protected Access 2 (WPA2) | A more secure version of WPA. With WPA2, a passphrase is used to create an encryption key.

Wi-Fi Protected Access 3 (WPA3) | An updated version of WPA2 providing additional security via 128-bit encryption among other features.

Wired Equivalent Privacy (WEP) | A security protocol for wireless networks defined by the 802.11 standard. Proven to be a weak method of security, WEP has been replaced by WPA and WPA2.

Wireless access point (WAP) | A device that allows wireless devices to connect to a wired network using Wi-Fi or related standards.

Wireless distribution system (WDS) | A system that enables the wireless interconnection of access points without the use of a backbone (wired) network.

Wireless extender | A device that takes a signal from a wireless access point and rebroadcasts it to create a second network. Also called a wireless repeater.

Wireless local area network (WLAN) | A local area network that allows access via radio waves rather than through cables. Access to the network is most often gained through a device known as an *access point*, which provides the physical connection to the network.

Wireless personal area network (WPAN) | A wireless network centered around a person's immediate workspace. Typically, it is used to wirelessly connect peripheral devices to a computer.

Wireless repeater | See *wireless extender*.

Wireless sensor network (WSN) | A group physically separated sensors used for monitoring status and collecting date. Data is passed through to a central location.

Worldwide Interoperability for Microwave Access (WiMAX) | A wireless standard that provides data rates of up to 1 Gbps. WiMAX is often viewed as a "last mile" broadband method, connecting home or small offices to the high-speed network backbone.

Worms: Similar to viruses | A worm replicates itself by spreading to other machines. Unlike viruses, worms are standalone pieces of code.

Xirrus Wi-Fi Inspector | Wireless network status monitoring software.

References

"4 New Features: First Wi-Fi Security Overhaul in 13 Years." SecureWorld. Accessed April 19, 2020. https://www.secureworldexpo.com/industry-news/new-wifi-wap3-features.

"10 Questions CISOs Should Ask About Mobile Security." Bitpipe.com, August 14, 2014. Accessed October 2, 2014. http://www.bitpipe.com.

"802.11 Network Security Fundamentals." Cisco. Accessed October 2, 2014. http://www.cisco.com/c/en/us/td/docs/wireless/wlan_adapter/secure_client/5-1/administration/guide/SSC _Admin_Guide_5_1/C1_Network_Security.html.

"802.1X: What Exactly Is It Regarding WPA and EAP?" SuperUser.com, January 12, 2012. http://superuser.com/questions/373453/802-1x-what-exactly-is-it-regarding-wpa-and-eap.

"90% of Unknown Malware Is Delivered via the Web." *Infosecurity*, March 26, 2013. http://www.infosecurity-magazine.com/news/90-of-unknown-malware-is-delivered-via-the-web.

"About the iOS Technologies." iOS Developer Library, Apple Inc. Accessed October 21, 2014. https://developer.apple.com/library/ios/documentation/miscellaneous/conceptual/iphoneostechoverview/Introduction/Introduction.html.

"A Brief History of Wi-Fi." *The Economist*, June 10, 2004. Accessed August 12, 2014. http://www.economist.com/node/2724397.

"Access Control and Authorization Overview." TechNet, Microsoft, February 20, 2014. http://technet.microsoft.com/en-us/library/jj134043.aspx.

"Advanced Persistent Threats: How They Work." Symantec. Accessed September 10, 2014. http://www.symantec.com/theme.jsp?themeid=apt-infographic-1.

"All About BYOD." *CIO*, June 24, 2014. Accessed August 5, 2014. http://www.cio.com/article/2396336/byod/all-about-byod.html.

"An Overview of the Sub-GHz ISM Bands." BehrTech Blog. Accessed April 10, 2020. https://behrtech.com/blog/an-overview-of-sub-ghz-ism-bands.

"Android Security Overview." Android Open Source Project. Accessed October 21, 2014. https://source.android.com/devices/tech/security/.

"Android Tools." Hackers Online Club. Accessed October 21, 2014. http://www.hackersonlineclub.com/android-tools.

"An Introduction to ISO 27001 (ISO27001)." *The ISO 27000 Directory*. Accessed August 30, 2014. http://www.27000.org/iso-27001.htm.

"ArubaOS User Guide." Aruba Networks, December 9, 2010. Accessed September 10, 2014. http://www.arubanetworks.com/techdocs/ArubaOS_60/UserGuide/.

Asadoorian, Paul. "Using Nessus to Discover Rogue Access Points." Tenable Network Security, August 27, 2009. Accessed September 10, 2014. http://www.tenable.com/blog/using-nessus-to-discover-rogue-access-points.

"AT&T Labs: Backgrounder." AT&T. Accessed August 15, 2014. http://www.corp.att.com/attlabs/about/backgrounder.html.

Beaver, Kevin. "How to Use Metasploit Commands for Real-World Security Tests." TechTarget, November 2005. Accessed November 26, 2014. http://searchsecurity.techtarget.com/tip/Using-Metasploit-for-real-world-security-tests.

"Best Practice Guide Mobile Device Management and Mobile Security." Kaspersky Lab, 2013. Accessed November 26, 2014. http://media.kaspersky.com/en/business-security/Kaspersky-MDM-Security-Best-Practice-Guide.pdf.

"Black Hat 2019: 5G Security Flaw Allows MiTM." Black Hat 2019. Accessed June, 2020. https://threatpost.com/5g-security-flaw-mitm-targeted-attacks/147073.

Blevins, Brandan. "Report: Backoff Malware Infections Spiked in Recent Months." TechTarget, October 24, 2014. Accessed November 26, 2014.

Bojinov, Hristo, Dan Boneh, Yan Michalevsky, and Gabi Nakibly. "Mobile Device Identification via Sensor Fingerprinting." Stanford University. Accessed November 11, 2014. https://crypto.stanford.edu/gyrophone/sensor_id.pdf.

Botelho, Jay. "Wireless in the Warehouse." Enterprise Networking Planet, February 10, 2014. Accessed August 12, 2014. http://www.enterprisenetworkingplanet.com/netsp/wireless-in-the-warehouse.html.

Bowers, Tom. "Finding the Balance Between Compliance & Security." *Information Week Dark Reading*, January 30, 2014. Accessed August 30, 2014. http://www.darkreading.com/compliance/finding-the-balance-between-compliance-and-security/d/d-id/1113620.

"Building Global Security Policy for Wireless LANs." Aruba Networks. Accessed October 2, 2014. http://www.arubanetworks.com/pdf/technology/whitepapers/wp_Global_security.pdf.

Carter, Jamie. "What Is NFC and Why Is It in Your Phone?" *TechRadar*, January 16, 2013. Accessed August 30, 2014. http://www.techradar.com/us/news/phone-and-communications/what-is-nfc-and-why-is-it-in-your-phone-948410.

Casey, Brad. "Identifying and Preventing Router, Switch and Firewall Vulnerabilities." TechTarget, December 2013. Accessed November 10, 2014. http://searchsecurity.techtarget.com/tip/Identifying-and-preventing-router-switch-and-firewall-vulnerabilities.

"CDMA/FDMA/TDMA: Which Telecommunication Service Is Better for You?" WINLAB, Rutgers, The State University of New Jersey. Accessed August 15, 2014. www.winlab.rutgers.edu/~crose/426_html/talks/foglietta_pres2.ppt.

"Cellebrite and Webroot Partner to Deliver Mobile Malware Diagnostics Capabilities to Cellular Retail Market." Cellebrite, 2014. Accessed November 26, 2014. http://www.cellebrite.com/pt/corporate/news-events/retail-press-releases/706-cellebrite-and-webroot-partner-to-deliver-mobile-malware-diagnostics-capabilities-to-cellular-retail-market.

"Cellular Networks." Northeastern University. Accessed August 15, 2014. http://www.ccs.neu.edu/home/rraj/Courses/6710/S10/Lectures/CellularNetworks.pdf.

Chandra, Praphul, Dan Bensky, Tony Bradley, Chris Hurley, Steve Rackley, John Rittinghouse, James F. Ransome, Timothy Stapko, George L. Stefanek, Frank Thornton, Chris Lanthem, and Jon S. Wilson. *Wireless Security: Know It All*. Amsterdam: Newnes, 2004.

Chebyshev, Victor. "Mobile malware evolution 2019." Kaspersky Feb 25, 2020. https://securelist.com/mobile-malware-evolution-2019/96280.

References

Chirillo, John, and Edgar Danielyan. *Sun Certified Security Administrator for Solaris 9 & 10 Study Guide*. New York: Osborne McGraw-Hill, 2005.

"Client Side Exploits." Metasploit Unleashed. Offensive Security Ltd. Accessed November 10, 2014. http://www.offensive-security.com/metasploit-unleashed/Client_Side_Exploits.

Cluley, Graham. "Revealed! The Top Five Android Malware Detected in the Wild." *Naked Security*. Sophos Ltd, June 14, 2012. Accessed November 26, 2014. http://nakedsecurity.sophos.com/2012/06/14/top-five-android-malware/.

"COBRA Risk Consultant." *The Security Risk Analysis Directory*, 2003. Accessed November 26, 2014. http://www.security-risk-analysis.com/riskcon.htm.

Coleman, David D. CWSP: *Certified Wireless Security Professional Official Study Guide*. Indianapolis, IN: John Wiley & Sons, 2010.

Coleman, David D., and David A. Westcott. *CWNA Certified Wireless Network Administrator Official Study Guide Exam PW0-105*. Hoboken: John Wiley & Sons, 2012.

Columbus, Louis. "IDC: 87% of Connected Devices Sales by 2017 Will Be Tablets and Smartphones." *Forbes*, September 12, 2013. Accessed August 15, 2014. http://www.forbes.com/sites/louiscolumbus/2013/09/12/idc-87-of-connected-devices-by-2017-will-be-tablets-and-smartphones/.

Compton, Stuart. "802.11 Denial of Service Attacks and Mitigation". Technical paper. SANS Institute, May 17, 2007. Accessed November 26, 2014. http://www.sans.org/reading-room/whitepapers/wireless/80211-denial-service-attacks-mitigation-2108.

Constantin, Lucian. "Dozens of Rogue Self-Signed SSL Certificates Used to Impersonate High-Profile Sites." *Computer World*, February 13, 2014. Accessed November 26, 2014. http://www.computerworld.com/article/2487761/encryption/dozens-of-rogue-self-signed-ssl-certificates-used-to-impersonate-high-profile-sites.html/02/13/2014.

Cooney, Michael. "10 Common Mobile Security Problems to Attack." *PCWorld*, September 21, 2012. Accessed November 26, 2014. http://www.pcworld.com/article/2010278/10-common-mobile-security-problems-to-attack.html.

"Cross-Site Scripting (XSS) Attack." Acunetix. Accessed October 15, 2014. https://www.acunetix.com/websitesecurity/cross-site-scripting.

Cruz, Benjamin, et al. "McAfee Labs Threats Report." McAfee Labs, June 2014. Accessed November 26, 2014. http://www.mcafee.com/hk/resources/reports/rp-quarterly-threat-q1-2014.pdf.

"Data Communications Milestones." Telecom Corner, Tampa Bay Interactive, Inc., October 25, 2004. Accessed August 5, 2014. http://telecom.tbi.net/history1.html.

"Delivering Enterprise Information Securely on Android, Apple IOS, and Microsoft Windows Tablets and Smartphones." Technical paper. Citrix, 2014. Accessed August 30, 2014. http://www.citrix.com/content/dam/citrix/en_us/documents/oth/delivering-enterprise-information-securely.pdf?accessmode=direct.

Dewan, Richard. "How to Install Free RADIUS Server in Kali Linux?" *Computer Trikes*, July 7, 2013. Accessed November 26, 2014. http://computertrikes.blogspot.com/2013_07_01_archive.html.

"Differences Between 802.11a, 802.11b, 802.11g and 802.11n." AT&T. Accessed August 12, 2014. http://www.wireless.att.com/support_static_files/KB/KB3895.html.

Drew, Jessica. "Mobile Phones Are Under Malware Attack." Top Ten Reviews, 2014. Accessed November 13, 2014. http://anti-virus-software-review.toptenreviews.com/mobile-phones-are-under-malware-attack.html.

Du, Hui, and Chen Zhang. "Risks and Risk Control of Wi-Fi Network Systems." *ISACA Journal*, Volume 4, 2008. Accessed October 15, 2014. http://www.isaca.org/Journal/Past-Issues /2006/Volume-4/Pages/Risks-and-Risk-Control-of-Wi-Fi-Network-Systems1.aspx.

Elliott, Christopher. "6 Wireless Threats to Your Business." Microsoft. Accessed October 2, 2014. http://www.microsoft.com.

"Enterprise Mobility Management: Embracing BYOD Through Secure App and Data Delivery." Technical paper. Citrix, 2013. Accessed August 30, 2014. http://www.citrixvirtualdesktops .com/documents/030413_CTX_WP_Enterprise_Mobility_Management-f-LO.pdf.

"Enterprise WLAN Market Grew 14.8% Year over Year in Second Quarter of 2013." IDC, August 26, 2013. Accessed August 12, 2014. http://www.idc.com/getdoc.jsp?containerId =prUS24278113.

"Facts about the Mobile. A Journey through Time." The Wayback Machine Internet Archive. Accessed August 15, 2014. http://web.archive.org/web/20100813122017/http://www .mobilen50ar.se/eng/FaktabladENGFinal.pdf.

"Fake AP Main." Wirelessdefence.org. Accessed October 15, 2014. http://www.wirelessdefence .org/Contents/FakeAPMain.htm.

Farley, Tom, and Mark Van Der Hoek. "Cellular Telephone Basics." Private Line, January 1, 2006. Accessed August 15, 2014. http://www.privateline.com/mt_cellbasics/.

Fitzpatrick, Jason. "HTG Explains: The Difference Between WEP, WPA, and WPA2 Wireless Encryption (and Why It Matters)." How-To Geek, LLC, July 16, 2013. Accessed September 16, 2014. http://www.howtogeek.com/167783/htg-explains-the-difference-between-wep-wpa -and-wpa2-wireless-encryption-and-why-it-matters/.

Fletcher, Grace. "Device Fingerprinting Methodology." Mobile App Tracking, June 18, 2013. Accessed November 26, 2014. http://support.mobileapptracking.com/entries/21771055 -Device-Fingerprinting-Methodology.

Forrest, Connor. "Wi-Fi is rebranding itself." TechRepublic. Accessed March 19, 2020. https://www .techrepublic.com/article/wi-fi-is-rebranding-itself-heres-how-to-understand-the-new-naming.

Forristal, Jeff. "Android Fake ID Vulnerability Lets Malware Impersonate Trusted Applications, Puts All Android Users Since January 2010 At Risk." Bluebox, July 29, 2014. Accessed November 26, 2014. https://bluebox.com/technical/android-fake-id-vulnerability/.

Frankel, Sheila, Bernard Eydt, Les Owens, and Karen Scarfone. *Special Publication 800-97: Establishing Wireless Robust Security Networks*. National Institute of Standards and Technology, February 2007. Accessed October 2, 2014. http://csrc.nist.gov/publications /nistpubs/ 800-97/SP800-97.pdf.

Gast, Matthew. 802.11 *Wireless Networks: The Definitive Guide*. Sebastopol, CA: O'Reilly, 2002.

Genig, Hannah. "There's an App for That." Benzinga.com. Accessed May 10, 2020. https:// www.benzinga.com/general/education/18/07/12001849/theres-an-app-for-that -apples-app-store-celebrates-10th-anniversary.

Georgiev, Martin, et al. "The Most Dangerous Code in the World: Validating SSL Certificates in Non-Browser Software." Association for Computing Machinery, October 16, 2012. Accessed November 16, 2014. http://www.cs.utexas.edu/~shmat/shmat_ccs12.pdf.

"Getting Started with Browser Exploitation Framework (BeEF) in Kali Linux." *Linux Digest*, July 22, 2014. Accessed November 13, 2014. http://sathisharthars.wordpress.com/2014 /07/22/getting-started-with-browser-exploitation-framework-beef-in-kali-linux/.

References

Gianchandani, Prateek. "KARMETASPLOIT, Pwning the Air!" InfoSec Institute, December 19, 2011. Accessed November 26, 2014. http://resources.infosecinstitute.com/karmetasploit.

Gibbs, Mark. "Top 10 Security Tools in Kali Linux 1.0.6." *Network World*, February 11, 2014. Accessed November 26, 2014. http://www.networkworld.com/article/2291215/security/139872-Top-10-security-tools-in-Kali-Linux-1.0.6.html.

Gilchrist, Alasdair. *Tackling Fraud: Behavioural Biometric Analysis*. Independently Published, 2017.

Gilman, Evan, and Doug Barth. *Zero Trust Networks: Building Secure Systems in Untrusted Networks*. Newton, MA: O'Reilly, 2017.

"Giving Business Travelers What They Want Is Both an Art and a Science." Hotel Managers Group, 2014. Accessed August 12, 2014. http://hmghotels.com/What-do-business-travelers-want-in-2014.html.

Gonen, Yoav, Kevin Fasick, and Bruce Golding. "NYPD to Get $160M Mobile Fingerprint Device." *New York Post*, October 23, 2013. Accessed November 26, 2014. http://nypost.com/2014/10/23/nypd-to-get-160m-mobile-fingerprint-device/.

Goodin, Dan. "Stealthy Technique Fingerprints Smartphones by Measuring Users' Movements." *Ars Technica*, October 14, 2013. Accessed November 26, 2014. http://arstechnica.com/security/2013/10/stealthy-technique-fingerprints-smartphones-by-measuring-users-movements.

Gopinath, K.N. "WiFi Rogue AP: 5 Ways to (Mis)use It." *AirTight Networks Blog*, July 28, 2009. Accessed September 10, 2014. http://blog.airtightnetworks.com/wifi-rogue-ap-5-ways-to-use-it/.

Greene, Tim. "Black Hat: Top 20 Hack-Attack Tools." *Network World*, July 19, 2013. Accessed November 26, 2014. http://www.networkworld.com/article/2168329/malware-cybercrime/black-hat--top-20-hack-attack-tools.html.

Gruman, Galen. "How Windows Phone 8 Security Compares to IOS and Android." *InfoWorld*, October 30, 2012. Accessed November 26, 2014. http://www.infoworld.com/article/2616016/windows-phone-os/how-windows-phone-8-security-compares-to-ios-and-android.html.

Guide for Conducting Risk Assessments. National Institute of Standards and Technology. September 2012. Accessed November 26, 2014. http://csrc.nist.gov/publications/nistpubs/800-30-rev1/sp800_30_r1.pdf.

Halasz, David. "IEEE 802.11i and Wireless Security." *EETimes*, August 25, 2004. Accessed November 26, 2014. http://www.eetimes.com/author.asp?section_id=36&doc_id=1287503.

"Hping Network Security—Kali Linux Tutorial." Ethical Hacking. Accessed November 10, 2014. http://www.ehacking.net/2013/12/hping-network-security-kali-linux.html.

Huadong, Chen. "LTE Network Design and Deployment Strategy–ZTE Corporation." ZTE Corporation, January 17, 2011. Accessed August 15, 2014. http://wwwen.zte.com.cn/endata/magazine/ztetechnologies/2011/no1/articles/201101/t20110117_201779.html.

"Installing Aircrack-ng from Source." Aircrack-ng, November 4, 2014. Accessed November 26, 2014. http://www.aircrack-ng.org/doku.php?id=install_aircrack.

"iOS Security." Apple Inc., February 2014. https://www.apple.com.

"iOS Technology Overview." *iOS Developer Library*, Apple Inc., September 19, 2014. Accessed November 26, 2014. https://developer.apple.com/library/ios/documentation/miscellaneous/conceptual/iphoneostechoverview/iOSTechOverview.pdf.

References

"IP Mobility." Aruba Networks. Accessed August 5, 2014. http://www.arubanetworks.com/techdocs/ArubaOS_60/UserGuide/Mobility.php.

"ISO/IEC 27001—Information Security Management." International Organization for Standardization. Accessed August 21, 2014. http://www.iso.org/iso/home/standards/management-standards/iso27001.htm.

"ISO/IEC 27002:2013 Information Technology—Security Techniques—Code of Practice for Information Security Controls." International Organization for Standardization. Accessed August 30, 2014. http://www.iso.org/iso/home/store/catalogue_ics/catalogue_detail_ics.htm?csnumber=54533.

Johnson, Linda A. "Bell Labs' History of Inventions." *USA Today*, December 1, 2006. Accessed August 15, 2014. http://usatoday30.usatoday.com/tech/news/2006-12-01-bell-research_x.htm.

Kabay, M. E. "Guidelines for Securing IEEE 802.11i Wireless Networks." *Network World*, February 19, 2009. Accessed November 16, 2014. http://www.networkworld.com/article/2263578/wireless/guidelines-for-securing-ieee-802-11i-wireless-networks.html.

"Kali Linux Evil Wireless Access Point." Offensive Security, June 10, 2014. Accessed November 26, 2014. http://www.offensive-security.com/kali-linux/kali-linux-evil-wireless-access-point.

Kalmes, Chad, and Greg Hedges. "Risk Assessment: Are You Overlooking Wireless Networks?" *CSO Online*, May 10, 2006. Accessed November 16, 2014. http://www.csoonline.com/article/2119881/security-leadership/risk-assessment--are-you-overlooking-wireless-networks-.html.

Kao, I-Lung. "Securing Mobile Devices in the Business Environment." Technical paper. IBM Corporation, October 2011. Accessed August 30, 2014. https://www-935.ibm.com/services/uk/en/attachments/pdf/Securing_mobile_devices_in_the_business_environment.pdf.

Karagiannidis, George. "App Security 101: A List of Top 10 Vulnerabilities and How to Avoid Them." Developer Economics, March 12, 2014. Accessed November 26, 2014. http://www.developereconomics.com/app-security-101-list-top-10-vulnerabilities/.

Kelly, Gordon. "Report: 97% Of Mobile Malware Is on Android. This Is the Easy Way You Stay Safe." *Forbes*, March 24, 2014. Accessed November 26, 2014. http://www.forbes.com/sites/gordonkelly/2014/03/24/report-97-of-mobile-malware-is-on-android-this-is-the-easy-way-you-stay-safe.

Kennedy, Susan. "Best Practices for Wireless Network Security." ISACA, 2004. Accessed September 16, 2014. http://www.isaca.org/Journal/Past-Issues/2004/Volume-3/Pages/Best-Practices-for-Wireless-Network-Security.aspx.

———. "Cell Phone Spy Software—The Complete Guide." AcisNI.com, 2014. Accessed November 11, 2014. http://acisni.com/cell-phone-spy-software-complete-guide.

Kirsch, Christian. "Introduction to Penetration Testing." Security Street, April 7, 2013. Accessed November 26, 2014. https://community.rapid7.com/docs/DOC-2248.

"Know the Risks of Ad Hoc Wireless LANs." *WLAN Watch Security Newsletter*, AirDefense, Inc., 2002. Accessed September 10, 2014. http://www.airdefense.net/eNewsletters/adhoc.shtm.

Koh, Rachel, et al. "Smartphones and Tablets: Economic Impacts." Accessed August 15, 2014. http://it1001tablet.blogspot.com/p/economic-impacts_25.html.

Lee, Timothy B. "What Killed BlackBerry? Employees Started Buying Their Own Devices." *Washington Post*, September 20, 2013. Accessed August 15, 2014. http://www.washingtonpost

References

.com/blogs/the-switch/wp/2013/09/20/what-killed-blackberry-employees-started-buying-their-own-devices/.

Legg, Gary. "The Bluejacking, Bluesnarfing, Bluebugging Blues: Bluetooth Faces Perception of Vulnerability." *EETimes*, August 4, 2005. Accessed September 10, 2014. http://www.eetimes.com/document.asp?doc_id=1275730.

Lessing, Marlese. "What is WiFI 6?" SDxCentral. Accessed April 10, 2020. https://www.sdxcentral.com/networking/wifi/what-is-wifi-6.

"Limiting or Removing Unwanted Network Traffic at the Client." The University of Iowa, 2014. Accessed October 2, 2014. http://its.uiowa.edu/support/article/3576.

Litten, David. "Qualitative and Quantitative Risk Analysis." *PMP Primer*. Accessed October 22, 2014. http://www.pm-primer.com/pmbok-qualitative-and-quantitative-risk-analysis/.

Malenkovich, Serge. "Is Your iPhone Already Hacked?" Kaspersky Lab, June 24, 2014. Accessed November 26, 2014. http://blog.kaspersky.com/iphone-spyware.

"Malware Delivery—Understanding Multiple Stage Malware." *Cyber Squared*. Accessed November 13, 2014. http://www.cybersquared.com.

Marin-Perianu, Raluca, Pieter Hartel, and Hans Scholten. "A Classification of Service Discovery Protocols." University of Twente (Netherlands), June 2005. Accessed November 26, 2014. http://doc.utwente.nl/54527/1/classification_of_service.pdf.

McNeil, Andrew. "Build Your Own WIFI Jammer." Instructables. Accessed October 15, 2014. http://www.instructables.com/id/Build-your-own-WIFI-jammer.

Mick, Jason. "The True Story: Two U.S. Nuclear Labs 'Hacked'" *DailyTech*, December 8, 2007. Accessed September 10, 2014. http://www.dailytech.com/article.aspx?newsid=9950.

Miessler, Daniel. "The Difference Between a Vulnerability Assessment and a Penetration Test." Danielmiessler.com. Accessed October 15, 2014. http://danielmiessler.com/writing/vulnerability_assessment_penetration_test.

Miliefsky, Gary. "SnoopWall Flashlight Apps Threat Assessment Report." SnoopWall, October 1, 2014. Accessed November 26, 2014. http://www.snoopwall.com/threat-reports-10-01-2014/.

Mills, Elinor. "On iPhone, Beware of That AT&T Wi-Fi Hot Spot." *CNET*, April 27, 2010. Accessed November 26, 2014. http://www.cnet.com/news/on-iphone-beware-of-that-at-t-wi-fi-hot-spot.

Minzsec. "Kali Linux—Get Control of Android Phone Using Armitage." Operating System Hacking & Security, July 1, 2014. Accessed November 26, 2014. http://operatin5.blogspot.com/2014/07/kali-linux-get-control-of-android-phone.html.

Mitchell, Bradley. "802.11 What? What Do These Different Wireless Standards Mean?" About.com. Accessed September 7, 2014. http://compnetworking.about.com/cs/wireless80211/a/aa80211standard.htm.

———. "What Is WPA2?" About.com. Accessed September 16, 2014. http://compnetworking.about.com/od/wirelesssecurity/f/what-is-wpa2.htm.

"Mobile Technology Fact Sheet." Pew Research Centers Internet & American Life Project, January 2014. Accessed August 13, 2014. http://www.pewinternet.org/fact-sheets/mobile-technology-fact-sheet/.

Moreau, Seán. "The Evolution of iOS." Computerworld. Accessed April 3, 2020. https://www.computerworld.com/article/2975868/the-evolution-of-ios.html

Murph, Darren. "Study: 802.11ac Devices to Hit the One Billion Mark in 2015, Get Certified in 2048." Engadget, February 8, 2011. Accessed September 8, 2014. http://www.engadget.com/2011/02/08/study-802-11ac-devices-to-hit-the-one-billion-mark-in-2015-get.

Murray, Jason. "An Inexpensive Wireless IDS Using Kismet and OpenWRT." SANS Institute, April 5, 2009. Accessed November 26, 2014. http://www.sans.org/reading-room/whitepapers/detection/inexpensive-wireless-ids-kismet-openwrt-33103.

Negus, Kevin J. "History of Wireless Local Area Networks (WLANs) in the Unlicensed Bands." George Mason University Law School Conference, Information Economy Project, April 4, 2008. Accessed August 12, 2014. http://iep.gmu.edu/wp-content/uploads/2009/08/WLAN_History_Paper.pdf.

Nerney, Chris. "Signs Your Android Device Is Infected with Malware (and What to Do about It)." CITEworld, October 16, 2013. Accessed November 26, 2014. http://www.citeworld.com/article/2114383/mobile-byod/android-malware-how-to-tell.html.

Quinn, Tim. "Non-Broadcast Wireless SSIDs: Why Hidden Wireless Networks Are a Bad Idea." Networking Blog, TechNet, Microsoft, February 8, 2008. Accessed September 5, 2014. http://blogs.technet.com/b/networking/archive/2008/02/08/non-broadcast-wireless-ssids-why-hidden-wireless-networks-are-a-bad-idea.aspx.

Okolie, C. C., F. A. Oladeji, B. C. Benjamin, H. A. Alakiri, and O. Olisa. "Penetration Testing for Android Smartphones." *IOSR Journal of Computer Engineering* 14, No. 3 (September/October 2013): 104–09. 2014. Accessed November 26, 2014. http://www.academia.edu/5320987/Penetration_Testing_for_Android_Smartphones.

Olifer, Natalia and Victor Olifer. *Computer Networks: Principles, Technologies and Protocols for Network Design*. Evolution of Computer Networks. Indianapolis, IN: John Wiley & Sons, 2005. Accessed August 5, 2014. http://czx.ujn.edu.cn/course/comnetworkarc/Reference/Evolution_of_Computer_Networks.pdf.

Park, Bok-Nyong, Wonjun Lee, and Christian Shin. "Securing Internet Gateway Discovery Protocol in Ubiquitous Wireless Internet Access Networks." Cham, Switzerland: Springer International Publishing AG. Accessed October 2, 2014. http://link.springer.com/chapter/10.1007/11802167_33#close.

Patil, Basavaraj. "IP Mobility Ensures Seamless Roaming." *Communication Systems Design*, February 2003, 11–19. Accessed August 5, 2014. http://m.eet.com/media/1094820/feat1-feb03.pdf.

Paul, Ian. "F-Secure Says 99 Percent of Mobile Malware Targets Android, but Don't Worry Too Much." CSO, April 29, 2014. Accessed November 26, 2014. http://www.csoonline.com/article/2148947/data-protection/f-secure-says-99-percent-of-mobile-malware-targets-android-but-dont-worry-too-much.html.

PCI Security Standards Council Website. Accessed August 30, 2014. https://www.pcisecuritystandards.org/.

Pearson, Dale. "Wireless Attack and Audit Tools…Recommendations List." Subliminal Hacking, February 7, 2013. Accessed October 15, 2014. http://www.subliminalhacking.net/2013/02/07/wireless-attack-and-audit-tools-recommendations-list.

Peltier, Thomas R., Justin Peltier, and John Blackley. *Information Security Fundamentals*. Boca Raton, FL: Auerbach Publications, 2005.

Phifer, Lisa. "Anatomy of a Wireless 'Evil Twin' Attack (Part 2: Countermeasures)." WatchGuard. Accessed September 10, 2014. http://www.watchguard.com/infocenter/editorial/27079.asp.

References

———. "Top Ten Wi-Fi Security Threats." *ESecurity Planet*, March 8, 2010. Accessed August 30, 2014. http://www.esecurityplanet.com/views/article.php/3869221/Top-Ten-WiFi-Security-Threats.htm.

Plaskett, Alex, and Dave Chismon. "Security Considerations in the Windows Phone 8 Application Environment." MWR InfoSecurity, August 8, 2013. Accessed December 1, 2014. https://www.mwrinfosecurity.com/articles/security-considerations-in-the-windows-phone-8-application-environment/.

Poole, Ian. "CDMA Technology Basics Tutorial." Radio-electronics.com. Accessed August 15, 2014. http://www.radio-electronics.com/info/rf-technology-design/cdma/what-is-cdma-basics-tutorial.php.

"Project Isizwe: Free Wi-Fi for Kids in Africa." ProjectIsizwe.org, June 2012. Accessed August 12, 2014. http://projectisizwe.org/downloads/socio-economic-impact.pdf.

"Projects/OWASP Mobile Security Project—Top Ten Mobile Risks." Open Web Application Security Project (OWASP), November 5, 2014. Accessed December 1, 2014. https://www.owasp.org/index.php/Projects/OWASP_Mobile_Security_Project_-_Top_Ten_Mobile_Risks.

PRNewswire. "Alcatel-Lucent Malware Report Reveals That More Apps Are Spying on Us, Stealing Personal Information and Pirating Data Minutes." *Yahoo! Finance*, September 4, 2014. Accessed November 26, 2014. http://finance.yahoo.com/news/alcatel-lucent-malware-report-reveals-140000872.html.

Qaissaunee, Michael, and Mohammad Shanehsaz. "Wireless LAN Auditing Tools." Brookdale Community College, Lincroft, NJ. Accessed October 15, 2014. http://www.ewh.ieee.org/r1/njcoast/events/WirelessSecurity.pdf.

Radack, Shirley, ed. "ITL Bulletin for August 2012." National Institute of Standards and Technology, August 2012. Accessed September 10, 2014. http://csrc.nist.gov/publications/nistbul/august-2012_itl-bulletin.pdf.

Raphael, J.R. "Android versions: A Living History from 1.0 to 11." Computerworld. Accessed April 4, 2020. https://www.computerworld.com/article/3235946/android-versions-a-living-history-from-1-0-to-today.html.

Reeves, Scott. "Try Kismet for Detecting Hidden 802.11 Wireless Networks." *TechRepublic*, December 2, 2011. http://www.techrepublic.com/blog/linux-and-open-source/try-kismet-for-detecting-hidden-80211-wireless-networks.

"Research in Motion Reports Third Quarter Fiscal 2013 Results." *Marketwire*, December 20, 2012. Accessed August 15, 2014. http://www.marketwired.com/press-release/research-in-motion-reports-third-quarter-fiscal-2013-results-nasdaq-rimm-1740316.htm.

Reynolds, Jake. "When 802.1x/PEAP/EAP-TTLS Is Worse Than No Wireless Security." *Depth Security*, November 19, 2010. http://blog.depthsecurity.com/2010/11/when-8021xpeapeap-ttls-is-worse-than-no.html.

Rouse, Margaret. "Enterprise Mobility Management (EMM)." TechTarget, July 2014. Accessed November 11, 2014. http://searchconsumerization.techtarget.com/definition/enterprise-mobility-management-EMM.

———. "Mobile Application Management (MAM)." TechTarget, June 2014. Accessed November 11, 2014. http://searchconsumerization.techtarget.com/definition/mobile-application-management.

Runnels, Tammie. "History of Wireless Networks." History of Wireless Networks, October 2005. Accessed August 12, 2014. http://www.arp.sprnet.org/default/inserv/trends/history_wireless.htm.

Russon, Mary-Ann. "19 Fake Mobile Base Stations Found Across US—Are They for Spying or Crime?" *International Business Times*, September 4, 2014. Accessed December 1, 2014. http://www.ibtimes.co.uk/19-fake-mobile-base-stations-found-across-us-are-they-spying-crime-1464008.

"Security for Windows Phone 8." Windows Dev Center, Microsoft, August 19, 2014. Accessed November 11, 2014. http://msdn.microsoft.com/en-us/library/windows/apps/ff402533%28v=vs.105%29.aspx.

"Security Risk Assessment and Audit Guidelines." The Government of the Hong Kong Special Administrative Region of the People's Republic of China Office of the Government Chief Information Officer, 2012. Accessed December 1, 2014. http://www.ogcio.gov.hk/en/infrastructure/methodology/security_policy/doc/g51_pub.pdf.

Segura, Jérôme. "A Cunning Way to Deliver Malware." *Malwarebytes Unpacked*, July 11, 2014. Accessed December 1, 2014. https://blog.malwarebytes.org/malvertising-2/2014/07/a-cunning-way-to-deliver-malware.

Seltzer, Larry. "Does IOS Malware Actually Exist?" *ZDNet*, June 13, 2014. Accessed December 1, 2014. http://www.zdnet.com/does-ios-malware-actually-exist-7000030518/.

"Single Sign-On Mythbusting." *Information Week Dark Reading*, January 10, 2013. Accessed October 2, 2014. http://www.darkreading.com/single-sign-on-mythbusting/d/d-id/ 1138961.

Six, Jeff. "An In Depth Introduction to the Android Permission Model." The OWASP Foundation, April 3, 2012. Accessed December 1, 2014. https://www.owasp.org/images/c/ca/ASDC12-An_InDepth_Introduction_to_the_Android_Permissions_Modeland_How_to_Secure_MultiComponent_Applications.pdf.

"Smartphone Sensors Leave Trackable Fingerprints." *ScienceDaily*, April 28, 2014. Accessed December 1, 2014. http://www.sciencedaily.com/releases/2014/04/140428121433.htm.

Snyder, Bill. "User Beware: That Mobile App Is Spying on You." *InfoWorld*, August 6, 2014. Accessed December 1, 2014. http://www.infoworld.com/article/2608494/mobile-apps/user-beware--that-mobile-app-is-spying-on-you.html.

Solomon, Sharon. "Top-10 Essential Challenges of Mobile Security." Checkmarx, November 29, 2013. Accessed December 1, 2014. https://www.checkmarx.com/2013/11/29/10-challenges-of-mobile-security/.

"SPF User Guide." Bulb Security. Accessed November 13, 2014. http://www.bulbsecurity.com/smartphone-pentest-framework/spf-user-guide/#Remote_Attack_Examples.

Srinivas. "Android Hacking and Security, Part 1: Exploiting and Securing Application Components." InfoSec Institute, March 27, 2014. Accessed December 1, 2014. http://resources.infosecinstitute .com/android-hacking-security-part-1-exploiting-securing-application-components/.

Stallings, William. "Security Comes to SNMP: The New SNMPv3 Proposed Internet Standards." *Internet Protocol Journal*, Cisco, Vol. 1, No 3, December 1998. Accessed October 2, 2014. http://www.cisco.com/web/about/ac123/ac147/archived_issues/ipj_1-3/snmpv3.html.

Stevenson, Alastair "Windows Phone 8.1 Review." *V3.co.uk*, May 26, 2014. Accessed December 1, 2014. http://www.v3.co.uk/v3-uk/review/2346443/windows-phone-81-review.

Svajcer, Vanja. "Sophos Mobile Security Threat Report." Sophos Ltd., 2014. Accessed October 21, 2014. http://www.sophos.com/en-us/medialibrary/PDFs/other/sophos-mobile-security-threat-report.pdf

References

Tarasenko, Nick. "iOS 7 Is Installed on 91% of Devices, Latest Android Version Only on 21%." iPhoneRoot.com, August 23, 2014. Accessed November 13, 2014. http://iphoneroot.com/ios-7-is-installed-on-91-of-devices-latest-android-version-only-on-21/.

Temple, James. "Stanford Researchers Discover 'Alarming' Method for Phone Tracking, Fingerprinting Through Sensor Flaws." *SFGate*, October 10, 2013. Accessed December 1, 2014. http://blog.sfgate.com/techchron/2013/10/10/stanford-researchers-discover-alarming-method-for-phone-tracking-fingerprinting-through-sensor-flaws/.

Tewson, Kathryn, and Steve Riley. "Security Watch: A Guide to Wireless Security." *TechNet Magazine*, Microsoft, December 2008. Accessed August 5, 2014. http://technet.microsoft.com/en-us/magazine/2005.11.securitywatch.aspx.

"The Rise of the Mobile Workforce and Deskless Workers." Skedulo. Accessed April 18, 2020. https://www.skedulo.com/the-rise-of-the-mobile-workforce-and-deskless-workers.

"Threat Report H1 2014." F-Secure Corporation, 2014. Accessed December 1, 2014. http://www.f-secure.com/documents/996508/1030743/Threat_Report_H1_2014.pdf.

"Top 10 Losing Warehouse Strategies and How to Avoid Them." Motorola Solutions, 2013. Accessed August 12, 2014. http://www.motorolasolutions.com/web/Business/Solutions/Manufacturing/_Documents/_staticFiles/Top%2010%20Losing%20Warehouse%20Strategies%20and%20How%20to%20Avoid%20Them.pdf.

"Understanding Encryption Types." Aruba Networks. Accessed August 5, 2014. http://www.arubanetworks.com/techdocs/Instant_40_Mobile/Advanced/Content/UG_files/Authentication/UnderstandingEncryption.htm.

"Understanding IEEE* 802.11 Authentication and Association." Intel Corporation, July 21, 2014. Accessed September 10, 2014. http://www.intel.com/support/wireless/wlan/sb/CS-025325.htm.

"Understanding WPA-PSK and WPA2-PSK Authentication." Juniper Networks, Inc., March 13, 2013. Accessed September 16, 2014. http://www.juniper.net/techpubs/en_US/junos-space-apps12.3/network-director/topics/concept/wireless-wpa-psk-authentication.html.

"Upgrade Cisco IOS on an Autonomous Access Point." Cisco, September 2, 2008. Accessed September 16, 2014. http://www.cisco.com/c/en/us/support/docs/wireless-mobility/wireless-lan-wlan/107911-ios-upgrade.html.

U.S. Department of Health and Human Services. "HITECH Act Enforcement Interim Final Rule," 2009. Accessed August 30, 2014. http://www.hhs.gov/ocr/privacy/hipaa/administrative/enforcementrule/hitechenforcementifr.html.

Vladimirov, Andrew A., Konstantin V. Gavrilenko, and Andrei A. Mikhailovsky. *Wi-Foo*. Boston: Addison-Wesley, 2004.

Warner, Jonathon. "The Complete Guide to Jailbreaking Windows Phone 7.8." *Windows Phone Hacker*, February 9, 2014. Accessed November 26, 2014. http://windowsphonehacker.com/articles/the_complete_guide_to_jailbreaking_windows_phone_7_and_7.5-09-24-11.

"Web Application Firewall Detection—Kali Linux Tutorial." Ethical Hacking. Accessed November 10, 2014. http://www.ehacking.net/2013/12/web-application-firewall-detection-kali.html.

Westervelt, Robert. "Droid Danger: Top 10 Android Malware Families." *CRN*, August 8, 2013. Accessed December 1, 2014. http://www.crn.com/slide-shows/security/240159651/droid-danger-top-10-android-malware-families.htm/pgno/0/3.

Westin, Ken. "Penetration Testing with Smartphones Part 1." *The State of Security*, November 30, 2012. Accessed December 1, 2014. http://www.tripwire.com/state-of-security/security-data-protection/penetration-testing-with-smartphones-part-1.

Wexler, Joanie. "Are All-Wireless Networks Vulnerable to Jamming?" *Network World*, August 27, 2007. Accessed December 1, 2014. http://www.networkworld.com/article/2294345/network-security/are-all-wireless-networks-vulnerable-to-jamming-.html.

"What Are Cookies in Computers?" All About Cookies. Accessed November 11, 2014. http://www.allaboutcookies.org/.

"What Is a Mobile Threat?" Lookout, Inc., 2013. Accessed August 30, 2014. https://www.lookout.com/resources/know-your-mobile/what-is-a-mobile-threat.

"What Is MU-MIMO and Why You Need It in Your Wireless Routers." NetworkWorld. Accessed April 10, 2020. https://www.networkworld.com/article/3250268.

"Why Choose WiFi as a Service?" Superloop. Accessed April 20, 2020. https://superloop.com/blog/why-choose-wifi-as-a-service.

"Wi-Fi Certified N: Longer-Range, Faster-Throughput, Multimedia-Grade Wi-Fi Networks (2009)" Wi-Fi Alliance, 2009. Accessed September 8, 2014. http://www.wi-fi.org/file/wi-fi-certified-n-longer-range-faster-throughput-multimedia-grade-wi-fi-networks-2009.

Wijayatunga, Champika. "Internet and Security Fundamentals." Asia Pacific Network Information Centre (APNIC). Accessed August 30, 2014. https://www.pacnog.org/pacnog10/track3/Security-Part-1.pdf.

Williams, Chris, Gabriel Solomon, and Robert Pepper. "What Is the Impact of Mobile Telephony on Growth?" Deloitte LLP, 2012. Accessed August 15, 2014. http://www.gsma.com/publicpolicy/wp-content/uploads/2012/11/gsma-deloitte-impact-mobile-telephony-economic-growth.pdf.

"Windows Phone Architecture Overview, Getting Started." Windows Dev Center, Microsoft, October 3, 2014. https://dev.windows.com.

Woods, John. "Fake AP on Kali Linux." *I'm Here to Protect You*, August 27, 2013. Accessed December 1, 2014. http://secjohn.blogspot.com/2013/08/fake-ap-on-kali-linu.html.

"Worldwide WLAN Market Reaches Nearly $6.4 Billion in 2011, According to IDC." *Reuters*, February 23, 2012. Accessed August 12, 2014. http://www.reuters.com/article/2012/02/23/idUS238517%2B23-Feb-2012%2BBW20120223.

Worth, Dan. "ATM Malware Thefts the 'Modern Day Bank Robbery' Raking in Millions for Crooks." *V3.co.uk*, November 3, 2014. Accessed December 1, 2014. http://www.v3.co.uk/v3-uk/analysis/2378908/atm-malware-thefts-the-modern-day-bank-robbery-raking-in-millions-for-crooks.

Index

Note: Page numbers followed by *f*, *t* denote figures and tables.

A

AAA. *See* authentication, authorization, and accountability
AAD. *See* additional authentication data
access control lists (ACLs), 187
access points. *See also specific access points*
  placement, 121, 155
  software-based, 73
access, vulnerabilities, 223
accountability, information security and, 69–70
ACLs. *See* access control lists
Acrylic Wi-Fi, 201–202
active fingerprinting, 305
active mode, coverage assessment, 122
active scanning, 96–97
ad hoc networks, 97, 148
  peer-to-peer hacking over, 148–149
  threats and, 73
additional authentication data (AAD), 167
Address Resolution Protocol (ARP), 134, 136
Adiantum, 262
administrative or management controls, risk-analysis, 234
ADSL. *See* asymmetric digital subscriber line
Advanced Encryption Standard (AES), 166, 261
Advanced Encryption Standard–Counter Mode Cipher Block Chaining Message Authentication Code Protocol (AES-CCMP), 184
Advanced Mobile Phone System (AMPS), 33
1G, 33–34
advanced persistent threat (APT), 127, 322
Advanced Research Projects Agency (ARPA), 7
advanced WLAN security measures
  authentication and access control implementation, 179–180, 180*f*
  authenticated Dynamic Host Configuration Protocol, 183
  Extensible Authentication Protocol, 180
  intrusion detection systems and intrusion prevention systems, 182
  protocol filtering, 182–183
  Remote Authentication Dial-In User Service, 180–182, 181*f*
comprehensive security policy
  centralized *vs.* distributed design and management, 176–177
  compliance considerations, 178
  employee training and education, 179
  guest policies, 177–178
  quarantining, 178
  remote access policies, 177
data protection
  Internet protocol security, 185
  malware and application security, 186
  virtual private networks, 186
  WPA2 personal and enterprise modes, 184
  WPA3, 184–185
network and user devices
  client security outside perimeter, 192–193
  coverage area and Wi-Fi roaming, 191–192
  device management and user logons, 193–194
  discovery protocols, 190
  hard drive encryption, 194
  IP Services, 190–191
  quarantining, 194–195
  simple network management protocol version 3, 189
  Wi-Fi as service, 195–196
user segmentation
  demilitarized zone segmentation, 188–189
  guest access and passwords, 188
  virtual local area networks, 187
Advertising Click Fraud, 321
advertising networks, 302
advertisingIdentifier, 306
AES. *See* Advanced Encryption Standard
AES-CCMP. *See* Advanced Encryption Standard–Counter Mode Cipher Block Chaining Message Authentication Code Protocol
AH. *See* Authentication Header
Airbase-ng, 209, 214
Aircrack-ng, 205, 209, 284
Airdeauth-ng, 209
Airdrop-ng, 290
Aireplay-ng, 209, 290

AirMagnet WiFi Analyzer, 201
Airmon-ng, 209
Airodump-ng, 209, 284
Airshark, 136, 143, 209–210
ALE. *See* annualized loss expectancy
Always on Listening, 263
AMPS. *See* Advanced Mobile Phone System
Android. *See also* Google Android
   malware on, 318–319
      criminal and developer collaboration, 320–323
      software fragmentation, 319–320
   sandbox, 261
   SDK, 248
   unique device identification, 306
   version evolution, 267–269
Android Framework for Exploitation, 248
Android Open Source Project (ASOP), 325
AndroRAT, 248
annual rate of occurrence (ARO), 234
annualized loss expectancy (ALE), 234
ANonce, 107
antennas
   wireless local area network, 211–212
   wireless. *See* wireless antennas
antimalware, 282
antispyware software, 282
APIs. *See* application programming interfaces
App Store, 46–47, 251, 252, 254–255
Apple iOS
   component-layered model, 252
   malware on, 325, 326t
   operating system, 252
   security challenges, 251–252
      App Store, 254–255
      architecture, 254
      exploits, 252–254
   security models
      Apple security model, 263
      application provenance, 264

encryption, 266
iOS sandbox, 264–265
jailbreaking iOS, 266
permission-based access, 265–266
security concerns, 265
unique device identification, 305–306
version evolution, 267–270
application-based threats. *See* mobile malware and application-based threats
application fingerprinting, 304
application identification, 304
Application Layer
   attacks, 147
   OSI Reference Model, 10
application programming interfaces (APIs), 49, 261, 291, 322
application provenance, 263, 264
application server network layer of defense, 80
APT. *See* advanced persistent threat
Armitage, 332
ARO. *See* annual rate of occurrence
ARP. *See* Address Resolution Protocol
ARPA. *See* Advanced Research Projects Agency
ARPANET project, 7
AR/VR mobile gaming, 36–37
ASOP. *See* Android Open Source Project
asset, 227
   identification and valuation, 229–230
   mapping, 232–233
   tracking, 18
association, authentication and, 160–161
asymmetric digital subscriber line (ADSL), 16
AT command codes, 142
AT&T Wi-Fi connection (ATTWIFI), 283
ATM. *See* automatic teller machine
attack surface, reducing, 280
attack tools and techniques
   denial of service, 213

   hijacking devices, 213–214
   hijacking session, 214–215
   radio frequency jamming, 212–213
attackers
   insiders *vs.* outsiders, 127–128, 128f
   skilled *vs.* unskilled, 127
   types of, 126–128
   unauthorized control, 149
audit tools, wireless local area network, 199–200
   antennas, 211–212
   attack tools and techniques
      denial of service, 213
      hijacking devices, 213–214
      hijacking session, 214–215
      radio frequency jamming, 212–213
   discovery tools, 200–201
      enterprise Wi-Fi audit tools, 201–202
      HeatMapper, 202–204
   hardware audit tools, 211
   network management and control tools
      Aircrack-ng, 209
      Airshark, 209–210
      network management system, 210–211
      wireless protocol analyzers, 208–209
   network utilities, 216–217
   password-capture and decryption tools, 205–207
   network enumerators, 208
   penetration testing tools
      Metasploit, 204
      Security Auditor's Research Assistant, 204–205
authenticated Dynamic Host Configuration Protocol, 183
authentication and access control
   advanced WLAN security measures, 179–180, 180f
   authenticated Dynamic Host Configuration Protocol, 183
   Extensible Authentication Protocol, 180

Index

intrusion detection systems and intrusion prevention systems, 182
protocol filtering, 182–183
Remote Authentication Dial-In User Service, 180–182, 181f
basic WLAN security measures, 158
authentication and association, 160–161
MAC filters, 160
SSID obfuscation, 159–160
virtual local area networks, 162, 163f
VPN over wireless, 161–162
authentication, authorization, and accountability (AAA), 80–82
Authentication Header (AH), 185
authentication server, 179
authenticator, 179
authority scam, 132
Auto Content Update, 263
automatic teller machine (ATM), 67
autonomous access points, 108
"autonomous" vehicles, 37
availability
  information security and, 69
  vulnerabilities, 223
Azure, 285

B

backhaul circuit, 29
Bank Trojans, 320
base controller station (BCS), 29
base transceiver station (BTS), 29
basic service set (BSS), 96–98
basic service set identification (BSSID), 136, 284
basic WLAN security measures
  authentication and access restriction, 158
  authentication and association, 160–161
  MAC filters, 160
  SSID obfuscation, 159–160
  virtual local area networks, 162, 163f
  VPN over wireless, 161–162

data protection, 163
  order of preference for Wi-Fi data protection, 167–168
  Wi-Fi protected access, 165
  Wired Equivalent Privacy (WEP), 164–165
  WPA2, 165–166
  WPA2 with AES, 166–167
  WPA2 with CCMP, 167
  WPA3, 168–169
design and implementation considerations
  best practices, 158
  equipment configuration and placement, 155, 156f
  interoperability and layering, 156–157
  radio frequency design, 154–155
  security management, 157–158
ongoing management security considerations
  firmware upgrades, 169–170
  Periodic Inventory, 170–171
  physical security, 170
  Rogue WLANs/wireless access points, 171
BCS. See base controller station
beacon transmission, 106–107
beam focusing, 102
Beck-Tews attack, 165
BeEF. See Browser Exploitation Framework
behavior analysis, 232
Bell Labs, 29
Bell Telephone System, 29
BES. See BlackBerry Enterprise Server
bidding down, 313
binary hardening. See binary protections
binary protections, 295
biometrics, 308–309
BlackBerry effect, 37–38
BlackBerry Enterprise Server (BES), 37
BleedingBit, 142
block ciphers vs. stream ciphers, 166
BlueBorne, 142

bluebugging, 140, 142–143
bluejacking, 140
bluesnarfing, 140, 141
Bluetooth 4.0. See Bluetooth Low Energy
Bluetooth, 9, 244
  bluebugging, 142–143
  bluejacking, 140–141
  bluesnarfing, 141
  functions, 262
  pairing, 139–140
  revisions compared, 138–139
  versions, 137–138
  vulnerabilities and threats, 137–143
  Wi-Fi hacks, 73
Bluetooth Low Energy, 55
Bluetooth Smart. See Bluetooth Low Energy
bots, 177
Bring Your Own Application (BYOA), 224, 271
Bring Your Own Cloud (BYOC), 49
Bring Your Own Device (BYOD), 154, 280, 330
  device management and user logons, 193
  other risks with, 77–78
  policy, steps, 110–111
  revolution, 37–38
  risk mitigation, 75–78
  security model, 271–272
  for small-to-medium businesses, 78
broadband fixed-line networks, 16
broadcast domains, 162
broken cryptography, 293
browser
  application, 289–290
  attacks, mobile, 328–329
  exploits, 72
Browser Exploitation Framework (BeEF), 290
brute-force attacks, 206
  protect against, 168
BSS. See basic service set
BSSID. See basic service set identification
BTS. See base transceiver station
business use cases for mobility, 40–42

BYOA. *See* Bring Your Own Application
BYOC. *See* Bring Your Own Cloud
BYOD. *See* Bring Your Own Device

C

C-I-A triad, 41
CA. *See* certificate authority
California Customer Privacy Act (CCPA), 86
CAPTCHA. *See* Completely Automated Public Turing Test to Tell Computers and Humans Apart
captive portal, 178, 183, 289–290, 327
care-of-address (CoA), 21
carriage return line feed (CRLF), 215
Carrier Sense Multiple Access with Collision Avoidance (CSMA/CA), 101
castle-and-moat model, 50
CCMP. *See* Counter Mode Cipher Block Chaining Message Authentication Code Protocol
CCPA. *See* California Customer Privacy Act
CDMA. *See* Code Division Multiple Access
CDP. *See* Cisco Discovery Protocol
cell towers, 29
cellular, 28
   coverage maps, 28–32, 30*f*, 33*f*
   design, 29
   Handoff, 32–33
   modulation, types of. *See specific types*
   networks, modern, 312–313
      man-in-the-middle (MITM) attack, 313
      MNmap, 313
certificate authority (CA), 288
chat-up scam, 131
chips, 104
churn, 16
circuit switching, 6
Cisco Discovery Protocol (CDP), 190
CLI. *See* command line interface

clickjacking, 328
client integrity control, 186
client security, 192–193
client-side exploits, 285–286
client-side injection, 293
client station, 96, 135
cloud apps *vs.* native mobile apps, 51
cloud computing, 50–51
cloud-managed wireless services. *See also* mobile cloud
   cost efficiency, 195
   deeper analytics and reporting, 196
   easy backup and real-time alerts, 196
   enhanced security, 196
   hands-off maintenance, 196
   on-demand scalability, 196
   simplified network management, 195
ClutchMod, 293
CoA. *See* care-of-address
Cocoa Touch Layer, 254
Code Division Multiple Access (CDMA), 32, 32*f*
   2G, 34–35
collocation model, 98
command line interface (CLI), 169
commercial mobile adware, 320. *See also* madware
common guest password, 188
communications. *See specific communications*
Completely Automated Public Turing Test to Tell Computers and Humans Apart (CAPTCHA), 207
compliance, 75
   considerations, 178
   regulatory. *See* regulatory compliance
component-layered model, 252
comprehensive security policy
   centralized *vs.* distributed design and management, 176–177
   compliance considerations, 178
   employee training and education, 179
   guest policies, 177–178

   quarantining, 178
   remote access policies, 177
confidentiality, information security and, 68
context-aware firewalls, 81
context-aware security devices, 81
"cookbook" approach, 169
cookies, 300, 301–302
COPE. *See* corporate owned personally enabled
Core OS Layer, 254
Core Service Layer, 254
corporate owned personally enabled (COPE), 74
Counter Mode Cipher Block Chaining Message Authentication Code Protocol (CCMP), 166
coverage analysis, site survey, 121
coverage area
   antennas, 118
   determining, 118, 119*f*
   omnidirectional, 119*f*
   semi-directional, 119*f*
   Wi-Fi roaming, 191–192
coverage assessment, 121–122
criminal and developer collaboration, 320–323
CRLF. *See* carriage return line feed
CRM. *See* customer resource management
cross-site profiling, 302–303
cross-site scripting (XSS), 215
crypto primitives, 261
cryptography, broken, 293
CryptoLocker, 194
cryptomining malware, 321
CSMA/CA. *See* Carrier Sense Multiple Access with Collision Avoidance
curiosity kills scam, 132
customer resource management (CRM), 42
customer satisfaction, 18–19
cybercrime, 39
cybercriminals, 320
   areas of interest for, 71
   and developer collaboration, 320–323
   goals of, 318
   Mobile IP security, 62

Index

mobile phone threats and vulnerabilities, 243–244
scanning corporate network for mobile attacks, 280
targets for, 247
Cypher, 284

D

D-AMPS. *See* Digital Advanced Mobile Phone System
data. *See also* information
  communication, 4
    early data networks, 5–7, 5f
    Internet revolution, 7
    personal computers, advances in, 7–8
  encoding, 104
  leakage, 292
  privacy, 168
    regulations, 178
  theft threats, 71–72
Data Link Layer, OSI Reference Model, 9, 12–13
data protection
  advanced WLAN security measures
    Internet Protocol Security, 185
    malware and application security, 186
    virtual private networks, 186
    WPA2 personal and enterprise modes, 184
    WPA3, 184–185
  basic WLAN security measures, 163
    order of preference for Wi-Fi data protection, 167–168
    Wi-Fi protected access, 165
    Wired Equivalent Privacy, (WEP) 164–165
    WPA2, 165–166
    WPA2 with AES, 166–167
    WPA2 with CCMP, 167
    WPA3, 168–169
Data Transport Layer Security (DTLS), 186
database server network layer of defense, 80
dead spots, 156
Debian Linux, 284
decryption tools, 205–207
deep packet inspection, 80
defense in depth, 78–80, 79f
defense, mobile malware, 330–331
delivery (drop off) loss mitigation, 41
demilitarized zone (DMZ), 80, 188–189
denial of service (DoS), 73–74, 130, 133, 147–148, 213
desktop virtualization, 78
developers
  criminal and, 320–323
  digital certificates, 289
device control threats, 72–73
device diversity, 303
device management, 193–194
  mobile, 330–331
devices without displays, connection process for, 168–169
DHCP. *See* Dynamic Host Configuration Protocol
Diameter, 36
dictionary password crackers, 206
Digital Advanced Mobile Phone System (D-AMPS), 34
digital certificate, 263, 264
digital communications. *See also specific communications*
  advantages of, 6
Digital Signature Algorithm (DSA), 261
digital subscriber line (DSL), 159
direct sequence spread spectrum (DSSS), 104–105
directed probe, 96
directional antennas, 211–212
discovery protocols, 190
discovery tools, 200–201
  enterprise Wi-Fi audit tools, 201–202
  HeatMapper, 202–204
distribution medium, 108
distribution network, 105
distribution service (DS), 108–109
distribution system service (DSS), 109
DMZ. *See* demilitarized zone
DNS. *See* Domain Name System
Dnsmasq tool, 287
Domain Name System (DNS), 80, 108, 188, 190, 287
DoS. *See* denial of service
dotted decimal, 11
drive-by attacks, 327–328
drive-by browser exploits, 290
driverless vehicles, 37
DroidBox, 248
DS. *See* distribution service
DSA. *See* Digital Signature Algorithm
DSL. *See* digital subscriber line
DSS. *See* distribution system service
DSSS. *See* direct sequence spread spectrum
DTLS. *See* Data Transport Layer Security
due diligence, practicing, 328
dumb terminal, 5, 5f
dwell time, 104
Dynamic Host Configuration Protocol (DHCP), 11, 108, 190–191, 304
  authenticated, 183

E

EAP. *See* Extensible Authentication Protocol
EAP over LAN (EAPoL), 180
EAP-TLS. *See* Extensible Authentication Protocol–Transport Layer Security
EAS. *See* Exchange Active Sync
easy-creds, 287
EDGE 2G+, 35
802.1Q tagging, 162
Ekahau Site Survey tool, 203
EMM. *See* enterprise mobility management
employee training and education, 179
Encapsulation Security Payload (ESP), 185
End User License Agreement (EULA), 76
endpoint attacks, 73

endpoint fingerprinting, 299
Enterprise App Store, 255
enterprise business management applications, 42
enterprise gateways, 111
enterprise mobility management (EMM), 281
   security using, 272
      mobile application management, 273, 275
      mobile device management, 273, 274–275t
enterprise modes, 184
environmental threats, 230
equipment configuration and placement, 155, 156f
ESP. *See* Encapsulation Security Payload
ESS. *See* extended service set`
ESSID. *See* extended service set identification
Ethernet, 12–16, 98, 110, 125, 171, 209
EtherType protocol filtering, 182–183
Ettercap, 205–206, 214
EULA. *See* End User License Agreement
evil twin, 72, 135–137, 214
   access points, 74
Evilgrade, 287
evolution of data and wireless networks, 3–4
   data communication, 4
      early data networks, 5–7, 5f
      Internet revolution, 7
      personal computers, advances in, 7–8
   data link layer, 12–13
   Internet of Things (IoT), 22–23
   IP addressing, 11–12
   network communication, 8–9, 11
   Open Systems Interconnection Reference Model, layers of, 8–10, 10f
   Physical Layer, 13–14
   Wi-Fi market, 19–20
      IP mobility, 20–22, 22f
   wired to wireless, 14–16

wireless networking
   business challenges addressed by, 16–19
   economic impact of, 16
   and way people work, 16–19
exabytes, 38
Exchange Active Sync (EAS), 271
exploitation stage, pentesting smartphones, 331
exploits, 245–246t
   Apple iOS, 252–254
   client and infrastructure, 285
   client-side, 285–286
   mobile software exploits and remediation, 290–291
      broken cryptography, 293
      client-side injection, 293
      data leakage, 292
      improper session handling, 294
      insufficient transport layer protection, 292
      lack of binary protections, 295
      poor authorization and authentication, 293
      security decisions through untrusted inputs, 294
      unsecure data storage, 291–292
      weak server-side security, 291
   network security protocol
      browser application and phishing exploits, 289–290
      developer digital certificates, 289
      drive-by browser exploits, 290
      public certificate authority exploits, 288–289
      RADIUS impersonation, 287–288
   other USB, 286
      network impersonation, 286–287
   tools, and techniques, 244
   Windows Phone OS, 255
extended service set (ESS), 97, 98, 191

extended service set identification (ESSID), 136, 284
Extensible Authentication Protocol (EAP), 164, 180, 181f
Extensible Authentication Protocol–Transport Layer Security (EAP-TLS), 285, 288
external network layer of defense, 80
external user segmentation, 187

F

FA. *See* foreign agent
FCC. *See* Federal Communications Commission
FDMA. *See* Frequency Division Multiple Access
Federal Communications Commission (FCC), 14, 95
FHSS. *See* frequency hopping spread spectrum
file-system permissions, 261
File Transfer Protocol (FTP), 169
fingerprint readers, 243
fingerprinting mobile devices, 297, 298
   advantages and disadvantages, 298
   defined, 298
   JavaScript, 307–308
   methods
      active fingerprinting, 305
      application identification, 304
      passive fingerprinting, 303
      TCP/IP headers, examining, 304
   modern cellular networks, 312–313
   man-in-the-middle (MITM) attack, 313
   MNmap, 313
   OS fingerprinting, 208
   spyware *vs.*, 309–312
   types of
      network access control and endpoint fingerprinting, 299

Index

network scanning and proximity fingerprinting, 299–300
  online or remote, 300–301
  unique device identification
    Android, 306
    Apple iOS, 305–306
    HTTP headers, 306–307
  users through biometrics, 308–309
firmware, upgrading, 169–170
5G, 36–37
foreign agent (FA), 21
4G, 36
four-way handshake, 107, 107f
  associated with authentication schemes, 161
free apps, 247, 309
free space optics (FSO), 57
FreeBSD, 263
FreeRADIUS, 287
Frequency Division Multiple Access (FDMA), 31, 31f
frequency hopping spread spectrum (FHSS), 104
frequency reuse, 29
frequency sharing, 31
FSO. *See* free space optics
FTP. *See* File Transfer Protocol
full duplex, 97
fully fledged stalkers, 321

G

GCHQ. *See* Government Communications Head Quarters
GCR. *See* general controls review
GDPR. *See* General Data Protection Regulation
general business and knowledge workers, wireless technology in, 19
general controls review (GCR), 228
General Data Protection Regulation (GDPR), 86
General Packet Radio Service (GPRS), 35
getDeviceID method, 306
GLBA. *See* Gramm–Leach–Bliley Act

Global Positioning System (GPS), 72, 133, 265
Global System for Mobile (GSM), 34–35, 305
Google Android
  security challenges, 244–247
    application architecture, 249–250
    criticism of, 247–248
    exploitation tools, 248–249
    security architecture, 249
    Google Play, 250
  security models
    file-system permissions, 261
    permission model, 262–263
    rooting and unlocking devices, 262
    sandbox, 261
    SDK security features, 261–262
    security model, 260
Google Cloud, 285
Google Play, 250
Government Communications Head Quarters (GCHQ), 247
GPRS. *See* General Packet Radio Service
GPS. *See* Global Positioning System
Gramm–Leach–Bliley Act (GLBA), 84–85
graphical user interface (GUI), 202, 331
green screen. *See* dumb terminal
grid antenna, 113, 115f
group temporal key (GTK), 107
Groupe Spécial Mobile (GSM). *See* Global System for Mobile
GSM. *See* Global System for Mobile
GTK. *See* group temporal key
guest access, 188
guest policies, 177–178
GUI. *See* graphical user interface

H

HA. *See* home agent
habits scam, 132
hack, 179
hackers, 32. *See also* attackers
  rogue access point vulnerabilities, 134

  targeted attack, 129–130
  using social engineering, 131–132
half duplex, 97
HaLow. *See* low-power Wi-Fi
Handoff, 32–33, 270
  -type features, security challenges of, 270
handover process, 33
hard drive encryption, 194
Hardware as a Service, 195
hardware audit tools, wireless local area network, 211
hash algorithms, 68–69
health care, wireless technology in, 17–18
Health Information Technology for Economic and Clinical Health (HITECH) Act, 85
Health Insurance Portability and Accountability Act (HIPAA), 85, 226
heartbeats, 289
Heartbleed, 248, 289
HeatMapper, 202–204
help desk scam, 132
high-level risk assessment, 225
High Speed Downlink Packet Access (HSDPA), 35
high throughput (HT), 166
highly directional antennas, 111, 113–114, 115f, 116f
hijacking
  devices, 213–214
  session, 214–215
HIPAA. *See* Health Insurance Portability and Accountability Act
HITECH Act. *See* Health Information Technology for Economic and Clinical Health (HITECH) Act
home address, 21
home agent (HA), 21
Homedale software, 202
hopping, 104
hosts, 11
Hotspot Shield, 217
Hotspotter, 136

HSDPA. *See* High Speed Downlink Packet Access
HT. *See* high throughput
HTTP. *See* Hypertext Transfer Protocol
HTTPS. *See* Hypertext Transfer Protocol Secure
human interface device, 286
Hypertext Transfer Protocol (HTTP), 188, 301
  headers, 306–307
Hypertext Transfer Protocol Secure (HTTPS), 50, 135, 188, 328

I

IAM. *See* identity and access management
IBSS. *See* independent basic service set
iCloud, 285
ICMP. *See* Internet Control Message Protocol
IdentifierForVendor, 306
identifying and implementing controls stage of security risk assessment, 234–235
identity and access management (IAM), 50, 193
IDSs. *See* intrusion detection systems
IEEE 802.11 standards, 98–99
  Wi-Fi Alliance naming system, 99–102
IEEE 802.15.4, 54
IEEE 802.1X standard, 179, 180f
IETF. *See* International Engineering Task Force
ifconfig, 217
IMEI. *See* International Mobile Station Equipment Identity
impact assessment, 233
IMSI. *See* International Mobile Subscriber Identity
independent basic service set (IBSS), 97, 98
individualized data encryption, 168

Industrial Internet of Things (IoT), 52
  IEEE 802.15.4, 54
  low-power device networks, 53–54
  low-power technologies, 53
information
  dissemination, 41–42
  gathering stage of security risk assessment, 228–229
  policy, 67
  security standards, 66
    ISO/IEC 27001:2013, 82, 82f
    ISO/IEC 27002:2013, 83
    NIST SP 800-53, 83
  theft, wireless networks and, 144–146
infrastructure exploits, 285–286
infrastructure mode, 97
infrastructure networks, 148
input validation, 215
inSSIDer Office tool, 160, 217, 282
insufficient transport layer protection, 292
integration service (IS), 108
integrity, information security and, 68–69
inter-process communication (IPC), 294
inter-stack communication, 10
interception, vulnerabilities, 222–223
interconnects, 56–58
interference, sources of, 120–121
internal network layer of defense, 80
internal user segmentation, 187
International Engineering Task Force (IETF), 21
International Mobile Station Equipment Identity (IMEI), 245t, 305
International Mobile Subscriber Identity (IMSI), 311
International Mobile Telecommunications-2000 (IMT-2000), 35
International Telecommunication Union (ITU), 35, 94
Internet Control Message Protocol (ICMP), 147

Internet of Things (IoT), 12, 22–23, 138
  industrial. *See* industrial Internet of Things
  wide area network technologies for
    LoRaWAN, 58
    low-power Wi-Fi (HaLow), 59
    millimeter radio, 59
    Sigfox, 58
Internet Protocol (IP), 108. *See also* Mobile IP
  addressing, 11–12
  mobility, 20–22, 22f, 177
  networking
    risk assessment. *See* risk assessment
    threat and vulnerability analysis. *See under* wireless local area network
  protocol filtering, 183
  services, 190–191
Internet Protocol Security (IPSec), 166, 186
Internet Protocol version 4 (IPv4), 11, 185
Internet Protocol version 6 (IPv6), 12, 185
Internet revolution, 7
Internet service providers (ISPs), 110
interoperability and layering, 156–157
intrusion detection systems (IDSs), 182
intrusion prevention systems (IPSs), 80, 182
inventory counts, 18
iOS malware, 325, 326t
IoT. *See* Internet of Things
IP. *See* Internet Protocol
IPC. *See* inter-process communication
ipconfig, 217
IPSec. *See* Internet Protocol Security
IPSs. *See* intrusion prevention systems
IPv4. *See* Internet Protocol version 4

Index

IPv6. *See* Internet Protocol version 6
IS. *See* integration service
island hopping, 73
ISM Unlicensed Spectrum, 94
    802.11 service sets, 97–98
    wireless client devices, 96–97
    WLAN anatomy, 96
ISO/IEC 27001:2013, 82, 82*f*
ISO/IEC 27002:2013, 83
ISPs. *See* Internet service providers
IT security management
    legal requirements, 226
    methodology, 225–226
    other justifications for risk assessments, 226
ITU. *See* International Telecommunication Union
IV collision attack, 164

jailbreaking, 73, 250, 266, 310
jamming, radio frequency, 212–213
JavaScript, 290, 307–308
    injection, 215, 293
jitter, 122

K

Kali Linux security platform, 283–284, 286
    Airodump-ng, scanning with, 284
Karmetasploit, 287
Key Negotiation of Bluetooth (KNOB), 142
key performance indicators (KPIs), 16
keylogger, 286
Kismet, 149, 160, 203, 204, 282
KNOB. *See* Key Negotiation of Bluetooth
knowledge workers, 17
KPIs. *See* key performance indicators

L

L0phtCrack, 206
LAN. *See* local area network
LanGuard, 208
latency, 122

LEAP. *See* Lightweight Extensible Authentication Protocol
least privilege, 67
Lightweight Extensible Authentication Protocol (LEAP), 180, 280
likejacking, 328
likelihood assessment. *See* probability assessment
lily padding, 73
Link Layer Discovery Protocol (LLDP), 190
LLDP. *See* Link Layer Discovery Protocol
local area network (LAN), 7. *See also* virtual local area networks, wireless local area network
long-range wide area network (LoRaWAN), 58
Long-Term Evolution (LTE), 36, 102
LoRaWAN. *See* long-range wide area network
loss control, 18
lost/stolen devices, data theft threats, 72
low-power device networks, 53–54
low-power technologies, 53
low-power Wi-Fi (HaLow), 59
LTE. *See* Long-Term Evolution
LTE networks, private, 59–60

M&M design, 79, 192
M7 coprocessor, 308
MAC Layer. *See* Media Access Control (MAC) Layer
MAC service data unit (MSDU), 164
machine-to-machine (M2M) communications, 52
macrocells, 30
madware, 244
    excessive application permissions, 323–325
    malware on Apple iOS devices, 325, 326*t*
    mobile malware delivery methods, 326–327

malicious data
    injection, 146
    insertion, on wireless networks, 146–147
malicious unnecessary functionality, 329
malware, 71–72
    and application security, 186
    mobile. *See* mobile malware and application-based threats
MAM. *See* mobile application management
man-in-the-middle (MITM) attack, 134, 313
Management as a Service, 195
Marshmallow (Android 6), 261
masquerading, 133
matrix approach, risk results analysis, 234
MBSS. *See* mesh basic service set
MCC. *See* mobile cloud computing
MDM. *See* mobile device management
Media Access Control (MAC) Layer, 105, 159
    address, 9, 160, 284, 304
    attacks, 148
    filters, 160
Media Layer, 254
mesh basic service set (MBSS), 97, 98, 191
message-digests algorithms, 68
message information base (MIB), 189
message integrity code (MIC), 107, 167
message modification, 133
Metasploit, 73, 204, 284
metro-style apps, 256
MIB. *See* message information base
MIC. *See* message integrity code
MITM. *See* man-in-the-middle attack
microcells, 30
millimeter radio, 59
MIMO. *See* multiple input/multiple output
MMS. *See* Multimedia Message Service
MN. *See* mobile node
MNmap, 313

mobile
  adware. *See* madware
  application
    attacks, 329–330
    convenience Trumps security, 47
    permissions, checking, 329
    removing unwanted, 329
    security, 76
    threat of, 47
    vendor-driven, 46–47
  attacks, 280
  browser attacks, 328–329
  communication, 28–33
  data usage, 38–39, 38f
  device providers, 303
  fingerprinting. *See* fingerprinting mobile devices
  networks
    AMPS 1G, 33–34
    GPRS and EDGE 2G+, 35
    GSM and CDMA 2G, 34–35
    3G Technology, 35–36
    4G and LTE, 36
    5G, 36–37
  security threats. *See* security threats
  threats and vulnerabilities, 242–244
  workforce, rise of, 48–49
mobile application management (MAM), 75, 76–77, 186
  security, 272, 273, 275
mobile cloud, 50–51
  cloud apps *vs.* native mobile apps, 51
  mobile cloud computing, 51
mobile cloud computing (MCC), 51
mobile communication security
  challenges, 241–242
  Apple iOS, 251–252
    App Store, 254–255
    architecture, 254
    exploits, 252–254
    exploits, tools, and techniques, 244
  Google Android, 244–247
    application architecture, 249–250
    criticism of, 247–248
    exploitation tools, 248–249
    security architecture, 249
    Google Play, 250
  mobile phone threats and vulnerabilities, 242–244
  Windows Phone
    architecture, 256
    OS exploits, 255
    security architecture, 255–256
    Windows Store, 256
mobile device management (MDM), 72, 75–76, 186
  malware and application-based threats, 330–331
  security using enterprise mobility management, 272, 273
  tasks, 274–275t
mobile device security models, 259–260
  Apple iOS
    Apple security model, 263
    application provenance, 264
    encryption, 266
    iOS sandbox, 264–265
    jailbreaking iOS, 266
    permission-based access, 265–266
    security concerns, 265
  BYOD and security, 271–272
  enterprise mobility management, 272
  mobile application management, 273, 275
  mobile device management, 273, 274–275t
  Google Android
    file-system permissions, 261
    permission model, 262–263
    rooting and unlocking devices, 262
    sandbox, 261
    SDK security features, 261–262
    security model, 260
  Handoff-type features, security challenges of, 270
  iOS and Android evolution, 267–268
    Android version evolution, 268–269
    Apple iOS, 269–270
    Windows Phone 8 security platform application security, 267
    security features, 267
Mobile IP, 21, 145
  arrival of, 40
  economic impact of, 38–39
  security, 62
mobile malware and application-based threats, 317–318
  madware
    excessive application permissions, 323–325
    malware on Apple iOS devices, 325, 326t
    mobile malware delivery methods, 326–327
    malware on Android devices, 318–319
    criminal and developer collaboration, 320–323
    software fragmentation, 319–320
  mitigating mobile browser attacks, 328–329
  mobile application attacks, 329–330
  mobile device management, 330–331
  mobile malware and social engineering
    captive portals, 327
    clickjacking, 328
    drive-by attacks, 327–328
    likejacking, 328
    plug-and-play scripts, 328
  mobile malware defense, 330–331
  penetration testing and smartphones, 331–332
mobile node (MN), 21
mobile Remote Access Trojans (mRATs), 253, 325
mobile revolution, 27
  BlackBerry effect and BYOD revolution, 37–38

business use cases, 40–42
cellular
  coverage maps, 28–32, 30f, 33f
  Handoff, 32–33
  Mobile IP, economic impact of, 38–39
  mobile networks
    AMPS 1G, 33–34
    GPRS and EDGE 2G+, 35
    GSM and CDMA 2G, 34–35
    3G Technology, 35–36
    4G and LTE, 36
    5G, 36–37
  mobility, business impact of, 40–42
mobile wireless attacks and remediation, 279
  client and infrastructure exploits, 285
  client-side exploits, 285–286
  corporate network for mobile attacks, scanning, 280
  Kali Linux security platform, 283–284
    Airodump-ng, scanning with, 284
  mobile software exploits and remediation, 290–291
    broken cryptography, 293
    client-side injection, 293
    data leakage, 292
    improper session handling, 294
    insufficient transport layer protection, 292
    lack of binary protections, 295
    poor authorization and authentication, 293
    security decisions via untrusted inputs, 294
    unsecure data storage, 291–292
    weak server-side security, 291
  network security protocol exploits
    browser application and phishing exploits, 289–290

developer digital certificates, 289
drive-by browser exploits, 290
public certificate authority exploits, 288–289
RADIUS impersonation, 287–288
other USB exploits, 286
network impersonation, 286–287
security awareness, 280–281
network, scanning, 281–282
vulnerabilities, scanning for, 282–283
mobility, business impact of, 40–42
modem, 6
modern cellular networks, 312–313
  man-in-the-middle (MITM) attack, 313
  MNmap, 313
monitoring stage of security risk assessment, 235
mRATs. *See* mobile Remote Access Trojans
MSDU. *See* MAC service data unit
MU-MIMO. *See* multi-user MIMO
multi-user MIMO (MU-MIMO), 100, 117
  with wireless devices, 117–118
Multimedia Message Service (MMS), 244, 262
multipath, 103–104, 104f
multiple input/multiple output (MIMO), 100, 201
  antennas, 114–118, 117f

N

NAC. *See* Network Access Control
narrowband, 103
NAT. *See* network address translation
National Security Agency (NSA), 247
native mobile apps, cloud apps *vs.*, 51
NDP. *See* Neighbor Discovery Protocol

near-field communication (NFC), 56, 73
near-far problem, 32
Neighbor Discovery Protocol (NDP), 190
Nessus, 205, 208, 223, 224
NetSpot, 203
netstat, 216
NetStumbler, 136, 204, 282
Network Access Control (NAC), 135, 299
  Port-based, 137, 179
network address translation (NAT), 11
network cloaking, 149
network communication, 8–9, 11
network effect, 4
network enumerators, 208
network impersonation, 286–287
network infrastructure program, 283
Network Layer
  attacks, 147
  OSI Reference Model, 9
network management and control tools
  Aircrack-ng, 209
  Airshark, 209–210
  network management system, 210–211
  wireless protocol analyzers, 208–209
network management system (NMS), 183, 210–211
network scanning, 299–300
network security protocol exploits
  browser application and phishing exploits, 289–290
  developer digital certificates, 289
  drive-by browser exploits, 290
  public certificate authority exploits, 288–289
  RADIUS impersonation, 287–288
Network Time Protocol (NTP), 190
network utilities, 216–217. *See also specific network utilities*

NFC. *See* near-field communication
NIST SP 800-53, 83
Nmap, 208, 304
NMS. *See* network management system
noise, sources of, 120–121
nomadic roaming, 98, 191
nonce, 107
nonrepudiation, information security and, 70
Nougat (Android 7), 261
NSA. *See* National Security Agency
NTP. *See* Network Time Protocol
null probe request, 96

O

OFDMA. *See* orthogonal frequency-division multiple access
Ohagi/Morii attack, 165
omnidirectional antennas, 111, 112, 211
One Drive, 285
online fingerprinting, 300–301
open access, 188
open share method, 206
open source model, 246
Open System Authentication (OSA), 160, 161
Open Systems Interconnection (OSI) Reference Model, 8–9
  layers of, 9–10, 10f. *See also specific layers*
Open Web Application Security Project (OWASP), 291
OpenSSL, 289
OpenVAS, 282
operational controls, risk-analysis, 235
orthogonal frequency-division multiple access (OFDMA), 101, 102
OS fingerprinting, 208
OSA. *See* Open System Authentication
OSI Reference Model. *See* Open Systems Interconnection (OSI) Reference Model
Outlook Web Access, 271
OWASP. *See* Open Web Application Security Project

P

packet analysis, 143–144
packet switching, 6
Packetforge-ng, 209
PacketManagerService, 250
pairing, Bluetooth, 139–140
  in public, 143
  short PINs, 143
PAN. *See* personal area network
Panopticlick, 304
parabolic antenna, 114, 116f
passive fingerprinting, 303
passive mode, coverage assessment, 122
passive scanning, 96, 97
password management system (PMS), 207
passwords, 188, 243
  -capture and decryption tools, 205–207
Patriot Act, 311
Payment Card Industry Data Security Standard (PCI DSS), 85
PBX. *See* Private Branch Exchange
PC. *See* personal computer
PCI DSS. *See* Payment Card Industry Data Security Standard
PDCA cycle, 82, 82f
peer-to-peer hacking over ad hoc networks, 148–149
penetration testing, 331–332
  tools
    Metasploit, 204
    Security Auditor's Research Assistant, 204–205
pentesting. *See* penetration testing
perimeter
  client security, 192–193
  network layer of defense, 80

periodic inventory, wireless local area network, 170–171
permission-based access control, 265–266
permission model, Android, 262–263
personal area network (PAN), 54, 94, 138
personal computer (PC), 5
  advances in, 7–8
personal identification numbers (PINs), 243
  short, 143
personally identifiable information (PII), 71
petabytes, 39
phablets, 319
phishing, 131
  exploits, 289–290
  wireless, 72
PhoneSheriff Investigator, 310
phreakers, 133
Physical Layer
  attacks, 148
  OSI Reference Model, 9, 13–14
physical security, 130–131, 170
picking efficiency, 18
picocells, 30
piconet, 55
PII. *See* personally identifiable information
Pineapple device, 211, 212
ping, 216
PINs. *See* personal identification numbers
planar antennas, 112, 114f
planar charts, 118
planning stage of security risk assessment, 227
platform application security, 267
plug-and-play scripts, 328
PMS. *See* password management system
Point-to-Point Protocol (PPP) networks, 180
poor authorization and authentication, 293

Index

pop-ups, blocking, 329
Port-based Network Access Control, 137, 179
port mirroring, 144
port scanning, 208
post-exploitation stage, pentesting smartphones, 331
potentially unwanted applications (PUAs), 244, 309, 320
PPP networks. *See* Point-to-Point Protocol (PPP) networks
premium SMS, 321
preproduction risk assessment, 225
Presentation Layer, OSI Reference Model, 9
Private Branch Exchange (PBX), 48
private LTE networks, 59–60
probability assessment, 233
process sandboxing, 260
ProPolice, 261
protocol analysis, 120–122
protocol filtering, 182–183
provisioned guest access, 188
proximity fingerprinting, 299–300
proximity hacking, 73
PSTN. *See* public switched telephone network
PUAs. *See* potentially unwanted applications
public certificate authority exploits, 288–289
public switched telephone network (PSTN), 4, 28
PwnSTAR, 287, 290

Q

QAM. *See* Quadrature Amplitude Modulation
QoS. *See* quality of service
Quadrature Amplitude Modulation (QAM), 101
qualitative risk assessment, 225, 234
quality of service (QoS), 107, 159, 162, 190
quantitative risk assessment, 225, 233–234
quarantining, 178, 194–195

R

radiation envelopes, 118
Radio Access Network (RAN), 313
radio frequency (RF)
　design, 154–155
　jamming, 212–213
radio-frequency identification (RFID), 17, 56
RADIUS. *See* Remote Authentication Dial-In User Service
rainbow table, 206
RAN. *See* Radio Access Network
ransomware, 321
RATs. *See* Remote Access Tools; Remote Access Trojans
recon stage, pentesting smartphones, 331
regulatory compliance
　California Customer Privacy Act, 86
　detrimental effects of, 86–87
　General Data Protection Regulation, 86
　Gramm–Leach–Bliley Act, 84–85
　Health Information Technology for Economic and Clinical Health Act, 85
　Health Insurance Portability and Accountability Act, 85
　Payment Card Industry Data Security Standard, 85
　Sarbanes–Oxley Act, 84
reliability, 223
remote access policies, 177
Remote Access Tools (RATs), 320
Remote Access Trojans (RATs), 244
Remote Authentication Dial-In User Service (RADIUS), 176, 180–182, 181*f*
　authentication, 158, 171
　impersonation, 287–288

remote fingerprinting, 299, 300–301
replay attack, 133
residential gateways, 110–111
retail industry, wireless technology in, 18–19
RF. *See* radio frequency
RFID. *See* radio-frequency identification
risk analysis stage of security risk assessment, 229–234, 234*f*
risk assessment, 158
　defined, 222
　IP networking, 221–224
　IT security management, 225–226
　other justifications for, 226
　other types of, 225
　security audits, 235–236
　security stages. *See* security, IP network risk assessment
　on wireless local area network, 224–225
risk map, 234
risk results analysis, 233–234
RiskInDroid, 248
robust antimalware software, 329
Robust Security Network (RSN), 159
rogue access points, 74, 134
　vulnerabilities, 134–315
rogue wireless local area networks, 171
rooting
　process, 250
　and unlocking devices, 262
RSA cryptographic protocol, 261
RSN. *See* Robust Security Network

S

SA. *See* Security Associations
SaaS. *See* Software as a Service
sandbox, 242
　Android, 260, 261
　Apple iOS, 264–265
　approach, 294
SARA. *See* Security Auditor's Research Assistant

Sarbanes–Oxley Act (SOX), 84, 226
SarbOx. *See* Sarbanes–Oxley Act
scanning stage, pentesting
    smartphones, 331
scream test, 170
SD-WAN. *See* software-defined
    wide area network
SDK. *See* software development kit
seamless roaming, 98
Secure Hash Algorithm (SHA), 261
Secure Shell (SSH), 331
Secure Simple Pairing (SSP), 139
Secure Sockets Layer (SSL),
    288–289
  certificate management
    service, 190
security, 271–272
  awareness, 280–281
    network, scanning, 281–282
    training, 179
    vulnerabilities, scanning for,
      282–283
  challenges. *See* mobile
    communication security
    challenges
  decisions through untrusted
    inputs, 294
  measures. *See* advanced WLAN
    security measures; basic
    WLAN security measures
  IP network risk assessment,
    226–227
    identifying and implementing
      controls, 234–235
    information gathering,
      228–229
    monitoring, 235
    planning, 227
    risk analysis, 229–234, 234*f*
    security audits, 235–236
    stages, 226–235
  IT security management
    legal requirements, 226
    methodology, 225–226
    other justifications for risk
      assessments, 226
  mobile apps, 47
  Mobile IP, 62
  models. *See* mobile device
    security models
  by obscurity, 267

policy, comprehensive. *See*
    comprehensive security
    policy
Security Associations (SA), 185
Security Auditor's Research
    Assistant (SARA), 204–205
security threats, 65–66
  authorization and access
    control, 80
  AAA, 80–82
  defense in depth, 78–80, 79*f*
  information security standards
    ISO/IEC 27001:2013, 82, 82*f*
    ISO/IEC 27002:2013, 83
    NIST SP 800-53, 83
  regulatory compliance
    California Customer Privacy
      Act, 86
    detrimental effects of
      regulations, 86–87
    General Data Protection
      Regulation, 86
    Gramm–Leach–Bliley Act,
      84–85
    Health Information
      Technology for Economic
      and Clinical Health Act, 85
    Health Insurance Portability
      and Accountability Act, 85
    Payment Card Industry Data
      Security Standard, 85
    Sarbanes–Oxley Act, 84
  risk mitigation, 74–75
    Bring Your Own Device,
      75–78
    BYOD for small-to-medium
      businesses, 78
  things to protect, 66
  threat categories, 67–68
    accountability, 69–70
    availability, 69
    confidentiality, 68
    integrity, 68–69
    nonrepudiation, 70
  to wireless and mobile devices,
    70–71
    data theft threats, 71–72
    device control threats, 72–73
    system access threats, 73–74
self-organizing wireless local area
    network, 122

semi-directional antennas, 111,
    112, 113*f*, 114*f*, 211–212
sensitive data disclosure, 322
service attacks, denial of, 147–148
service level agreements
    (SLAs), 223
service-orientated
    architecture, 49
service set identifier (SSID), 74,
    96, 159–160, 283
service sets, 97. *See also specific*
    service sets
session handling, improper, 294
session hijacking, 214–216
Session Layer, OSI Reference
    Model, 9
session side-jacking, 214–216
SHA. *See* Secure Hash Algorithm
Shared Key Authentication
    (SKA), 161
Short Message Service (SMS), 34,
    244, 262
  premium, 321
Sigfox system, 58
Signaling System 7 (SS7), 36
SIM. *See* subscriber identity
    module
Simple Mail Transfer Protocol
    (SMTP), 188
Simple Network Management
    Protocol (SNMP), 183,
    210, 305
Simple Network Management
    Protocol version 3
    (SNMPv3), 189
simplex communication, 97
Simultaneous Authentication of
    Equals, 185
single loss expectancy (SLE), 233
single sign-on (SSO), 193
single-user MIMO (SU-MIMO),
    115–116
site surveys, 118–119
  spectrum and protocol analysis
    access point placement, 121
    coverage analysis, 121
    coverage assessment, 121–122
    noise and interference,
      sources of, 120–121
    self-organizing WLANs, 122
6LoWPAN, 56

Index

SKA. *See* Shared Key Authentication
SLAs. *See* service level agreements
SLE. *See* single loss expectancy
small office/home office (SOHO), 60
small-to-medium business (SMB), 78
Smartphone Pentest Framework (SPF), 331
smartphones, 11, 37. *See also* mobile
  penetration testing and, 331–332
SMB. *See* small-to-medium business
SMS. *See* Short Message Service
SMTP. *See* Simple Mail Transfer Protocol
sniffer, 133
sniffing, 71
SNMP. *See* Simple Network Management Protocol
SNMP traps, 210
SNMPv3. *See* Simple Network Management Protocol version 3
SNonce, 107
snooping. *See* sniffing
social engineering, 131–132
  mobile malware and
    captive portals, 327
    clickjacking, 328
    drive-by attacks, 327–328
    likejacking, 328
    plug-and-play scripts, 328
social threats, 230
Software as a Service (SaaS), 195, 271
software-defined wide area network (SD-WAN), 56, 57–58
software development kit (SDK), 47, 260
  security features, 261–262
software fragmentation, 319–320
SOX. *See* Sarbanes–Oxley Act
specific targets, targets of opportunity *vs.*, 128–129
spectrum analysis/analyzer, 120–122

SPF. *See* Smartphone Pentest Framework
spread spectrum, 103
spy cells, 311–312
spy software, 310–311
spyware, 71, 301, 309–310
  fingerprinting *vs.*, 309–312
  spy cells, 311–312
  spy software, 310–311
  Stingray, 311–312
SQL. *See* Structured Query Language
SS7. *See* Signaling System 7
SSH. *See* Secure Shell
SSID cloaking, 159–160
SSL. *See* Secure Sockets Layer
SSLsplit, 287
SSO. *See* single sign-on
SSP. *See* Secure Simple Pairing
STA. *See* station
stack, 8, 9
stalkerware, 321
star trek, 132
station (STA), 96
Stingray, 311–312
stream ciphers, block ciphers *vs.*, 166
Structured Query Language (SQL), 293
  injection, 215
SU-MIMO. *See* single-user MIMO
subnets, 20
subscriber identity module (SIM), 34
supplicant, 179
SweynTooth, 142–143
switching
  circuit, 6
  dominance of, 13
  packet, 6
Symbian OS, 244
SYN flood attack, 69
system access threats, 73–74
system review, 229
system/device takeover, 72

T

T1/E1 trunks, 29
tablets, 319
tailgating, 132
Tamper Data, 215

targeted attack, scouting for, 129–130
targets of opportunity, *vs.* specific targets, 128–129, 129f
TCP SYN flood attack, 147
TCP/IP. *See* Transmission Control Protocol/Internet Protocol
TDM. *See* Time Division Multiplexing
TDMA. *See* Time Division Multiple Access
technical controls, risk-analysis, 234
technical threats, 230
telegraphy, 4
telephony, 4
Telnet, 169
Temporal Key Integrity Protocol (TKIP), 165
thin access points, 108
Thread, 56
threats, 227. *See also specific categories*
  analysis, 230
  of loss and theft, 78
  mapping, 232–233
  security. *See* security threats and vulnerability. *See* wireless local area network
3G Technology, 35–36
Time Division Multiple Access (TDMA), 31, 32f
Time Division Multiplexing (TDM), 312
TKIP. *See* Temporal Key Integrity Protocol
TLS. *See* Transport Layer Security
toothing, 140
towers. *See* cell towers
traceroute and tracert, 216
trackers, 321
Transmission Control Protocol/ Internet Protocol (TCP/IP), 7
  headers, examining, 304
Transmit Beamforming technology, 201
Transport Layer
  attacks, 147
  OSI Reference Model, 9

Transport Layer Security
(TLS), 180
  hijacking session, 216

U

ubiquitous mobile data access, 16
UDI. *See* unique device identifier
UDID. *See* universal device
  identifier
unauthorized and modified
  clients, threats and, 73
unique device identification
  Android, 306
  Apple iOS, 305–306
  HTTP headers, 306–307
unique device identifier
  (UDI), 305
universal device identifier
  (UDID), 305
Universal Plug and Play
  (UPnP), 190
unlicensed bands, 102–103
  direct sequence spread
    spectrum, 104–105
  frequency hopping spread
    spectrum, 104
  multipath, 103–104, 104*f*
  narrowband and spread
    spectrum, 103
unsecure data storage, 291–292
untrusted inputs, security
  decisions through, 294
UPnP. *See* Universal Plug and Play
USB exploits, 286–287
user-based permission
  model, 260
user logons, 193–194
user segmentation
  demilitarized zone
    segmentation, 188–189
  guest access and
    passwords, 188
  virtual local area networks, 187

V

version analysis, 232
very high throughput (VHT), 103
very-small-aperture terminal
  (vSAT), 57
VHT. *See* very high throughput

virtual local area networks
  (VLANs), 159, 162,
  163*f*, 187
virtual machine (VM), 260
virtual private network (VPN),
  186, 190, 271
  cloud, 56–58
  over wireless, 161–162
VLAN. *See* virtual local area
  networks
VM. *See* virtual machine
Voice over Internet Protocol
  (VoIP), 13, 48, 177
Voice over WLAN (VoWLAN),
  157, 177
VoIP. *See* Voice over Internet
  Protocol
VoWLAN. *See* Voice over WLAN
VPN. *See* virtual private network
vSAT. *See* very-small-aperture
  terminal
vulnerabilities, 227, 242–244
  assessment, 231–232
  aware of, 319
  categories of, 222–223
  mapping, 232–233
  scanning for, 282–283
  threat and. *See under* wireless
    local area network

W

WaaS. *See* Wi-Fi as a Service
walled garden, 178, 250
WAN. *See* wide area network
WAP. *See* wireless access point
wardialing, 133
wardriving, 133–134
warehousing and logistics
  industry, wireless
  technology in, 18
WDS. *See* wireless distribution
  system
weak server-side security, 291
Web Proxy Autodiscovery
  Protocol (WPAD), 190
WEP. *See* Wired Equivalent
  Privacy
white list, 294
Wi-Fi, 9, 244
  roaming, 191–192

virtual private network for, 162
Wi-Fi 1 (802.11b), 99
Wi-Fi 2 (802.11a), 99
Wi-Fi 3 (802.11g), 100
Wi-Fi 4 (802.11n), 100
Wi-Fi 5 (802.11ac), 99, 100
Wi-Fi 6 (802.11ax), 99, 100–102
Wi-Fi 6E, 102
Wi-Fi 802.11ah, 59
Wi-Fi Alliance, 165
Wi-Fi Cracker, 284
Wi-Fi Inspector, 202
Wi-Fi market, 19–22, 22*f*
Wi-Fi as a Service (WaaS),
  195–196
Wi-Fi Protected Access (WPA), 61,
  165, 290
Wi-Fi Protected Access 2 (WPA2),
  61, 107, 161, 165–166, 280
  with Advanced Encryption
    Standard, 166–167
  with Counter Mode Cipher
    Block Chaining Message
    Authentication Code
    Protocol, 167
  personal and enterprise
    modes, 184
Wi-Fi Protected Access 2–
  preshared key mode
  (WPA2-PSK), 166, 184
Wi-Fi Protected Access 3
  (WPA3), 168
  data privacy over public
    networks, 168
  data protection
  advanced WLAN security
    measures, 184–185
  basic WLAN security
    measures, 168–169
  higher security for government,
    defense, and industrial
    applications, 169
  protection against brute-force
    attacks, 168
wide area network (WAN), 7,
  56–58
  technologies for Internet of
    Things (IoT)
  long-range wide area
    network (LoRaWAN), 58

low-power Wi-Fi (HaLow), 59
millimeter radio, 59
Sigfox, 58
WiMAX. *See* Worldwide Interoperability for Microwave Access
Win Sniffer, 205–206
Windows Phone 8 (WP8)
  security challenges
    architecture, 256
    OS exploits, 255
    security architecture, 255–256
    Windows Store, 256
  security models
    platform application security, 267
    security features, 267
Windows Store, 256
WIPS. *See* wireless intrusion prevention system
Wired Equivalent Privacy (WEP), 61, 136, 157, 164–165, 280
wired network. *See also* Ethernet
  security threats, 65–88
  to wireless, 14–16
wireless access point (WAP), 106f, 171
  architecture, 108–109
  working of, 105–108, 107f
wireless anonymity, 300
wireless antennas, 111, 156f
  coverage area, determining, 118, 119f
  highly directional antennas, 113–114, 115f, 116f
  MIMO antennas, 114–118, 117f
  omnidirectional antennas, 112
  semi-directional antennas, 112, 113f, 114f
wireless attacks and remediation, mobile. *See* wireless attacks and remediation
wireless bridges
  enterprise gateways, 111
  residential gateways, 110–111
  wireless workgroup bridges, 109–110
wireless broadcast, 154

wireless client devices, 96–97
wireless communication technologies, 51–52, 54
  Bluetooth Low Energy, 55
  near-field communication, 56
  radio-frequency identification (RFID), 56
  6LoWPAN, 56
  Thread, 56
  Z-Wave, 55
  Zigbee IP, 55
wireless distribution system (WDS), 109
wireless extender, 156
wireless intrusion prevention system (WIPS), 135, 182, 208, 282, 287, 300
  host-based, 182
  network-based, 182
wireless local area network (WLAN), 13, 15, 16, 93–94
  audit tools. *See* audit tools, wireless local area network
  IEEE 802.11 standards, 98–99
    Wi-Fi Alliance naming system, 99–102
  and IP network risk assessment. *See* risk assessment, IP networking
  and IP networking threat and vulnerability analysis, 125–126
    attacker gaining unauthorized control, 149
    attackers, types of, 126–127
    Bluetooth vulnerabilities and threats. *See* Bluetooth
    denial of service attacks, 147–148
    evil twins, 135–137
    malicious data insertion on wireless networks, 146–147
    packet analysis, 143–144
    peer-to-peer hacking over ad hoc networks, 148–149
    physical security and wireless networks, 130–131
    rogue access points, 134–135

    social engineering, 131–132
    targeted attack, scouting for, 129–130
    targets of opportunity *vs.* specific targets, 128–129
    wardriving, 133–134
    wireless networks and information theft, 144–146
  ISM Unlicensed Spectrum
    802.11 service sets, 97–98
    wireless client devices, 96–97
    WLAN anatomy, 96
  security measures
    advanced. *See* advanced WLAN security measures
    basic. *See* basic WLAN security measures
  self-organizing, 122
  site surveys, 118–119
  spectrum and protocol analysis, 120–122
  topologies, 94
  unlicensed bands, 102–103
    direct sequence spread spectrum, 104–105
    frequency hopping spread spectrum, 104
    multipath, 103–104, 104f
    narrowband and spread spectrum, 103
  wireless access points, 106f
    architecture, 108–109
    working of, 105–108, 107f
  wireless antennas, 111
    coverage area, determining, 118, 119f
    highly directional antennas, 113–114, 115f, 116f
    MIMO antennas, 114–118, 117f
    omnidirectional antennas, 112
    semi-directional antennas, 112, 113f, 114f
  wireless bridges
    enterprise gateways, 111
    residential gateways, 110–111
    wireless workgroup bridges, 109–110

wireless network. *See also specific types*
   business challenges addressed by, 16–19
   economic impact of, 16
   evolution of data and. *See* evolution of data and wireless networks
   and information theft, 144–146
   malicious data insertion on, 146–147
   physical security and, 130–131
   scanning, 281–282
   security, 60–61, 61*f*
      lingering security issues, 62
      threats. *See* security threats
   virtual local area networks, 162
   and way people work, 16–17
      general business and knowledge workers, 19
      health care, 17–18
      retail, 18–19
      warehousing and logistics, 18
   from wired to, 14–16
wireless personal area network (WPAN), 137
wireless phishing, 72
wireless protocol analyzers, 208–209
wireless repeater, 156
wireless sensor network (WSN), 53
wireless wide area network (WWAN), 56, 94
wireless workgroup bridges, 109–110
Wireshark, 147
WLAN. *See* wireless local area network
Worldwide Interoperability for Microwave Access (WiMAX), 9, 36, 57
WPA. *See* Wi-Fi Protected Access
WPA2-Personal. *See* Wi-Fi Protected Access 2-preshared key mode
WPA2-PSK. *See* Wi-Fi Protected Access 2-preshared key mode
WPA2. *See* Wi-Fi Protected Access 2
WPA3. *See* Wi-Fi Protected Access 3
WPAD. *See* Web Proxy Autodiscovery Protocol
WPAN. *See* wireless personal area network
WSN. *See* wireless sensor network
WWAN. *See* wireless wide area network

Xirrus Wi-Fi Inspector, 202
XSS. *See* cross-site scripting
Xsser, 325

Yagi antennas, 112, 113*f*

Z-WAVE, 55
Zero Trust, 50
Zigbee, 54, 55